D0151649

CONCEPTS IN SYSTEMS AND SIGNALS

John D. Sherrick
Rochester Institute of Technology

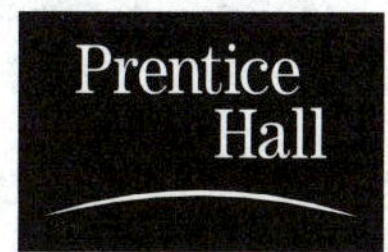

Upper Saddle River, New Jersey
Columbus, Ohio

Library of Congress Cataloging-in-Publication Data

Sherrick, John D.
Concepts in systems and signals / John D. Sherrick.
p. cm.
ISBN 0-13-084115-3
1. Signal processing--Mathematics. 2. System analysis--Data processing.
3. MATLAB. I. Title.

TK5102.9. S5325 2001
621.382′2–dc21 00-029816

Vice President and Publisher: Dave Garza
Editor in Chief: Stephen Helba
Assistant Vice President and Publisher: Charles E. Stewart, Jr.
Production Editor: Alexandrina Benedicto Wolf
Production Coordination: Clarinda Publication Services
Design Coordinator: Robin G. Chukes
Cover Designer: Tom Mack
Cover Image: Marjorie Dressler
Production Manager: Matthew Ottenweller
Marketing Manager: Barbara Rose

This book was set in New Century Schoolbook by The Clarinda Company. It was printed and bound by R. R. Donnelley & Sons Company. The cover was printed by Phoenix Color Corp.

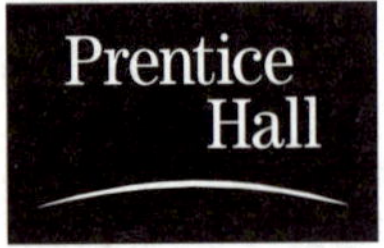

10 9 8 7 6 5 4 3 2 1
ISBN 0-13-084115-3

PREFACE

This text was written for a 40-lecture-hour, junior-year course in the Department of Electrical, Computer, and Telecommunications Technology at Rochester Institute of Technology. It is a required upper-division course for all three of the department's engineering technology programs. Considering the wealth of diverse topics competing for students' attention, the faculty action establishing this common upper-division course requirement is significant. The goals of the course are to:

1. Review the foundations of continuous-time systems and introduce, with equal emphasis, the "new circuit theory" of discrete-time systems.
2. Introduce the concepts and analysis tools associated with signal spectra, with emphasis on periodic signals and the discrete Fourier transform.
3. Make students aware of the capabilities of MATLAB, the general-purpose mathematical software available in our department computer facilities.

Typical electrical engineering technology programs in our region start students off with an ac circuits course in their freshman year before they have the mathematical background to really understand where the phasor comes from. Just when they are getting comfortable working with phasors and complex numbers, they are sent off to electronics courses, where capacitors and inductors are open or short circuits, and their ac circuits skills deteriorate. This text attempts to recapture and extend those skills in preparation for such upper-division specialty courses as communications theory and transmission lines. The elements of linear, discrete-time systems parallel exactly those of the continuous-time domain. Filtering is a subject well within the mathematical capabilities of an undergraduate technology student, and serves as a common focus for the two domains.

The sampling theorem, a key issue in converting between the continuous and discrete domains, is presented at two levels of mathematical sophistication. It is deduced from sampled sinusoids, but it is also derived more formally using the definition of an ideal sampler. Sampling schemes that minimize the need for analog filtering conclude the course.

Among department faculty who use a general mathematical software package in their courses, MATLAB is the package of choice. It is introduced systematically in this text, and used both to support the concepts presented and as a filter design handbook. The student edition of MATLAB v.5 was used for all MATLAB applications in this text. The latest student versions, which also contain several features not used in this text, may be purchased on the Web from The MathWorks, Inc. (http:// www.mathworks.com) of Natick, Massachusetts, at a reasonable price.

The discrete-time portion of the course benefits significantly from some demonstrations of sampling and real-time filtering. Any digital scope having an FFT math option can be used to demonstrate most of the sampling concepts and aliasing. A modest microcontroller, such as the Motorola 68HC11, contains all the features necessary to demonstrate real-time processing using a finite averager program. An 8-term averager has an interesting frequency response easily compared to theory. Lately, we have had the benefit of demonstrations provided by a few students pursuing an independent study of Analog Devices' EZ Lite 16-bit DSP kit. Even in the absence of these demonstrations, filtering a signal that has been saved to a file can be an impressive, eye-opening experience for most students.

ACKNOWLEDGMENTS

This text is a collection of concepts that I have had, in some cases, 40 years to digest. In that time many of these ideas have blended together, and it would be impossible to pinpoint where they all originated. All I can do is acknowledge that few, if any, of the ideas presented are originally mine. This is an attempt to recognize and thank those sources who are the most likely to have contributed to my background as represented in this text.

s Domain

Most electrical engineers of my age were introduced to the *s* domain by M. E. Van Valkenburg in his classic text *Network Analysis* (Prentice Hall, 1955). The form of the Routh–Hurwitz test I have used in this text comes from *Introduction to the Design of Servomechanisms,* by Bower and Schultheiss (Wiley, 1958). Modern texts favor a different form of the test, although I fail to see its advantages.

B. P. Lathi set the standard for continuous-time signals in *Signals, Systems, and Communication* (Wiley, 1965). I understand this text is used in some technology programs today, although it is really an engineering-level text.

Graduate school brought Louis Weinberg's *Network Analysis and Synthesis* (McGraw-Hill, 1962) to my collection. This theoretical treatise on network synthesis concludes with practical filter-design information that includes the classical filters discussed in this text.

Over the years I have used many different texts that may have contributed to my knowledge of the *s* domain. My favorite text to teach from was *Linear Circuits* by Ronald E. Scott (Addison-Wesley, 1960).

z Domain

In this area, both the texts and my experiences are more recent and my sources more accurately identified. My first experience with DSP came from an excellent 3-day seminar conducted in 1981 by Edward R. Salem and W. F. Walker of the RIT electrical engineering department.

A tremendous number of engineering-level texts in DSP made an appearance in and since the mid-1980s. I am sure some of them were exceptional, but I paid little attention to those that were not at the mathematical level I was seeking. *Digital Signal Processing* by Stanley, Dougherty, and Dougherty (Reston Publishing, 1984), although written for engineers, was the first text I found mathematically suitable for my senior-level technology elective in DSP. I use its explanation of the sequential design steps in the development of the window design of FIR filters. When it went out of print, I was forced to begin preparing these notes for my course.

In its initial version, my DSP elective contained a lab practicum. It did not seem possible to teach both a special processor and the required DSP theory in the allotted time. As a result, we performed some simple DSP using the Motorola 68HC11 microcontroller, which students learn in a required course. Doing this made the problems of coefficient accuracy clear and led to some results we could not, at that time, explain (limit cycling). These experiences also raised questions about the value of digital filtering when high-order analog anti-aliasing or reconstruction filters still seemed to be needed. These and other questions were addressed in two recent publications: a draft manuscript of *Understanding Digital Signal Processing* by Richard G Lyons (Addison-Wesley, 1997) and a Motorola University publication, *Digital Signal Processing and the Microcontroller* by Dale Grover and John R. Deller, with illustrations by Jonathon Roth (Prentice Hall, 1998). I am sure these issues were presented in other publications and texts, but this is where I found them and understood them.

Finally, Version 4 of the *User's Guide to MATLAB,* by The MathWorks Inc. (Prentice Hall, 1995), is as well done as the software it supports.

Thanks also to my students who shared the adventure of exploring DSP with me and who contributed ideas and corrections to the manuscript. I would also like to acknowledge the valuable feedback from the following reviewers: John Cmelko, Bryant & Stratton; Judy Serwatka, Purdue University–Calumet; Donald Stelzer, DeVry Institute–Phoenix; and Omar Zia, Southern Polytechnic State University.

JOHN D. SHERRICK

CONTENTS

1

NUMBERS, ARITHMETIC, AND MATHEMATICS

OUTLINE

OBJECTIVES

1. Discuss the origin of complex numbers.
2. Represent complex numbers in polar form and rectangular form.
3. Define and interpret Euler's identity.
4. Conduct numerical calculations with complex numbers.
5. Describe properties of a function of a complex variable.
6. Interpret and resolve indeterminate values.
7. Use MATLAB for arithmetic operations.

INTRODUCTION

Electrical, mechanical, hydraulic, and other systems contain components that obey calculus-based laws. Determining the behavior of such a system in response to an input signal involves solving differential equations. There is, however, one particular type of signal, e^{st}, for which the linear differential equations reduce to simple algebraic equations. Fortunately, this exponential function can represent the kinds of signals of interest to us. While this signal is a simple function of time, it unfortunately returns values that are complex numbers.

Since complex numbers are central to this text, a brief review of how they arise and the arithmetic they obey is likely to be helpful. A few other mathematical topics related to functions, and especially complex functions, will also be discussed. Otherwise, we will generally prefer to introduce mathematical concepts as they are needed.

1.1 THE NUMBER SYSTEM

The names that have been given to different types of numbers show that it is a common reaction to view new concepts with astonishment and possibly skepticism. After reading this section you should be able to:

- Discuss the origin of complex numbers.
- Explain why "imaginary" numbers are necessary.
- Recognize that you are not alone in finding new concepts difficult.

Much speculation has occurred on the origins of our numbering system, but it most certainly started with counting. A prehistoric hunter might have recorded the success of his efforts by the number of skins obtained, or the length of his efforts by the number of sunrises that occurred while he was away. Some of his skins might have been bartered away for roots and berries or for tools and weaponry. "I'll give you two of these for one of those" types of transactions. All of these processes required positive integers, a concept that is both natural and easy for us to appreciate.

Manuscripts dating from about 2000 B.C. show that simple fractions were also understood early on. Apparently, questions of how a lord could divide his territory uniformly among his subjects so that they might be equally and fairly taxed were among the most pressing early mathematical issues. Some priorities never change.

If a man had 5 skins and traded 3 of them for a new spear, he could determine how many skins he had left by counting out those needed for the trade and then counting those that remained. No new skills or concepts needed to be developed. It was unthinkable that 5 skins would be traded if only 3 were available. Such a transaction would be impossible! Several millennia would pass before that changed.

It is difficult, today, to fully appreciate some of the discoveries of the early mathematicians. The concepts with which they struggled are now so ingrained in our experience that they are second nature to us. The notations and algebraic laws that allow complicated questions to be addressed systematically were not available to them. Yet, once the philosophers of the day postulated that no transaction should be impossible, the existence of an entirely new set of numbers was required. Thus negative numbers were discovered, the number system doubled in size, and debt became an element in some transactions.

The concept of the number *zero* was especially difficult to establish. After all, why create a number to count nothing? It is only when man began to philosophize about numbers that meaning could be attached to such an abstraction. The discovery that there were numbers that could not be represented as a ratio of integers, numbers like those special ones we denote today as π or e, did not come from everyday experience. Indeed, the claim that there were such numbers was greeted with much skepticism and disbelief. Such numbers were inconceivable, irrational! We know now that the set of *irrational* numbers is infinitely larger than all those that were known previously.

Finally, but not until around the 16th century, it was recognized that there had to be yet again an entirely new set of numbers, in fact a new kind of number, if equations like $x^2 = -1$ were to have solutions. These new numbers were called *imaginary,* a regrettable though understandable label in order to distinguish them from the *real* numbers known previously. A j (mathematicians prefer i) is used to indicate the imaginary part of a number, where $j = +\sqrt{-1}$.

Today we conveniently think of the real number system as being represented by a continuous line of numbers running from $-\infty$ to $+\infty$, with every point along the line being a unique number. We think of the imaginary numbers in much the same way, with two exceptions: (1) Each imaginary number is tagged with a j to distinguish it from the real numbers, and (2) the line of imaginary numbers is independent of, or perpendicular to, the line of real numbers. Together the real and imaginary numbers form a *complex-number plane,* each point of which represents a generalized number that could be used in, or result from, a mathematical calculation. Every equation conceived of by man has yielded a result somewhere in this number plane . . . so far.

It is not necessary to introduce complex numbers into the study of electrical technology. All of our variables—voltage, current, resistance, inductance, etc.—are physical quantities that are described by ordinary real numbers. Ultimately, complex numbers gain their meaning through how we apply and interpret them, not through the arbitrary tags of "real" or "imaginary" that mathematicians use to distinguish between their two component parts. The fact that these components are independent allows a single complex number to hold information about two separate physical properties. We introduce complex numbers because they can make many of our calculations easier.

1.2 RECTANGULAR/POLAR CONVERSIONS

Complex numbers may be represented in rectangular or polar form. After reading this section you will be able to:

- Represent complex numbers in polar form and rectangular form.
- Convert from one form to the other.

A complex number can be identified by giving its real and imaginary parts to identify its location in the complex-number plane of Figure 1.1. This is called the *rectangular* form of the number. If $z = a + jb$, where a and b are real numbers, a is called *the real part* of z, and b is called *the imaginary part* of z. To emphasize this terminology, whenever we speak of the imaginary part of a complex number, we are talking about a real number that is being multiplied by j.

Some people like to think of j as a rotational *operator,* since it converts a real number into an imaginary number by rotating it 90° to the j axis in the complex-number plane. We prefer simply to define j to be a number, namely, the positive

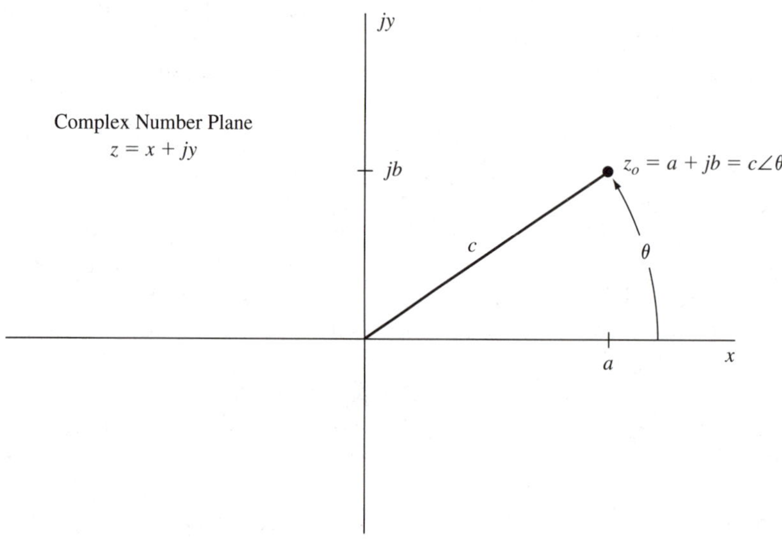

Figure 1.1 A specific complex number may be described either by its rectangular or its polar coordinates in a complex number plane.

square root of -1, and to treat it like any other number. To us, there is no difference between the numbers $j2$ and $2j$. Some of the products of j are:

$$j = +\sqrt{-1} \qquad j^2 = -1 \qquad j^3 = -j \qquad j^4 = 1 \qquad j^5 = j$$

A complex number may alternatively be specified by giving its distance from the origin along a radius line having a specified angle relative to the positive real axis. This is called the *polar* form of the number, and is often stated using the notation $c\angle\theta$, read as "a magnitude of c at an angle of θ." By convention, θ is taken as positive in the counterclockwise direction. Both c and θ are real numbers.

It is essential to be able to convert between polar and rectangular forms. The required relationships are easily deduced from Figure 1.1. The conversion to polar form is complicated slightly by the fact that the inverse tangent function must be evaluated in the proper quadrant. A sketch is often useful in establishing the proper angle.

Polar/Rectangular Conversions

$P \rightarrow R$ $\qquad a + jb = c\angle\theta \qquad$ $P \leftarrow R$

$$a = c\cos\theta \qquad \textbf{(1.1a)}$$

$$b = c\sin\theta \qquad \textbf{(1.1b)}$$

$$c = \sqrt{a^2 + b^2} \qquad \textbf{(1.1c)}$$

$$\phi = \arctan(|b/a|) \qquad \textbf{(1.1d)}$$

quadrant	***a***	***b***	θ
1st	+	+	ϕ
2nd	−	+	$\pi - \phi$
3rd	−	−	$\pi + \phi$
4th	+	−	$-\phi$

EXAMPLE 1.1A

Find the rectangular form of the complex number $z_o = 2\angle 140° = 2\angle 2.444$.

Solution

$$a = 2\cos 140° = 2(-0.7664) = -1.533$$

$$b = 2\sin 140° = 2(0.6428) = 1.286$$

$$\therefore z_0 = -1.533 + j1.286$$

EXAMPLE 1.1B

Find the polar form of the complex number $z_o = 1.5 - j0.5$.

Solution

A sketch is usually helpful:

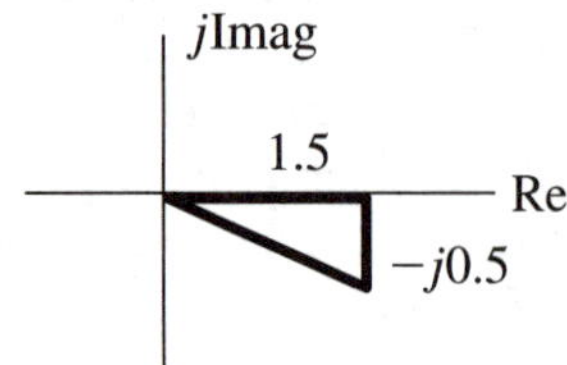

$$c = \sqrt{1.5^2 + (-0.5)^2} = \sqrt{2.25 + .25} = 1.581$$

$$\phi = \arctan\left(\left|\frac{-0.5}{1.5}\right|\right) = \arctan\left(\frac{1}{3}\right) = 0.322 = 18.43^\circ$$

A z_o with a positive real part and a negative imaginary part is in the 4th quadrant, so $\theta = -18.43^\circ$

$$\therefore z_o = 1.581\angle -18.43^\circ$$

Although angles in radians are, strictly speaking, correct, either degrees or radians are generally accepted. It is important, however, to make sure you enter numerical values corresponding to the units your calculator is expecting!

The conversion procedures of Equations 1.1 and Examples 1.1a–b are rarely used in practice. Modern technical calculators have built-in rectangular/polar functions that take care of making these conversions for you, including finding the proper quadrant angle information. It is essential that you consult your calculator manual and become proficient in making these conversions.

1.3 EULER'S IDENTITY

In preparing to investigate arithmetic operations involving complex numbers, we need to establish the proper mathematical equivalence of our polar notation $c\angle\theta$. After completing this section you will be able to:

- State and interpret Euler's identity.
- Describe a polar-form complex number using standard math functions.
- Express a cosine as a sum of complex exponentials.
- Express a sine as a sum of complex exponentials.

A very important relationship, known as *Euler's identity,* is

$$e^{\pm j\theta} \equiv \cos\theta \pm j\sin\theta \tag{1.2}$$

If Euler's identity is converted from its rectangular form to polar form, it becomes

$$e^{j\theta} = \cos\theta + j\sin\theta = \sqrt{\cos^2\theta + \sin^2\theta}\,\angle\arctan(\tan\theta) = 1\angle\theta$$

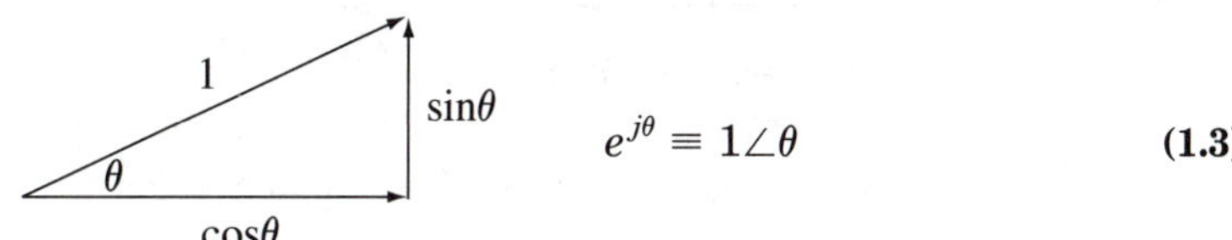

$$e^{j\theta} \equiv 1\angle\theta \tag{1.3}$$

Thus, the true mathematical interpretation of $A\underline{/\theta}$ is $Ae^{j\theta}$, and the algebraic rules for exponents apply to the angle information of polar-form complex numbers (Table 1.1).

Table 1.1 A Few Laws of Exponents

$e^{jx}e^{jy} = e^{j(x+y)}$	$(e^{jx})^y = e^{jxy} = (e^{jy})^x$	$1/e^{jx} = e^{-jx}$

A few additional special cases follow by direct substitution into Equation 1.2:

$$e^{\pm j0} = 1 \qquad e^{\pm j\pi} = -1 \qquad e^{\pm j2\pi} = 1 \qquad e^{\pm j\pi/2} = \pm j$$

The two signs in Euler's identity give us two equations, which may be solved for the cosine or sine function. For instance,

$$\begin{aligned} e^{j\theta} &= \cos\theta + j\sin\theta \\ e^{-j\theta} &= \cos\theta - j\sin\theta \\ e^{j\theta} + e^{-j\theta} &= 2\cos\theta \qquad \text{(adding the 2 equations)} \\ e^{j\theta} - e^{-j\theta} &= 2j\sin\theta \qquad \text{(subtracting the 2 equations)} \end{aligned}$$

$$\cos\theta = \frac{e^{j\theta} + e^{-j\theta}}{2} \tag{1.4}$$

$$\sin\theta = \frac{e^{j\theta} - e^{-j\theta}}{2j} \tag{1.5}$$

Using these relationships, any sinusoidal signal may be expressed in terms of the even more elementary functions of complex exponentials. This is, in fact, the origin of the *phasor* concept, which is fundamental to our study of alternating current (a-c) circuits. It is difficult to overstate the importance of Euler's identity; we will spend the rest of this text making use of it.

EXAMPLE 1.2

Express $f(x)$ in terms of cosine functions:

$$f(x) = 2e^{j2x} + 4e^{-jx} + 4e^{jx} + 2e^{-j2x}$$

Solution

Regrouping the terms to compare with Equation 1.4, we get

$$f(x) = 2[e^{j2x} + e^{-j2x}] + 4[e^{jx} + e^{-jx}]$$

and

$$f(x) = 4\cos 2x + 8\cos x$$

1.4 COMPLEX-NUMBER ARITHMETIC

Complex numbers are "complex" only in the sense that their arithmetic is more involved than that for real numbers. Adding two complex numbers involves the same amount of effort as adding two sets of two real numbers. This doubles the chances for error as well as the time required to complete the operation. After completing this section you will be able to:

- Conduct numerical calculations with complex numbers.
- Calculate the sum, difference, product, or ratio of complex numbers.
- Raise a complex number to a power.

- Find the conjugate of a complex number.
- Combine a number with its conjugate to find its real or imaginary part, its magnitude, or its angle.

1.4.1 Addition/Subtraction

If $z_1 = x_1 + jy_1$ and $z_2 = x_2 + jy_2$, then $z_1 + z_2 = (x_1 + x_2) + j(y_1 + y_2)$

Similarly,

$$z_1 - z_2 = (x_1 - x_2) + j(y_1 - y_2)$$

The j term serves to remind us that two completely independent types of numbers are present and that they must be handled separately. We also like the final results to be recombined into a single real part and a single imaginary part.

If complex numbers are given in polar form, the usual way to add them is to immediately convert both of them to rectangular form. Most calculators will accept polar-form numbers, convert and add them, and put the result back in polar form for you, all in a single operation. Of course, if two complex numbers have exactly the same angle, their magnitudes may be added directly as if they were vectors.

 **EXAMPLE 1.3A**

Find $z_1 - z_2$ if $z_1 = 2 + j5$ and $z_2 = 2\angle 140°$.

Solution

Since a subtraction is called for, the numbers must be in rectangular form. From Example 1.1a, $z_2 = 2\angle 140° = -1.533 + j1.286$. Then $z_1 - z_2 = 2 + j5 - (-1.533 + j1.286) = 3.533 + j3.714$.

1.4.2 Multiplication

Complex numbers may be multiplied in either rectangular or polar form. The procedure used depends primarily on what form the numbers are in initially and what form we want for the result. Starting with rectangular-form numbers, we may simply multiply them out, remembering that $j^2 = -1$:

If $z_1 = x_1 + jy_1$ and $z_2 = x_2 + jy_2$, then

$$z_1z_2 = (x_1 + jy_1)(x_2 + jy_2) = x_1(x_2 + jy_2) + jy_1(x_2 + jy_2)$$
$$z_1z_2 = x_1x_2 + jx_1y_2 + jy_1x_2 + j^2y_1y_2$$
$$z_1z_2 = (x_1x_2 - y_1y_2) + j(x_1y_2 + y_1x_2)$$

You are not expected to memorize these results, but rather, when faced with making a hand calculation, to carry out this multiplication process in an expeditious fashion. It is really no different from carrying out the multiplication of real factors like $(x + a)(y + b)$ except that with complex numbers we like to finish things off by regrouping the results into their real and imaginary parts.

The multiplication of two complex numbers is particularly easy if they are in polar form. Their magnitudes are multiplied and their angles are added:

$$\text{If } z_1 = c_1\angle\theta_1 \text{ and } z_2 = c_2\angle\theta_2, \text{ then } z_1z_2 = c_1e^{j\theta_1}c_2e^{j\theta_2} = c_1c_2e^{j(\theta_1+\theta_2)}$$

Raising a complex number to a power may be regarded as a special case of repeated multiplication. Powers are most easily found in polar form, since

$$z^n = (ce^{j\theta})^n = c^ne^{jn\theta} = c^n\angle n\theta$$

1.4.3 Conjugation

The conjugate of a complex number z, denoted with the symbol z^*, is defined as follows:

$$\text{If } z = a + jb, \text{ then } z^* = a - jb; \text{ or if } z = ce^{j\theta}, \text{ then } z^* = ce^{-j\theta}$$

Notice that the conjugate is formed by simply replacing j by $-j$ in these basic complex-number forms. This continues to be true when the numbers are not fully simplified, and may be generalized in the following theorem.

The Conjugation Theorem

If z results from evaluating an arithmetic expression involving the addition (subtraction), multiplication, and/or division of complex numbers, z^* may be formed simply by replacing every j with a $-j$ in that expression.

Notice that if $z = a + jb$, then adding or subtracting its conjugate will separate out the real and imaginary parts:

$$z + z^* = 2a = 2\text{Real}(z) \qquad \text{(Read as "twice the real part of } z\text{")}$$

and

$$z - z^* = 2jb = 2j\text{Imag}(z)$$

Multiplying a number by its conjugate separates out the magnitude information:

$$zz^* = (a + jb)(a - jb) = a^2 + b^2 \qquad \text{or} \qquad zz^* = (ce^{j\theta})(ce^{-j\theta}) = c^2$$

1.4.4 Division

Division of rectangular-form numbers is converted to a multiplication problem by *rationalizing*. This involves multiplying the numerator and denominator by the conjugate of the denominator, which converts the denominator to a real number and moves all of the angle information to the numerator. Here again, it is the process that should be remembered, not the results.

$$\text{If } z_1 = x_1 + jy_1 \text{ and } z_2 = x_2 + jy_2, \text{ then}$$

$$\frac{z_1}{z_2} = \frac{x_1 + jy_1}{x_2 + jy_2} = \frac{(x_1 + jy_1)(x_2 - jy_2)}{(x_2 + jy_2)(x_2 - jy_2)} = \frac{x_1x_2 + y_1y_2 + j(y_1x_2 - x_1y_2)}{x_2^2 + y_2^2}$$

Division is easiest if the numbers are initially in polar form. The procedure is evident immediately from the rules for exponentials (see Table 1.1, page 7):

$$\frac{z_1}{z_2} = \frac{c_1\angle\theta_1}{c_2\angle\theta_2} = \frac{c_1e^{j\theta_1}}{c_2e^{j\theta_2}} = \frac{c_1}{c_2}e^{j(\theta_1 - \theta_2)}$$

Dividing a complex number by its conjugate allows us to sort out the number's angle information, since

$$\text{If } z_2 = z_1^*, \text{ then } \frac{z_1}{z_1^*} = e^{j2\theta_1} = 1\angle 2\theta_1$$

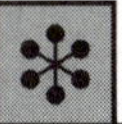

EXAMPLE 1.3B

Find

$$\frac{z_1z_2}{z_2^*} + z_1^*$$

$$\text{if } z_1 = 1 - j3 \text{ and } z_2 = 2 - j1.$$

Solution

Since the final operation will be the addition of z_1^*, we will retain rectangular form throughout the calculations.

$$\frac{(1-j3)(2-j1)}{2+j1}+(1+j3)=\underbrace{\frac{(1-j3)(2-j1)^2}{5}}_{\text{rationalizing}}+(1+j3)$$

$$=\frac{(1-j3)(3-j4)}{5}+(1+j3)=-\frac{9}{5}-j\frac{13}{5}+\left(\frac{5}{5}+j\frac{15}{5}\right)$$

$$=-\frac{4}{5}+j\frac{2}{5}$$

EXAMPLE 1.3C

Find the polar form of

$$\frac{(z_1+z_2^2)}{z_2{}^*}$$

if $z_1 = 1 - j3$ and $z_2 = 2 - j1$.

Solution

In this case there is a variety of ways to proceed. Since z_2 is only squared, we will retain rectangular form in the numerator until it has been fully simplified. (If z_2 had been raised to a higher power, we would be tempted first to convert it to polar form, then to raise it to the required power, and finally to convert it back to rectangular form for the addition to z_1.)

$$\frac{z_1+z_2^2}{z_2{}^*}=\frac{(1-j3)+(2-j1)^2}{\sqrt{5}\angle 0.4636}=\frac{(1-j3)+(3-j4)}{\sqrt{5}\angle 0.4636}=\frac{4-j7}{2.236\angle 0.4636}$$

$$=\frac{8.062\angle -1.0517}{2.236\angle 0.4636}=3.606\angle\underbrace{-1.5153}_{-86.82^\circ}$$

Each of the numerical examples of this section might have been accomplished on your calculator without worrying about the procedures to be used. However, these procedures are also important in manipulating complex variables that have not been assigned specific values.

1.5 FUNCTIONS OF A COMPLEX VARIABLE

Throughout this text we will encounter functions consisting of polynomials of a complex variable. The theory of functions of complex variables can be a rather challenging mathematical investigation, but our concerns are fairly basic. After completing this section you will be able to:

- Describe properties of a function of a complex variable.
- Properly interpret questions related to complex functions.
- Identify a function's poles and zeros.
- Represent the function with a pole-zero diagram.

Real functions have, at most, a sign and an amplitude. Questions related to the properties of such functions are usually unambiguous as a result. We can seek to find where the real function has a particular value, or look for its maximum or minimum values. A function of a complex variable, however, is usually itself complex, so it is important to exercise due care in asking questions about it. A complex function, just like a complex number, has a real part, an imaginary part, a magnitude, and an angle. Our questions must clearly reflect these properties. Asking, for instance, if a function F meets the condition that $F = 1$ is really asking two questions: Is there a value of the independent variable for which the real part of F equals $+1$ while the imaginary part is simultaneously zero? It is definitely not the same as asking if $|F| = 1$ or even if $\text{Real}(F) = 1$. We must ask the question correctly to be able to answer it correctly.

EXAMPLE 1.4

If $F(z) = 6/z$, where $z = x + jy$, under what conditions, if any, does Real $F = 3$?

Solution

The first step is to decide whether it will be easier to address the issue if F is expressed in rectangular or polar form. In this case rectangular form is appropriate.

$$F = \frac{6}{x + jy} = \frac{6(x - jy)}{x^2 + y^2} = \frac{6x}{x^2 + y^2} - j\frac{6y}{x^2 + y^2}$$

Separating out the real part, we require that

$$\text{Real } F = \frac{6x}{x^2 + y^2} = 3 \qquad \therefore x^2 + y^2 = 2x$$

Completing the square, you might recognize the result as the equation of a circle centered at $z = 1 + j0$ and having unit radius:

$$(x - 1)^2 + y^2 = 1$$

We also have a problem displaying information about complex functions. If a variable, z, is complex, it can take on any value in the two-dimensional complex-number plane. If $F(z)$ is also complex, two additional dimensions would be required to characterize it. Since four-dimensional graphs are impossible to visualize, we would probably break the function into its constituent parts, real and imaginary or magnitude and phase, and plot each separately over the z plane. Figure 1.2 shows one such graph for the real part of the function $F = \sin(z)$. It is not all that easy getting useful numerical information from even this three-dimensional plot.

It may have surprised you that the sine function can take on complex values, but that is only because you have never allowed it to have a complex argument before. The sine function may be defined by Equation 1.5, where θ is allowed to be complex. Any function you have ever encountered is likely to produce complex values if allowed to take on a complex argument. Just because it can be done does not mean we will find

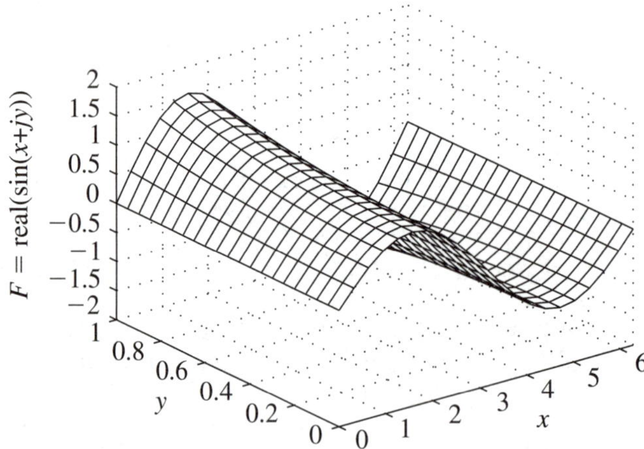

Figure 1.2 A three-dimensional plot of the real part of the function $F = \sin(z)$, where $z = x + jy$ using the MATLAB **mesh** command.

it useful. When we evaluate the sine function, the only values we will ask for are those along the path consisting of the real axis, just as we have always done.

A similar situation will exist for all of our complex functions. While they may be defined in terms of a complex variable, our interest in them will be focused primarily on their behavior along a specific path in the complex plane. Once a path has been specified, we can plot the magnitude and phase of the function in standard two-dimensional plots, with which we are more familiar.

On those occasions when we do investigate a function over the entire complex-number plane, a *pole-zero plot* will provide the kind of information we will be looking for. The pole-zero plot of a function identifies the extreme points in its magnitude. Poles are locations (values of z) where the function is infinite, while zeros are locations where the function is zero. Zeros are indicated with an "o" and poles with an "x." If a function contains a repeated root, such as would result from a term like $(z - a)^n$, the *multiplicity* of the root is indicated by placing the value of n next to it on the pole-zero plot. The pole-zero diagram (Figure 1.3) is like a topographical map showing points of elevation extremes, but without the added detail provided by intermediate contour lines of constant elevation.

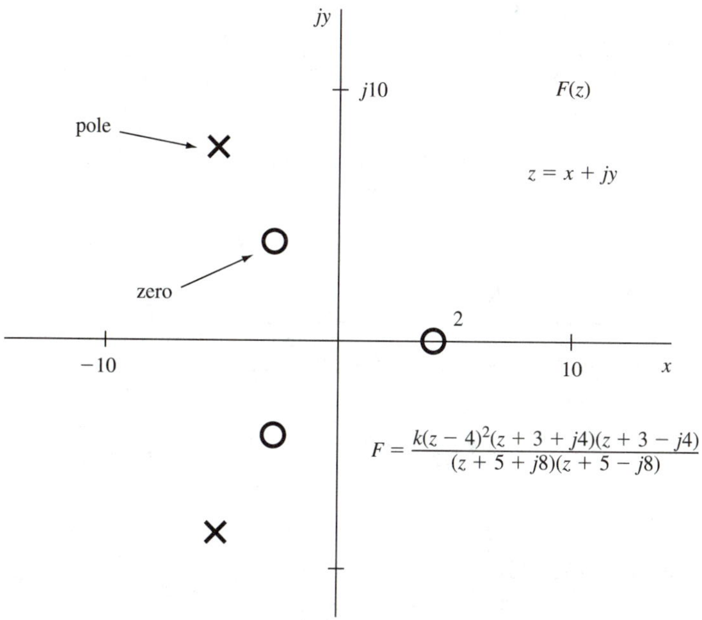

Figure 1.3 A *pole-zero diagram* indicates locations in the z plane where the function has a magnitude of infinity or zero, respectively. For the functions encountered in this text, complex poles or zeros will always occur as conjugate pairs. In this figure, a double zero is located at $z = 4 + j0$. The original function can be recreated from its pole-zero diagram to within a multiplicative constant.

Most of the complex functions we will encounter will consist of a ratio of polynomials. For such functions, the finite zeros are simply the zeros of the numerator polynomial, and the finite poles are the zeros of the denominator polynomial. Functions having this form may be represented by Equation 1.6, where N is the degree of the denominator polynomial whose roots represent poles, p_n, of the function. M is the degree of the numerator polynomial whose roots, z_m, are the zeros of the function. If the multiplier, k, is added to the pole-zero plot, the plot describes F as completely as does the equation.

$$F(z) = \frac{k\prod_{m=1}^{M}(z - z_m)}{\prod_{n=1}^{N}(z - p_n)} = \frac{k(z - z_1)(z - z_2)\cdots(z - z_M)}{(z - p_1)(z - p_2)\cdots(z - p_N)} \tag{1.6}$$

If the number of finite poles and zeros are not equal ($M \neq N$), then the function will have poles or zeros at infinity. At very large values of z, only the highest powers of z need be considered. Then

$$\lim_{|z|\to\infty} F(z) = \lim_{|z|\to\infty} \frac{k\prod_{m=1}^{M}(z - z_m)}{\prod_{n=1}^{N}(z - p_n)} = \lim_{|z|\to\infty} kz^{M-N} = \begin{cases} \infty & M > N \\ 0 & N > M \end{cases}$$

Every function has an equal number of poles and zeros if those at infinity are included.

EXAMPLE 1.5

Provide a pole-zero plot for the following function:

$$F(z) = \frac{2(z^2 + 4)}{(z^2 - z)(z + 2)}$$

Solution

Placing F in factored form gives

$$F(z) = \frac{2(z + j2)(z - j2)}{z(z - 1)(z + 2)}$$

which shows that the numerator of F goes to zero if $z = \pm j2$. Similarly, the denominator of F goes to zero if $z = 0, +1$, or -2. A zero also exists at $|z| = \infty$, since the total number of zeros must equal the number of poles. See Figure 1.4.

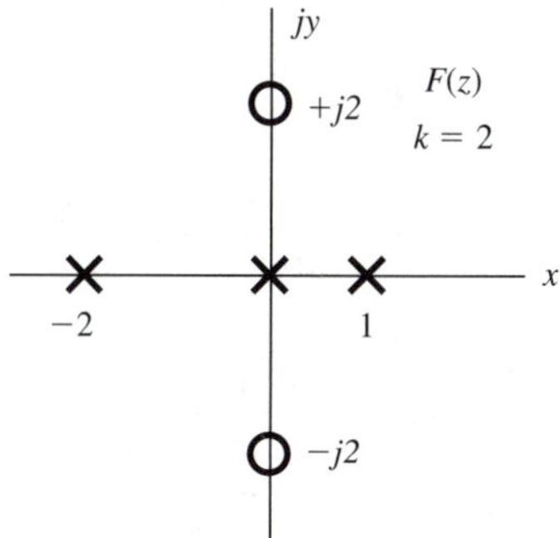

Figure 1.4 The pole-zero diagram for the function of Example 1.5.

EXAMPLE 1.6

Find the magnitude of

$$F(z) = \frac{2(z + 4)}{(z - 1)}$$

along the path $z = x + j2$.

Solution

The pole-zero diagram for this function (Figure 1.5) shows that the indicated path will take us close to a zero and a pole as we move from $z = -\infty + j2$ to $+\infty + j2$. It seems reasonable that the function will dip as we pass the zero, and peak as we pass the pole. The zero affects the landscape in the vicinity of the pole, and vice versa, so minimum and maximum points do not necessarily occur at points on the evaluation path closest to the poles or zeros.

To check this, we evaluate F along the specified path:

$$F(z)\big|_{z=x+j2} = \left.\frac{2(z + 4)}{(z - 1)}\right|_{z=x+j2} = 2\frac{(x + j2 + 4)}{(x + j2 - 1)} = 2\frac{[(x + 4) + j2]}{[(x - 1) + j2]}$$

$$F(x + j2) = 2\frac{[(x + 4) + j2][(x - 1) - j2]}{(x - 1)^2 + 4} = \frac{2x^2 + 6x - j20}{x^2 - 2x + 5}$$

The result is clearly a complex function of the real variable x. Evaluating the magnitude of the function over a range of x where the pole and zero exercise their greatest influence confirms our predictions and augments them with specific numerical

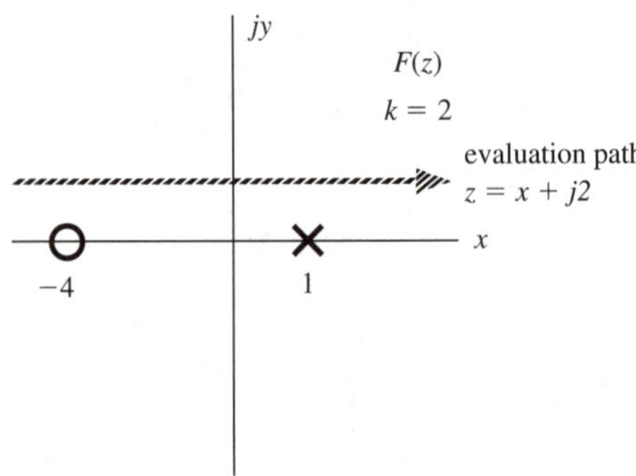

Figure 1.5 Pole-zero diagram and evaluation path for Example 1.6.

information (Figure 1.6). Notice that a relatively simple function of the complex variable z has become a quite messy function of the real variable x. Having a calculator or software package that routinely handles complex variables would allow the simpler complex function to be evaluated along a specified path without complicating the problem with a lot of unnecessary algebra.

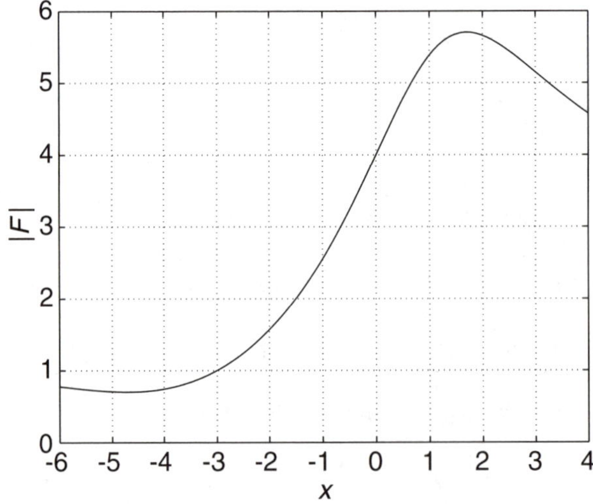

Figure 1.6 The magnitude of the function in Example 1.6 is shown evaluated along the path $z = x + j2$. Notice that the maximum and minimum points do not correspond to the points of closest approach to the pole and the zero respectively. The equation of this curve is $|F(x + j2)| = \dfrac{\sqrt{4x^2(x+3)^2 + 400}}{x^2 - 2x + 5}$

1.6 INDETERMINATE VALUES

Division by zero is undefined in mathematics and can also create havoc with some computer operating systems. However, important physical events can be associated with this mathematical condition, and measurable properties can be calculated by resolving indeterminate values. After completing this section you will be able to:

- Interpret and resolve indeterminate values.
- Recognize common indeterminate forms.
- Use L'Hopital's rule to define a value for an indeterminate form.
- Predict conditions under which special events may occur.

If a class of students is asked the value of the function $F(m) = (\sin am)/m$ at $m = 0$, about half of them would first notice the numerator and declare the result to be zero. The rest would first notice the denominator, and declare the result to be *infinite* or *undefined.* This very substantial discrepancy occurs because the value of the function is, in fact, *indeterminate* at $m = 0$. Three kinds of situations arise.

Situation 1. If the numerator of the function is not zero, we say the function becomes *infinite,* denoted by the symbol ∞, as the divide-by-zero condition is approached. We interpret this symbol as indicating that the value of the function increases without bound as the divide-by-zero condition is approached, but we are forbidden to try and further quantify or evaluate it. We have called these points poles in our complex functions. They are also called *singular points,* or *singularities.* In investigating properties of complex functions, we are permitted to get as close to a pole as we wish, as long as we do not actually reach it.

Situation 2. If a function takes the form 0/0 at a point (∞/∞, $0 \cdot \infty$, and $\infty - \infty$ are a few related forms), we say that the value of the function is *indeterminate* at that point. We will interpret this as meaning that the function is hiding its true value from us at this point and that further investigation is required to assign it the appropriate value.

It is not uncommon to obtain different functions as a result of solving simultaneous equations, depending on the procedure used. One method might lead to a result $f(x)$, for instance, whereas using a rote procedure involving determinants might lead to the result $g(x) = (x - 1)f(x)/(x - 1)$. In this case f and g are identical everywhere, except that g is indeterminate at $x = 1$. This situation arises if opportunities to cancel common factors are not noticed during a derivation. Clearly, the physical property being represented by these functions is correctly given by f, and the indeterminacy of g at a particular point should not be allowed to cloud the issue.

A function like

$$F(m) = \int_0^a (\cos mx)\, dx = \frac{\sin ma}{m}$$

becomes indeterminate at $m = 0$ after the integration, but does not if the special case of $m = 0$ is entered before the integral is evaluated:

$$F(0) = \int_0^a (\cos 0x)\, dx = \int_0^a 1\, dx = a$$

For the most part, indeterminate forms result from a failure to simplify expressions when the opportunity presents itself or from trying to generalize a complicated result in a single formula. In these circumstances we assume that the correct value of the function at the indeterminacy can be found by taking the limit of the function as we approach arbitrarily close to the indeterminacy. A procedure routinely used for resolving this value is *L' Hopital's rule.*

L'Hopital's Rule

If

$$f(x) = \frac{n(x)}{d(x)}$$

becomes indeterminate at $x = a$, then

$$f(a) = \lim_{x \to a} f(x) = \frac{n'(a)}{d'(a)} \tag{1.7}$$

where $n'(a)$ and $d'(a)$ are the respective derivatives of $n(x)$ and $d(x)$ evaluated at $x = a$.

If the indeterminacy is still not resolved, the second derivatives of numerator and denominator are taken, and so on until it is resolved. Remember, L' Hopital's rule requires you to take the derivative of the numerator and denominator *separately,* which is much easier than taking the derivative of the whole function!

EXAMPLE 1.7

Evaluate $F(z)$ at $z = 1, 2,$ and 3, where

$$F(z) = 3\frac{z^2 - 3z + 2}{z^2 - 5z + 6}$$

Solution

Substituting directly we find

$$F(1) = \frac{0}{2} = 0, \qquad F(2) = \frac{0}{0} = ?, \qquad \text{and} \qquad F(3) = \frac{6}{0} = \infty$$

Neither $F(1)$ nor $F(3)$ is indeterminate, but note that both the numerator and the denominator values must be found before that fact can be ensured.

Factoring the polynomials will resolve the indeterminacy at $z = 2$:

$$F(z) = 3\frac{z^2 - 3z + 2}{z^2 - 5z + 6} = 3\frac{(z-1)(z-2)}{(z-2)(z-3)} = 3\frac{z-1}{z-3} \qquad \therefore F(2) = -3$$

The resulting simplified equation for F applies at all values of z and would be used in place of the original expression.

Alternatively, we could have applied L'Hopital's rule to resolve the issue at the single indeterminate point:

$$\lim_{z \to 2} F(z) = 3\frac{2z - 3}{2z - 5}\bigg|_{z=2} = -3$$

Situation 3. Sometimes our only interest in a function is to determine the condition under which it becomes indeterminate. Suppose, for example, a function is given by $F(z) = 0/D(z)$. Since its numerator is identically zero, the function is zero . . . unless the denominator also happens to be zero. Then there is at least a possibility that F will not be zero at locations where it normally has poles. In these cases it will be sufficient for us to know that something may be going on, even if we cannot immediately have more detailed information about it.

This final interpretation of an indeterminate form probably seems bazaar at this point, but it eventually will prove to have great physical significance.

1.7 INTRODUCTION TO MATLAB

MATLAB is a mathematical software package that has gained considerable favor in the engineering community. By using its function calls (subroutines) related to signals and systems, we can easily and quickly investigate many of the concepts introduced in this text. In this section you will:

- Use MATLAB for arithmetic operations.
- Observe MATLAB's response to typical numerical calculations.

- Observe MATLAB's handling of complex numbers.
- Create arrays and make calculations with arrays.
- Use some plotting functions to create and label a graph.
- Identify your workplace variables and array dimensions.
- Use the MATLAB help facility.

To complete this section, you need to be seated at a computer with the MATLAB Command Window active. Type the commands indicated after the prompt and follow with a carriage return. The standard MATLAB prompt is » but some past student versions have also used >, and EDU>. If you are using a networked computer that restricts writing to its hard drive, you will need to have a floppy disk handy should you want to save any of your results.

You cannot modify lines in the Command Window after the carriage return has been executed. If you make an error in entering a command, just retype the command correctly. Usually Control-C will abort an apparent infinite loop.

MATLAB EXAMPLES

We will start with some ordinary arithmetic operations. The symbols +, −, *, /, and ^ indicate the matrix operations of addition, subtraction, multiplication, division, and exponentiation, respectively. Operations are performed left to right. The highest-priority operation is exponentiation, then multiplication/division, and finally addition/subtraction. Use parentheses to group operations to produce the result intended.

```
>2*3/4              % calculations do not need an "=" if you do not want to
                      save the result
>y=2+4/3            % saves the result as y
>Y=(2+4)/3          % names are case sensitive, both y and Y now have values
>y=2^3/4            % 2 cubed and divided by 4 (original y was overwritten)
>y=2^(3/4)          % 2 raised to the 3/4 power, use parentheses to make
                      the interpretation clear
>y1=2*(1+j)/(1-j)   % complex numbers are OK too
>b=2j               % MATLAB recognizes this as an imaginary number
>c=j2               % but not this
>c=j*2              % but you can always think of j (or i) as a number by
                      itself
>y2=2*exp(j*pi/4)   % the usual standard math functions are available, a
                      few are:
```

sin cos tan asin acos atan exp log log10 sqrt

```
>y3=tan(pi*(1+j)/2)  % angles are in radians, complex angles allowed but
                       have no meaning to us
>y4=(1+j)^2          % complex numbers are presented in rectangular form.
```

real imag conj abs angle

These and all other MATLAB functions consist of a name and one or more arguments in parentheses, and separated by commas.

```
>y5=real(y4)       % select out the real part of a previous calculation
>mag=abs(y4)       % find the magnitude of the complex number y4
>phase=angle(y4)   % get the angle of y4
>y_star=conj(y4)   % take the conjugate of a complex number
>y=1/0             % infinity is recognized
>y=0/0             % indeterminate results are marked with NaN: Not a
                     (known) Number
```

Variable names must start with a letter. They can be longer than you would want to type. Use good computer programming practice for variable names. MATLAB will let you know if it does not like your choice.

In MATLAB, variables are held in arrays. Our applications will involve mostly single-row or single-column arrays, which are called *vectors*. Arithmetic operations can be performed on two or more vectors, provided the vectors have the same number of rows or columns. Vector arithmetic is done on an element-by-element basis and is identified by a dot (period) preceding the operation symbol. Exception: the dot is omitted in an addition or subtraction of arrays.

```
>x=[0 1 2 ;3 4 5]      % creates a 2-row, 3-column array; extra spacing is
                         optional
>y = [1 2 3; 1 2 3]    % creates another one
>z1=x.*y               % each element of x is multiplied by the
                         corresponding element of y
>z2=x./y               % each element of x is divided by the corresponding
                         element of y
>z3=y.\x               % an alternate way to express z2
>z4=x.^y               % each element of x is raised to the power of the
                         corresponding element in y
>z5=x.+y               % the dot is inappropriate in addition or subtraction
>2.^y                  % operations by constants apply to all elements of y
>y.^2
>y+2
```

[ ]   linspace   logspace   colon notation   help FunctionName

One of the most common applications we will have for a row vector is to represent an independent variable in an equation. The equation is used to calculate a second array containing the values of the dependent variable for each point in the first array. Then we will plot the resulting curves. Suppose we want an independent variable running from -20 to $+30$. Any of the following commands can be used to create it:

```
>x1=[-20 -10 0 10 20 30]   % manually creates a six-element row vector
>x2=linspace(-20,30, 6)    % uses the linspace function: x2 is identical to
                             x1
```

```
>x3=linspace(-20,30)      % gives 100 points (default value) over the
                            specified range
>x4=-20:0.2:30            % colon notation: starting
                            value:increment:ending value
>x5=(-100:150)/5          % x5=x4 The default increment is unity in colon
                            notation
```

The **logspace** function would be inappropriate for this independent variable specification, but demonstrates the help facility. If you know a MATLAB function name but are not sure what it does or what its argument options are, you can be reminded using the help facility.

```
>help logspace   % help function_name
```

Up to now we have wanted to see the calculation results at each step. As we start dealing with large arrays and more involved command sequences, we will usually want to suppress having the results printed to the screen until we specifically ask to see them. *Concluding a command with a semicolon causes the calculation to be made but not printed to the screen.*

Suppose we want to calculate the formula $y = x^2/4$ for the set of points in x3. The dependent variable statement is

```
>y1 =(x3.*x3)/4;   % the parentheses are only for clarity in expressions
                     y1, y2, and y3
>y2=(x3.^2)/4;     % y1, y2, and y3 are all the same result
>y3=(x3.^2)./4 ;   % no dot is needed for division by the scalar, but it
                     would be accepted
```

Let's reward ourselves by plotting the result.

```
>plot(x3,y1)   % plots y1 versus x3
```

plot grid title xlabel ylabel gtext axis hold

```
>grid                  % add an optional grid and proper documentation for
                         your graph
>title('anything you want to say at the top of the graph')
>xlabel('x axis variable and units')
>ylabel('y axis variable and units')
>gtext('puts this text wherever you click on the graph')
>axis([0 20 0 400])    % sets graph axes: axis([xstart xend ystart yend])
>hold                  % holds the current graph while more curves are added
                         to it
>plot(x3,y2,'r+')      % same plot using just data points marked by a red +
>hold                  % the hold command cycles the hold action on and off
```

The plot command interpolates between actual calculated points in order to produce a smooth curve. Sometimes that interpolation is not wanted. To plot data points only, specify a data point marker without any line type. Try some of the line/data point options suggested by the help facility.

```
>help plot
```

Commands are available to remind you what variable names you have used, the dimensions of your arrays, or both, and to allow you to clear all or some of your variables from the command window.

who size whos clear

```
>who                 % identifies existing variable names
>size(variable_name) % identifies the row and column dimensions of the
                       variable_name array
>whos                % does both of the above, including bytes allocated
>clear x y Y         % clears variables, x, y, and Y
                     % clear without a qualifier clears all variables;
                       watch out for this!
```

diary load workplace save workplace

In future MATLAB sessions, or when doing MATLAB problem assignments, you may wish to keep a record of your Command Window activities. If you do not understand MATLAB's response to your commands, obtaining a printout of the Command Window will allow you to seek an explanation from your instructor or from other experienced MATLAB users. Your instructor may also require such a record for graded homework or test problems. The **diary** command will duplicate the Command Window activities in a text file. You could also just cut and paste the Command Window to a text editor.

```
>diary filename  % on a networked Microsoft Windows system the filename
                   usually must include an A: prefix.
```

If you have filled your Command Window with unnecessary array elements because you forgot the semicolon at the end of an instruction, your instructor would probably appreciate your editing those sections from your text file before obtaining hardcopy.

If you must interrupt a MATLAB session, and want to be able to pick up where you left off you can select **save workplace** under the **file** menu and be prompted for a filename and destination. The workplace variable names and values are stored in a digital format *.mat* file. You cannot **open** this file, but you can **load** it while in MATLAB. If you want to try this out, do the following:

```
>whos                       % shows the current workplace variables
save workplace A: saver.mat % (assumes a Microsoft Windows operating
                              system)
>clear
>whos                       % verify that the workplace is now empty
>load A: saver.mat          % or select load workplace from the file menu
>whos                       % verify that the workplace variables have
                              been restored
```

EXAMPLE 1.8

Use MATLAB to plot the phase of

$$F(z) = \frac{2(z + 4)}{(z - 1)}$$

along the path $z = x + j2$

Solution

This is the same function evaluated in Example 1.6. If we select the same range for x as in that example, the programming is identical no matter what property of F is desired. See Figure 1.7.

```
> x=linspace(-6,4);          % create a 100-point independent variable
> z=x+2j;                    % now the range of z has been defined
> F=2*(z+4)./(z-1);          % calculate the function for each value of z
> plot(x,180*angle(F)/pi)    % plot phase in degrees
> grid
> xlabel('x')
> ylabel('phase in degrees')
```

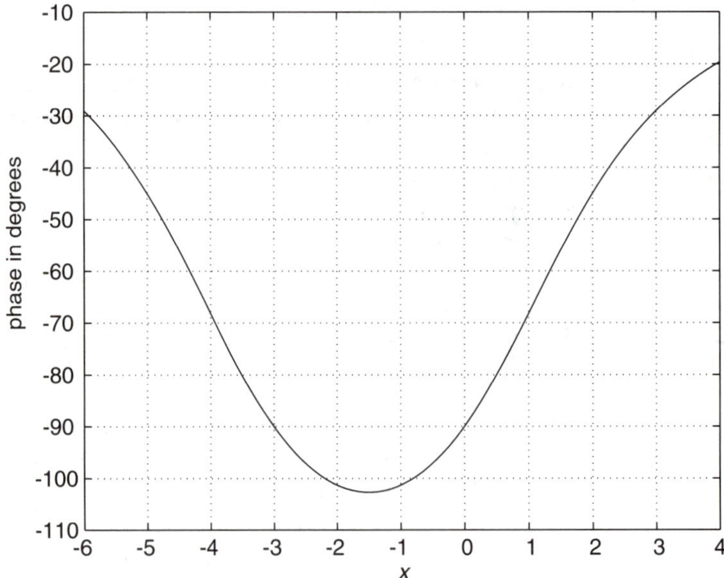

Figure 1.7 The phase of the function of Examples 1.6 and 1.8 evaluated along the path $z = x + j2$.

In this example, F was just specified in a formula, as it might have been in any general-purpose programming language. A much simpler way of describing this type of function is available with MATLAB; it will be introduced in the next chapter.

SUMMARY

Complex numbers are like aliens from the planet of Mathematics. They are foreign to our everyday experience, but they have many beneficial characteristics once we become friendly with them. This chapter has attempted to review how to make arithmetic calculations with complex numbers and how to convert between their polar and rectangular forms, and it has introduced a software package that routinely works with them. We have also encountered Euler's identity, which establishes the mathematically correct form for our shorthand polar-number notation. Euler's identity also shows us how a sinusoid can be expressed in terms of a more elementary function, namely, the exponential.

Complex variables lead to complex functions. Unlike real functions, which have only a magnitude and a sign, complex functions have a magnitude and an angle, or a real part and an imaginary part. It becomes important to be clear which of these parts of the function are the desired features. We have considered ways to present information about a complex function graphically, and have settled on a simple method called a pole-zero diagram.

Finally, we have reviewed the significance of a division by zero in functions, and distinguished between infinite values and indeterminate values. We will avoid evaluating a function at its poles. We will use the methods of L'Hopital to assign functions values at points of indeterminacy.

PROBLEMS

Section 1.2

1. Convert the following numbers to polar form with angles in degrees.
   a. $-2 + j1$ b. $2 - j4$ c. $1 - j3$
2. Convert the following numbers to polar form with angles in radians.
   a. $-1 + j3$ b. $3 + j2$ c. $3 - j4$
3. Convert the following numbers to rectangular form.
   a. $\pi\underline{/2\text{ rad}}$ b. $2\underline{/-4\text{ rad}}$ c. $5\underline{/18\text{ rad}}$
4. Convert the following numbers to rectangular form.
   a. $4\underline{/27^\circ}$ b. $2\underline{/-120^\circ}$ c. $5\underline{/180^\circ}$

Section 1.3

5. Determine the magnitude and angle of the following complex numbers.
   a. $3e^{j\pi}$ b. $-2e^{j2}$ c. $0.25e^{-j5}$
6. Express these numbers in rectangular form.
   a. $e^{-j\pi/4}$ b. $3e^{j4}$ c. $2\pi e^{j1}$
7. Express the given sum as a sum of sine or cosine functions.
   a. $-3e^{-j2} + 2e^{-j} + 2e^{j} - 3e^{j2}$ b. $4e^{-j3} - e^{-j\pi/2} + e^{j\pi/2} - 4e^{j3}$

Section 1.4

8. Evaluate the expressions and leave the result in rectangular form for $z_1 = 1 + j2$, $z_2 = 2 - j1$, and $z_3 = -1 + j$.
   a. $z_1 + \dfrac{z_2}{z_3}$ b. $\dfrac{z_1 z_2{}^*}{z_3}$ c. $\dfrac{z_1 - z_3{}^*}{z_2}$
9. Evaluate the expressions and leave the result in rectangular form for $z_1 = 2 + j3$, $z_2 = 1 - j3$, and $z_3 = 0 + j2$.
   a. $z_1 z_2 z_3$ b. $\dfrac{z_1 + z_2}{z_3{}^*}$ c. $(z_1/z_2)^3$
10. Evaluate the expressions of Problem 9 and put the result in polar form for $z_1 = \pi\underline{/35^\circ}$, $z_2 = 2\underline{/-140^\circ}$, and $z_3 = 1\underline{/-50^\circ}$.
11. Evaluate the indicated expressions and give the answer in rectangular form for $z_1 = 0.5\angle 0.24$, $z_2 = 4\angle 1.20$, and $z_3 = 3\angle -4.0$.
    a. $\dfrac{z_1 z_2}{z_2 + z_3}$ b. $\left(\dfrac{z_1}{z_2 + 1}\right)^2$ c. $z_1 + z_2{}^* + z_3$

Section 1.5

12. For the function $F(z) = z^2$, where $z = x + jy$:
    a. Determine the locations where Real $F = 1$.
    b. Determine the locations where $F = 1$.
13. For the function $F(z) = 1/(z - 1)$, where $z = x + jy$:
    a. Determine the conditions under which $|F| = 2$.
    b. Determine the conditions under which $F = j2$.
14. a. Sketch a fully labeled pole-zero diagram for the following complex function:

$$F(z) = \frac{2}{z + 2}$$

    b. Obtain expressions for the real and imaginary parts of $F(z)$ along the path $z = -1 + jy$, and sketch each for $0 < y < \infty$. Identify actual values at $y = 1$.

15. a. Sketch a fully labeled pole-zero diagram for the following complex function:

$$F(z) = \frac{3z}{z + 2}$$

b. Obtain expressions for the magnitude and angle of $F(z)$ along the path $z = 0 + jy$, and sketch each for $0 < y < \infty$. Identify actual values at $y = 2$.

Section 1.6

16. a. Evaluate $F(x)$ at $x = -1$ and $+1$:

$$F(x) = \frac{x^2 - 1}{x - 1}$$

b. Evaluate $G(x)$ at $x = 0$:

$$G(x) = \frac{1 - \cos 2x}{2x}$$

17. a. Evaluate $F(z)$ at $z = 0$, 2, and 10:

$$F(z) = \frac{z^3 - 6z^2 - 40z}{z^2 - 12z + 20}$$

b. Evaluate $G(z)$ at $z = 2$:

$$G(z) = \frac{\sin[2(z - 2)/5]}{(z - 2)}$$

18. In numerical problems a calculator can often resolve indeterminate values. Simply calculate the function at a point extremely close to the value causing the indeterminate result. For example, if a function is indeterminate at $z = 1$, calculate the function at $z = 1 \pm 10^{-4}$. Use this technique to resolve any indeterminate values in Problem 17.

Section 1.7

19. Show two ways of producing a MATLAB row vector that starts at 0 and ends at π, in 13 equally spaced points.

20. Given the row vectors x = [3 6 9] and y = [5 3 0], what are the results of the following MATLAB operations.
a. x+y b. x.*y c. x.^y d. x./y

21. Describe the result of the following MATLAB command sequence:

```
>t=linspace(0,pi/2,123);
>y=sin(2*t);
>plot(t,y,'r')
```

Advanced Problems

22. a. Sketch a fully labeled pole-zero diagram for the following complex function:

$$F(z) = \frac{z^2 + 1}{z^2 - 1}$$

b. Based on your pole-zero diagram, sketch roughly how you would expect the magnitude of F to vary if the function were evaluated along the path $z = 1.1\underline{/\theta}$. Don't try to specify any actual magnitudes, just where the peaks and valleys are expected for $-\pi \geq \theta \geq \pi$.

c. Use MATLAB to plot $|F|$ along the path indicated, and compare to your expectations.

d. Does the real part of F pass through zero over this path?

23. Use MATLAB to demonstrate that if $F = 6/z$, the real part of $F = 3$ everywhere on a circle of unit radius centered at $z = 1$ (Example 1.4, page 13).

2

CONTINUOUS-TIME SYSTEMS

OUTLINE

OBJECTIVES

1. Identify a linear differential equation.
2. Interpret variations of the exponential signal.
3. Establish an appropriate phasor transformation for each problem.
4. Transform between differential equations and transfer functions.
5. Determine aspects of the natural response of a linear system.
6. Use MATLAB for efficient polynomial operations.

INTRODUCTION

Any complete set of mathematical relationships between input and output variables constitutes a system. Continuous-time systems obey laws of physics expressed in the form of differential equations. Given the initial status of the system, these equations prescribe the subsequent behavior of the system at any instant in time.

The general solution of differential equations involves processes significantly different from those of algebraic equations. However, linear differential equations with constant coefficients reduce to algebraic equations for an exponential time signal of the form e^{st}, where s is a constant. Such a function repeats itself under differentiation: $d/dt(e^{st}) = se^{st}$. The interpretation of this signal for complex s leads to the definition of the phasor.

The solutions of linear differential equations consist of the superposition of a forced response and a natural response. The natural response represents the system's reaction to any disturbance; the forced response is the system's reaction to a specific applied input signal. The forced response is generally the desired feature, but an unacceptable natural response may make the system unusable.

2.1 SYSTEM EQUATIONS

Physical systems obey the laws of physics. The exact formulation of these laws involves calculus-based relationships. This section will demonstrate how these relationships result in differential equations and will discuss the problem of solving such equations. After reading this section you will be able to:

- Identify a linear differential equation.
- Write differential equations for simple mechanical and electrical systems.
- State the characteristics that define a linear system.
- Appreciate the problem of solving a differential equation.

Most systems obey a central law or laws and have system components whose variables are related to the central law. In electrical systems, the central laws are those of Kirchhoff:

The algebraic sum of the voltages around any closed path equals zero:

$$\sum v(t) = 0 \tag{2.1a}$$

The algebraic sum of the currents entering a junction equals zero:

$$\sum i(t) = 0 \tag{2.1b}$$

The term *algebraic sum* means that voltage rises must be distinguished from voltage drops by giving one a plus sign and the other a minus sign. Similarly, currents entering a junction must be distinguished by their sign from those leaving. The criterion governing which type of terms is given the plus sign is immaterial.

The passive devices used to construct a circuit are shown in Figure 2.1. A circuit results when some of these devices are connected together and attached to a voltage or current source.

$+ \; v(t) \; -$, $i(t)$, R	$+ \; v(t) \; -$, $i(t)$, L	$+ \; v(t) \; -$, $i(t)$, C
$v(t) = R\, i(t)$	$v(t) = L\dfrac{di(t)}{dt}$	$v(t) = \dfrac{1}{C}\int_{-\infty}^{t} i(t)dt$
$i(t) = \dfrac{v(t)}{R}$	$i(t) = \dfrac{1}{L}\int_{-\infty}^{t} v(t)dt$	$i(t) = C\dfrac{dv(t)}{dt}$

Figure 2.1 The integro-differential relationships between currents and voltages in linear passive circuit elements. The equations are positive for currents entering the positive voltage end of the circuit element.

EXAMPLE 2.1

Write the differential equation governing the behavior of the circuit shown in Figure 2.2.

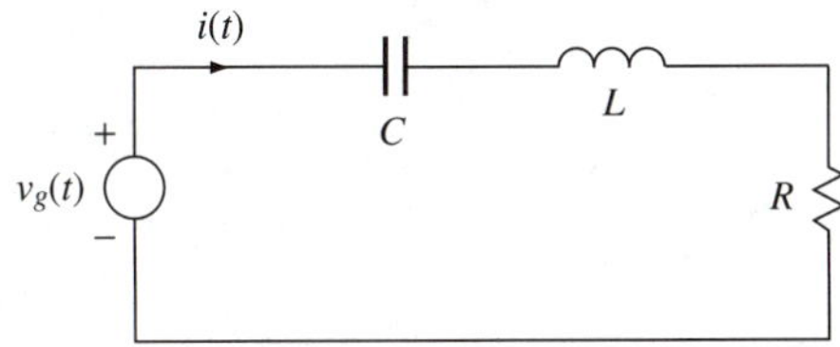

Figure 2.2

Solution

In this case, Kirchhoff's voltage law is needed (Eq. 2.1a):

$$v_g(t) = \frac{1}{C}\int i(t)\,dt + L\frac{di(t)}{dt} + i(t)R$$

This is know as an integro-differential equation, due to the presence of the integral term. If the equation is differentiated, a pure differential equation results:

$$L\frac{d^2i(t)}{dt^2} + R\frac{di(t)}{dt} + \frac{i(t)}{C} = \frac{dv_g(t)}{dt} \quad \textbf{(2.2)}$$

A mechanical system is governed by Newton's laws. In a purely rotational system, with J representing the system's moment of inertia, θ the angular displacement, and T the torques present, Newton's second law states that

$$J\frac{d^2\theta}{dt^2} = \sum T \quad \textbf{(2.3)}$$

where those torques that contribute to the motion must be algebraically distinguished from those opposing the motion. Figure 2.3 shows the most common elements. The differential equation of this system is summarized as

$$J\frac{d^2\theta(t)}{dt^2} + k_f\frac{d\theta(t)}{dt} + k_s\theta(t) = T_i(t) \quad \textbf{(2.4)}$$

While electrical and mechanical systems are important of themselves, systems often involve an interaction of the two. Suppose, for instance, that we wish to predict the response of the permanent magnet d-c motor shown in Figure 2.4 to an input voltage. Both Newton's and Kirchhoff's laws provide equations.

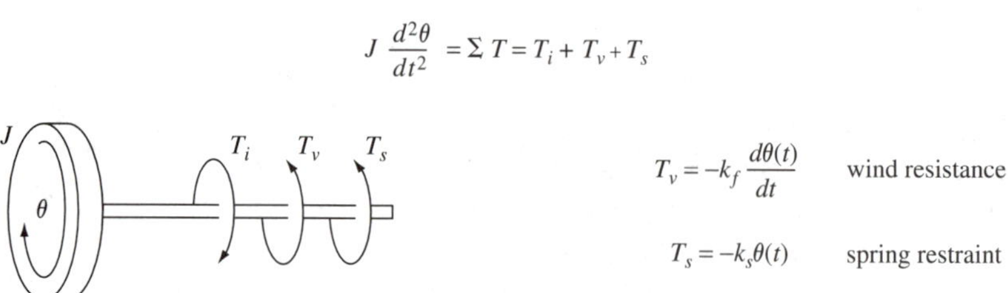

Figure 2.3 In a rotational system, the torque, T_i, causing the motion defines the positive direction of θ. Opposing the motion is a frictional force proportional to the velocity of rotation. In some systems, such as in D'Arsonval meter movements, the system may also be restrained by a spring, which produces a countertorque proportional to the angular displacement.

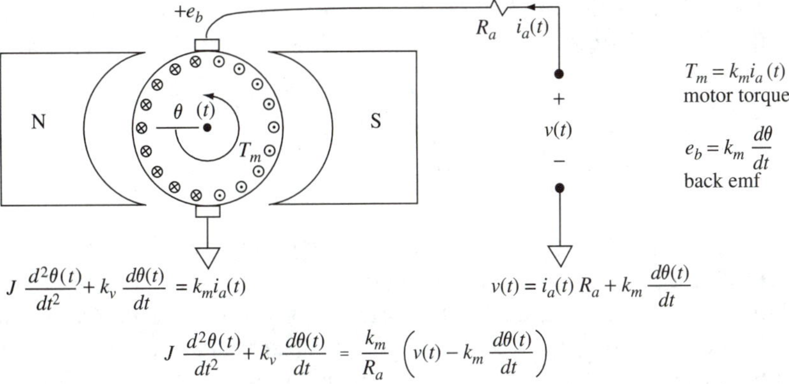

Figure 2.4 In a permanent magnet d-c motor, the action of the commutator and brushes combine to keep the armature-current spatial distribution constant (⊗ represents a current entering the paper, ⊙ represents a current coming up out of the paper). For the currents shown, the armature looks like a coil producing a north magnetic pole downward. Since like poles repel and unlike poles attract, the motor will rotate counterclockwise. The equations for the two systems are coupled by the *Blv-Bli* laws of physics, where *B* is the magnetic flux density, and *l* is the effective length of the conductor in the magnetic field. These values, as well as geometrical factors, have been lumped into the motor constant, k_m.

A current will flow in the armature (rotor) circuit of the d-c motor if a voltage is applied to it. The resulting current in the armature wires, which are in the field produced by the permanent magnets, produces a torque, $T_m = k_m i_a$, that causes the motor to begin turning. As these same wires move through the magnetic field, they also experience a counterelectromotive force opposing the buildup of current, $e_b = k_m(d\theta/dt)$. These two effects couple the mechanical and electrical equations. Rearranging them so that input and output variables are on opposite sides of the equation gives the following result:

$$J\frac{d^2\theta(t)}{dt^2} + \left[k_v + \frac{k_m^2}{R_a}\right]\frac{d\theta(t)}{dt} = \frac{k_m}{R_a}v(t) \tag{2.5a}$$

Using ω to represent the motor's angular velocity, where $\omega = d\theta/dt$, we could also write this differential equation as

$$J\frac{d\,\omega(t)}{dt} + \left[k_v + \frac{k_m^2}{R_a}\right]\omega(t) = \frac{k_m}{R_a}v(t) \tag{2.5b}$$

All of the systems examined so far have been described in terms of a linear differential equation with constant coefficients. The general form for such an equation is

$$\frac{d^M y(t)}{dt^M} + \cdots + b_1\frac{dy(t)}{dt} + b_0 y(t) = a_N\frac{d^N x(t)}{dt^N} + \cdots + a_1\frac{dx(t)}{dt} + a_0 x(t) \tag{2.6}$$

where $x(t)$ is the input signal or forcing function and $y(t)$ is the resulting output. One significance of linearity is that an increase in the input results only in a corresponding increase in the output. With nonlinear systems, the shape of the output waveform depends not only on the shape of the input waveform but also on the size of the input waveform. Solving the differential equations of nonlinear systems requires a numerical integration approach, such as is produced by a SPICE transient analysis; the nature of the results will vary as the input amplitude changes. It is, with few exceptions, impossible to describe the behavior of such systems with an algebraic equation.

The most important result of linearity is that it allows us to break complicated input waveforms into their components and to apply the simpler components individually to the system. The individual results may then be superimposed to obtain the full response. Breaking a complex system into smaller, simpler pieces is always an attractive analysis option. For the purposes of this text, and all introductory a-c circuit texts, linearity/superposition is essential.

Even with a linear system, the task of solving a differential equation is not easy.

EXAMPLE 2.2

Find $v_g(t)$ in the circuit of Figure 2.5, given that the capacitor is initially uncharged and $i(t) = 6t\,e^{-t}$ for $t > 0$.

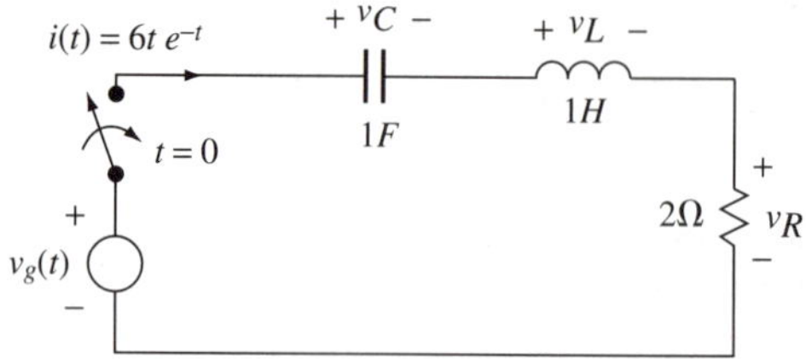

Figure 2.5

Solution

Since the current is known everywhere in the series circuit for $t > 0$, the voltages across each circuit element may be found immediately:

$$v_R = iR = 12te^{-t} \qquad v_L = L\frac{di}{dt} = [-6te^{-t} + 6e^{-t}]$$

$$v_C = \frac{1}{C}\int_{-\infty}^{t} i\,dt = \int_0^t 6te^{-t}\,dt + v_C(0) = 6e^{-t}(-t-1)\Big|_0^t = 6 - 6e^{-t} - 6te^{-t}$$

$$\therefore v_g(t) = v_R + v_L + v_C = \underbrace{12te^{-t}}_{v_R} + \underbrace{-6te^{-t} + 6e^{-t}}_{v_L} + \underbrace{6 - 6e^{-t} - 6te^{-t}}_{v_C} = 6$$

The result, perhaps surprising, is that a 6-V battery is the input source that causes the current specified.

The typical differential equation problem is just the opposite of Example 2.2. Given that the input source is a 6-V d-c battery, determine the resulting current. Obviously this is not a simple algebraic problem, but rather a search for a special function whose integrals or derivatives meet just the right conditions to satisfy Kirchhoff's laws for the circuit. Since the derivative or integral of a function is usually not the same shape as the function itself, this search can become complicated. There is, however, one function that does repeat its shape under differentiation and integration: the exponential.

2.2 THE EXPONENTIAL SIGNAL

A time function of the form Ae^{st}, where A and s are constants, repeats itself under differentiation and integration with respect to time. Linear differential equations are easily solved for such a function. We will specialize in this type of signal. After completing this section, you will be able to:

- Interpret variations of the exponential signal.
- Define the Euler phasor.
- Use a superposition of exponentials to represent a sinusoid.
- Solve linear differential equations for their forced response to an exponential signal.

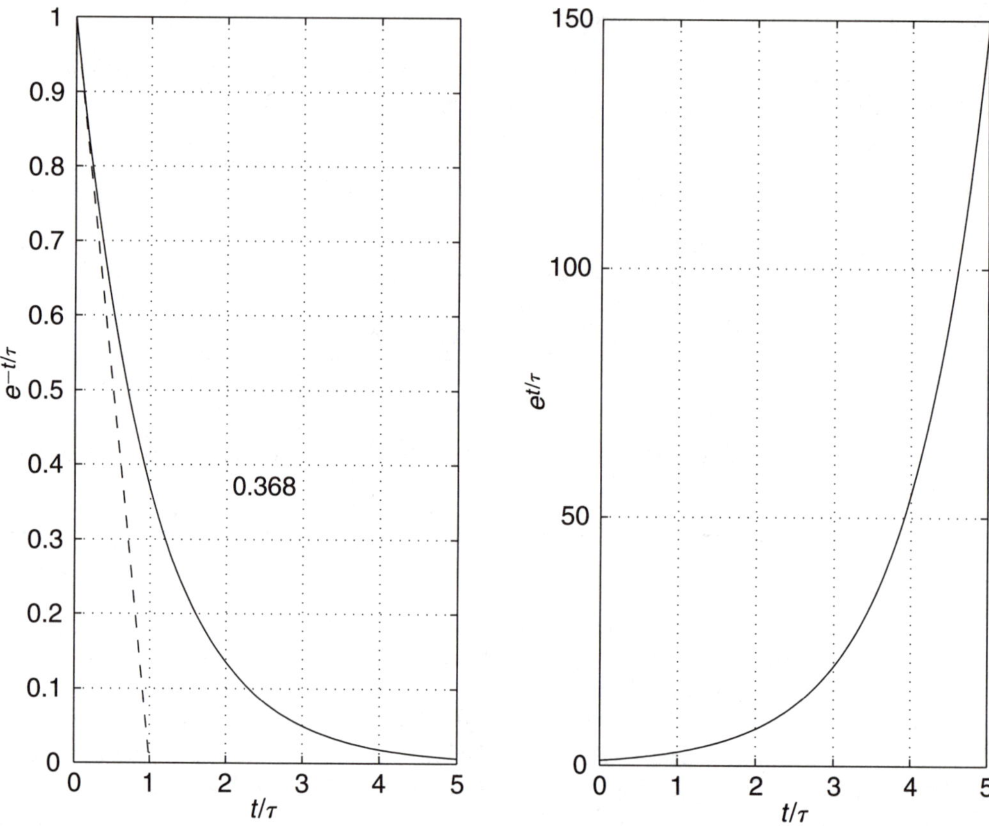

Figure 2.6 The real decreasing exponential is often expressed in terms of its time constant, τ. It decreases to 36.8% of its initial value after 1 time constant. If it were to continue decreasing at its initial rate, it would be down to zero after 1 time constant. It is essentially negligible after 5 time constants. The increasing exponential also starts at unity, but grows to about 150 times its initial value in 5 time constants.

If s is real, the exponential signal takes one of the forms shown in Figure 2.6. The degenerate case of $s = 0$ results in a signal of constant amplitude, as encountered in d-c circuits. Finding the forced response of a differential equation with a real exponential forcing function is very straightforward.

EXAMPLE 2.3

Find the forced response of the differential equation

$$\frac{dy(t)}{dt} + 3y(t) = 3e^{-4t}$$

Solution

Since exponential functions repeat themselves under differentiation, the forced response must have the same s as the forcing function. The solution must therefore be $y(t) = Ke^{-4t}$. Substituting this into the differential equation requires that

$$-4Ke^{-4t} + 3Ke^{-4t} = 3e^{-4t} \qquad \text{or} \qquad K = -3$$

The solution for the forced response is consequently

$$y(t) = -3e^{-4t}$$

With a minor modification, the procedure of Example 2.3 may be extended to sinusoidal sources. Using Euler's identity, we can express a cosine function of time in terms of exponentials with imaginary exponents

$$f(t) = A_m \cos(\omega t + \theta) = \frac{A_m}{2} e^{j(\omega t+\theta)} + \frac{A_m}{2} e^{-j(\omega t+\theta)}$$

Grouping the constants separately from the time dependent part of the signal gives

$$f(t) = A_m \cos(\omega\, t + \theta) = \underbrace{\left[\frac{A_m}{2} e^{j\theta}\right]}_{\vec{A}} e^{j\omega t} + \underbrace{\left[\frac{A_m}{2} e^{-j\theta}\right]}_{\vec{A}^*} e^{-j\omega t} \qquad \textbf{(2.7)}$$

We will call the bracketed constant that is multiplying the positive exponential time function the *Euler phasor.* It has a magnitude that is half the peak value of the sinusoid, and an angle equal to the angle of the cosine function. To emphasize that the phasor is a complex number, and therefore somewhat removed from reality, it is common practice to place an arrow over it in introductory courses. We will adhere to that practice.

In principle, we will solve a differential equation for a sinusoidal forcing function by applying the $e^{+j\omega t}$ term by itself, then applying the $e^{-j\omega t}$ signal, and superimposing the results. In practice, however, we will never bother to apply the $e^{-j\omega t}$ signal. Conjugate inputs produce conjugate outputs in linear systems. The conjugate outputs superimpose to provide a result that is twice the real part of the $e^{+j\omega t}$ term.

EXAMPLE 2.4

Find the forced response of the linear differential equation

$$\frac{dy(t)}{dt} + 3y(t) = 6\cos 2t$$

Solution

Replacing the cosine function using Euler's identity, we will actually solve the differential equation for the exponential forcing function

$$\frac{dy_1(t)}{dt} + 3y_1(t) = 3e^{j2t} \qquad \text{where} \qquad y_1(t) = \vec{Y}e^{j2t}$$

A subscript has been added to the y variable to remind us that we are using superposition to find one component of the full solution.

Substituting the proposed solution into the differential equation gives

$$j2[\vec{Y}e^{j2t}] + 3[\vec{Y}e^{j2t}] = 3e^{j2t}$$

$$[3 + j2]\vec{Y} = 3$$

$$\therefore \vec{Y} = \frac{3}{3 + j2} = 0.832\angle - 33.7^\circ$$

This makes $y_1(t) = 0.832\, e^{j(2t-33.7^\circ)}$, and the solution to the original problem is

$$y(t) = 2\text{Re}\, y_1(t) = 1.664\cos(2t - 33.7^\circ)$$

While sinusoidal sources are very important to us, we can actually further generalize the exponential class of signals. Consider the $x(t)$ of Equation 2.8:

$$x(t) = X_m e^{\sigma t}\cos(\omega t + \theta) \qquad \textbf{(2.8)}$$

where all of the constants are real numbers. Again using Euler's identity this becomes

$$x(t) = X_m e^{\sigma t} \cos(\omega t + \theta) = \frac{X_m}{2} e^{\sigma t}(e^{j(\omega t + \theta)} + e^{-j(\omega t + \theta)})$$

$$x(t) = \left[\frac{X_m}{2} e^{j\theta}\right] e^{(\sigma + j\omega)t} + \left[\frac{X_m}{2} e^{-j\theta}\right] e^{(\sigma - j\omega)t} \tag{2.9}$$

$$x(t) = \tilde{X}e^{st} + \tilde{X}^* e^{s^* t} \tag{2.10}$$

where we have defined a complex frequency variable for exponentials as

$$s = \sigma + j\omega \tag{2.11}$$

The damping constant, σ, has the units of nepers/second. It is also the inverse of the exponential's time constant. ω is the frequency of the sinusoid in radians/second. Both nepers and radians are actually dimensionless. Lab instruments are calibrated to indicate frequency, f, in hertz, while ω is the frequency variable of choice in theoretical work. These frequency variables are related by

$$\omega = 2\pi f \tag{2.12}$$

Notice that the definition of the Euler phasor is not affected by this more generalized signal. Only the value of s has changed. The procedure for handling this type of signal is, therefore, exactly the same as that for the pure sinusoid.

EXAMPLE 2.5

Solve for the forced response of the differential equation

$$\frac{dy(t)}{dt} + 3y(t) = 2e^{-3t} \cos(2t + 50°)$$

Solution

The input forcing function for this problem can be represented by an Euler phasor of $1\underline{/50°}$, and its complex frequency is $s = -3 + j2$. We therefore need to solve the equation

$$\frac{dy_1(t)}{dt} + 3y_1(t) = e^{j50°} e^{(-3+j2)t}$$

A solution of the form $y_1(t) = \vec{Y}e^{(-3+j2)t}$ is expected. Substituting this into the differential equation gives

$$(-3 + j2)\,\vec{Y}e^{(-3+j2)t} + 3\vec{Y}e^{(-3+j2)t} = e^{j50^\circ}e^{(-3+j2)t}$$
$$j2\vec{Y} = e^{j50^\circ}$$
$$\vec{Y} = \frac{1\angle 50^\circ}{j2} = 0.5\angle -40^\circ$$
$$y_1(t) = 0.5e^{-3t}e^{j(2t-40^\circ)}$$

The forced response of the original problem is consequently

$$y(t) = e^{-3t}\cos(2t - 40^\circ)$$

2.3 PHASOR TRANSFORMATIONS

The Euler phasor has been defined from the cosine function. It is not necessary to retain a rigorous mathematical link between the differential equation forcing function and the phasor. After reading this section you will:

- Establish an appropriate phasor transformation for each problem.

So far we have retained the mathematical equality between a sinusoidal time function and its complex components defined from

$$X_m e^{\sigma t}\cos(\omega t + \theta) = \underbrace{\left[\frac{X_m}{2}e^{j\theta}\right]}_{\vec{X}} e^{st} + \underbrace{\left[\frac{X_m}{2}e^{-j\theta}\right]}_{\vec{X}^*} e^{s^*t}$$

We have, however, never actually solved for the conjugate component, but simply recognized what its contribution would be. By replacing the equality with a bidirectional arrow, $\Leftrightarrow$, we can establish a one to one correspondence between a sinusoid and an e^{st} signal that will represent it. This mathematically less formal procedure releases us from always using the Euler phasor. Instead of halving X_m to get the magnitude of the Euler phasor, and then doubling the result to get back to the sinusoid, we can just define the phasor to have an amplitude X_m, and skip the doubling step. This phasor transformation would be stated as

$$X_m e^{\sigma t}\cos(\omega t + \theta) \Leftrightarrow \underbrace{[X_m e^{j\theta}]}_{\vec{X}}\, e^{st}$$

Suppose further that the forcing function for a differential equation is a sine wave rather than a cosine. Since $\sin\beta = \cos(\beta - 90°)$, we could use the following phasor transform:

$$X_m e^{\sigma t}\sin(\omega t + \theta) \Leftrightarrow \underbrace{[X_m e^{j(\theta-90°)}]}_{\vec{X}}\, e^{st}$$

Here again, there is no advantage to carrying the extra $-90°$ along in the phasor angle just because the Euler phasor is referenced to a cosine function. Instead, we will choose one of the following transformations, depending on whether the forcing function is given as a sine or a cosine function:

$$\begin{aligned} X_m e^{\sigma t}\sin(\omega t + \theta) &\Leftrightarrow \underbrace{[X_m e^{j(\theta)}]}_{\vec{X}}\, e^{st} \qquad s = \sigma + j\omega \\ X_m e^{\sigma t}\cos(\omega t + \theta) &\Leftrightarrow \overbrace{[X_m e^{j(\theta)}]}\, e^{st} \end{aligned} \tag{2.13}$$

As long as the transformation establishes the relationship between a signal and its phasor, any linear relationship is acceptable. The only restriction is that whatever definition we use for one phasor must be applied to all phasors in a given problem. If a problem involves both sine and cosine forcing functions, the cosine is the preferred reference.

EXAMPLE 2.6

Find the forced response of the differential equation

$$\frac{d^2y(t)}{dt^2} + 3y(t) = 4\sin(2t + 20°)$$

Solution

The equation we will solve is

$$\frac{d^2y(t)}{dt^2} + 3y(t) = \vec{X}e^{st}$$

where we choose the phasor transformation to be

$$X_m e^{\sigma t} \sin(\omega t + \theta) \Leftrightarrow [X_m e^{j(\theta)}]\, e^{st}$$

so

$$\tilde{X} = 4e^{j20^\circ} \qquad \text{and} \qquad s = 0 + j2$$

The expected solution is $y(t) = \tilde{Y}e^{j2t}$; substituting it in the differential equation gives

$$s^2\tilde{Y}e^{j2t} + 3\tilde{Y}e^{j2t} = 4e^{j20^\circ}e^{j2t}$$

$$\tilde{Y} = \frac{4\angle 20^\circ}{s^2 + 3} = \frac{4\angle 20^\circ}{(j2)^2 + 3} = -4\angle 20^\circ$$

Reversing the phasor transformation gives the forced response as

$$y(t) = -4\sin(2t + 20^\circ) = 4\sin(2t - 160^\circ) = 4\sin(2t + 200^\circ)$$

Although we have decided not to use the Euler phasor when solving differential equations, it is still the mathematically correct link between sinusoidal and exponential signals. It will surface again in Fourier analysis.

2.4 TRANSFER FUNCTIONS

If the s value of an exponential forcing function is left general, it is still possible to find the relationship for the ratio of a system's output and input phasor from the system differential equation. The result is a transfer function describing the system as a function of s. After completing this section you will be able to:

- Transform between differential equations and transfer functions.
- Distinguish between the s domain and the time domain.
- Provide a pole-zero diagram of a transfer function.

Starting with the general linear differential equation with constant coefficients, we may propose an arbitrary input signal $x(t) = \tilde{X}e^{st}$ and a corresponding forced output $y(t) = \tilde{Y}e^{st}$:

$$\frac{d^M y(t)}{dt^M} + \cdots + b_1\frac{dy(t)}{dt} + b_0 y(t) = a_N\frac{d^N x(t)}{dt^N} + \cdots + a_1\frac{dx(t)}{dt} + a_0 x(t)$$

$$s^M\tilde{Y}e^{st} + \cdots + b_1 s\tilde{Y}e^{st} + b_0\tilde{Y}e^{st} = a_N s^N\tilde{X}e^{st} + \cdots + a_1 s\tilde{X}e^{st} + a_0\tilde{X}e^{st}$$

Canceling the common e^{st} terms and arranging as a phasor ratio gives

$$\frac{\tilde{Y}}{\tilde{X}} = \frac{a_N s^N + \cdots + a_1 s + a_0}{s^M + b_{M-1}s^{M-1} + \cdots + b_1 s + b_0} \tag{2.14}$$

This transformation between the differential equation and the transfer function is readily made by inspection.

When a system is described in terms of real, physical constants and the forcing functions are real functions of time, the system description is said to be in the *time domain.* When the system description is in terms of phasors, or transfer functions of the complex frequency s, the description is said to be in the *s domain,* or the *frequency domain,* or the *phasor domain.* These three terms are used interchangeably. The only instances when the t and s variables should appear in the same equation is in the definition of a phasor, $x(t) = \tilde{X}e^{st}$ or temporarily while making the transition between domains.

The transfer function depends both on the differential equation coefficients and on the value of s for the forcing function. We may consequently explore the response of any differential equation for a whole range of signals by plotting its transfer function's pole-zero diagram over the complex s plane. At zeros of the transfer function, an input will produce no output. At s values close to poles of the transfer function, a small input may produce very large outputs. If we are particularly interested in the response to sinusoids, we need to evaluate the transfer function along the path $s = j\omega$, which gives what is generally meant by the term *frequency response.*

EXAMPLE 2.7

Determine the transfer function of the differential equation

$$\frac{d^2y(t)}{dt^2} + 3\frac{dy(t)}{dt} + 2y(t) = 3\frac{dx(t)}{dt} + x(t)$$

and briefly discuss the system's forced response.

Solution

By inspection, the transfer function is

$$\frac{\tilde{Y}}{\tilde{X}} = \frac{3s + 1}{s^2 + 3s + 2} = \frac{3(s + 1/3)}{(s + 1)(s + 2)}$$

A forcing function like $e^{-t/3}$ will produce no output, while forcing functions close to e^{-t} or e^{-2t} should produce a very large output. We can only speculate about what

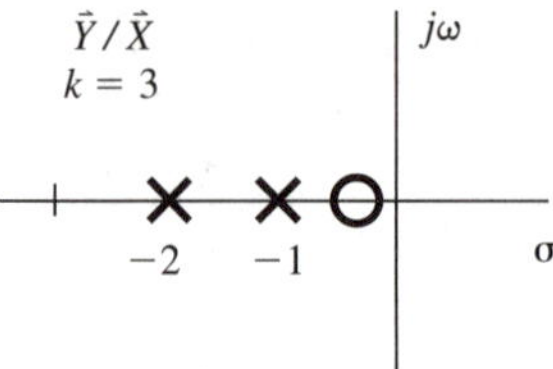

Figure 2.7 The pole-zero diagram for the system of Example 2.7.

happens to signals equal to e^{-t} or e^{-2t}. The pole-zero diagram of Figure 2.7 offers a more complete visualization of the situation.

In the future, we will spend more time working with transfer functions than with the differential equations from which they are derived. While working with the algebraic transfer function has the advantage of simplicity, questions that are unanswerable from the vantage point of the transfer function can always be answered in the more fundamental differential equation. It can be shown, for instance, that if a forcing function has exactly the same frequency as a transfer function pole, the assumption of a solution in the form e^{st} is too restrictive. An additional term of the form te^{st} also satisfies the differential equation under this condition.

In other words, the problems posed in this text will be carefully framed so that they may be correctly answered from the transfer function viewpoint. This will not present an important limitation to the systems we can consider, but those who wish to know all the details should eagerly anticipate taking a formal course in differential equations or Laplace transforms.

2.5 THE NATURAL RESPONSE

When a system is disturbed (something changes), it reacts in a way that is characteristic of the system. Any forcing function is irrelevant. After completing this section you will be able to:

- Determine aspects of the natural response of a linear system.
- Distinguish between the forced and natural response of the system.
- Characterize the stability of the system.

In addition to its forced response, every linear differential equation has a natural response. This is found by solving the equation with the forcing functions set to zero. Mathematicians call this the *homogeneous* equation:

$$\frac{d^M y(t)}{dt^M} + \cdots + b_1 \frac{dy(t)}{dt} + b_0 y(t) = 0 \tag{2.15}$$

We can accomplish the same thing from the point of view of the transfer function. If $\tilde{X} = 0$, $\tilde{Y}$ must normally also be zero. However, at its poles the expression for $\tilde{Y}$ becomes indeterminate, and a nonzero output is possible:

$$\frac{\tilde{Y}}{\tilde{X}} = \frac{a_N s^N + \cdots + a_1 s + a_0}{s^M + b_{M-1} s^{M-1} + \cdots + b_1 s + b_0}$$

The denominator of the transfer function is called the *characteristic equation* of the system, and the s values obtained from that equation are the system's *natural frequencies*, s_n. Signals of the form $e^{s_n t}$ can and will be generated anytime the forcing function is turned on or off or the system is otherwise disturbed.

It is not possible to establish magnitudes of the natural response terms from these observations, only the natural s_n values. The magnitudes of the natural response terms are affected by the conditions existing in the system when the source is turned on. It is possible that these conditions could happen to cause some of the natural frequency signals to appear with zero amplitude, but that is an exception rather than the rule.

EXAMPLE 2.8

Determine the full response of the system whose differential equation is

$$\frac{d^2 y(t)}{dt^2} + 2\frac{dy(t)}{dt} + 4y(t) = 2\frac{dx(t)}{dt} + x(t)$$

when $x(t) = 2 \cos t$.

Solution

The transfer function for this system is

$$\frac{\tilde{Y}}{\tilde{X}} = \frac{2s + 1}{s^2 + 2s + 4}$$

For the forced response, the phasor transform gives

$$\tilde{X} = 2\angle 0^\circ \qquad \text{and} \qquad s = j1$$

$$\therefore \tilde{Y} = \frac{1 + 2j}{3 + 2j} 2\angle 0^\circ = 1.2403\angle 29.74^\circ \Leftrightarrow y_f(t) = 1.2403 \cos(t + 29.74^\circ)$$

For the natural response, the transfer function poles are at

$$s^2 + 2s + 4 = 0 \qquad s = -1 \pm j\sqrt{3}$$

The natural response terms will consequently be

$$y_n(t) = K_1 e^{-t} e^{j\sqrt{3}t} + K_2 e^{-t} e^{-j\sqrt{3}t}$$

where K_1 and K_2 are, in general, complex conjugates. If K_1 is thought of as $(K/2)e^{j\theta}$, the natural response can be written as

$$y_n(t) = Ke^{-t}\cos(\sqrt{3}t + \theta)$$

The full response of this system is

$$y(t) = 1.2403\cos(t + 29.74°) + Ke^{-t}\cos(\sqrt{3}t + \theta)$$

where K and θ are arbitrary real constants.

Designating the system poles by $s_n = \sigma_n + j\omega_n$, there are three possible system categories based on the pole locations in the complex s plane (Figure 2.8):

1. If all poles have $\sigma_n < 0$, the natural response dies out with time and is said to be transient. The forced response then eventually dominates the system performance. Such systems are *stable.*

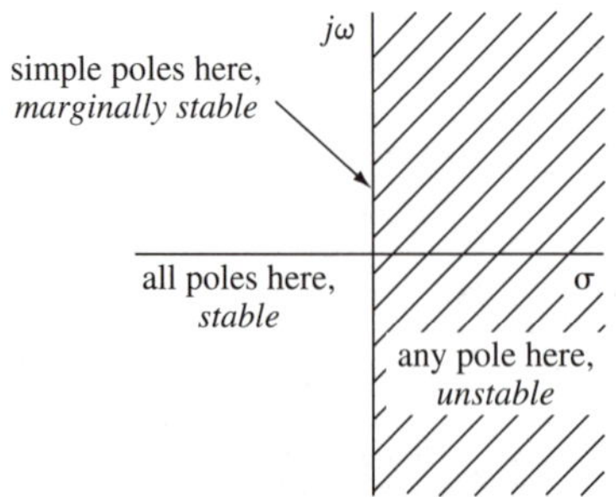

Figure 2.8 Transfer functions having poles in the right-hand side of the s plane (RHP) have growing natural responses and are *unstable.* Simple poles on the $j\omega$ axis produce a natural response that is present but neither growing nor decaying. Such a system, called *marginally stable,* is used to create oscillators. Higher-order poles on the $j\omega$ axis also produce an unstable natural response.

2. If any one pole has a $\sigma_n = 0$, the natural response of that pole does not die out with time. Most often this happens with a conjugate pair of poles, and the result is a sinusoidal oscillator. The system is described as *marginally stable.*
3. If any one pole has a $\sigma_n > 0$, its natural response grows exponentially with time and completely overwhelms any forced response. Such a system is *unstable* and unusable. This also occurs if a multiple (repeated) pole has $\sigma_n = 0$. In short, for a system to be stable, all of its poles must be in the left half of the s plane.

2.6 MATLAB LESSON 2

This chapter has shown that linear systems may be described by a ratio of polynomials in s called a transfer function. After completing this section you will be able to:

- Use MATLAB for efficient polynomial operations.
- Correct command typo errors easily.
- Modify MATLAB's numerical display format

We start with a reminder that you will get nothing out of this section unless you are seated at a computer and entering the MATLAB commands indicated. Feel free to experiment with them beyond the instructions.

MATLAB EXAMPLES

MATLAB allows a polynomial to be described simply by entering its coefficients into an array. The final coefficient must be the coefficient of the s^0 term. It also provides a variety of commands that recognize this simplified notation:

roots **poly** **polyval** **polyder** **conv** **deconv**

```
>p1=[1 0 0 2]     % defines the polynomial s^3+2. The last entry must be the
                    s^0 term
>r1=roots(p1)     % creates a vector containing the roots of polynomial p1
>p2=poly(r1)      % given the roots, find the polynomial form, p2 = p1
>r2=[-1 -1 -1]    % defines three identical roots at s = -1
>p2=poly(r2)      % p2 = (s + 1)^3 =s^3 + 3s^2 + 3s + 1
```

If you make a mistake, hit the up-arrow key to bring up the last command ready for editing at the current > cursor position. Hit it again to bring up the next earlier command, and so forth. The down-arrow key moves you back down the list. The side-arrow keys move you along the line being edited. Try using the up-arrow key now, and you will cycle through the five preceding statements on your active command line. You can also cut and paste previous commands to the active line.

```
>z=[-2 -1 0 1 j];          % defines a set of 5 points
>polyval(p2,z)             % calculates p2 at the set of points in z. We did
                             the same thing in Chapter 1, but had to
                             describe the polynomial in a formula. With
                             polyval we can do it with this simpler
                             notation.
>p3=[1 -1]                 % defines the polynomial p3 = s - 1
>p4=conv(p1,p3)            % takes the product of polynomials p1 and p3
>[p5,rem]=deconv(p1,p3)    % divides p1 by p3, result is p5 with remainder
                             rem
>[p6,re]=deconv(p4,p3)     % divides p4 by p3, should get back p1 with no
                             remainder
>r1                        % let's see the r1 roots (roots of p1) again
```

The elements of any array may be addressed individually. In an array called X, individual elements can be addressed by giving their row/column position as X(row,column). For row or column vectors, only the column or row number is needed. We will use this technique to select the pair of conjugate roots of p1 and find the quadratic polynomial they come from:

```
>q1=[r1(2) r1(3)]   % form an array consisting of the 2nd and 3rd elements
                      of r1 (these should be the conjugate roots)
>p7=poly(q1)        % forms the quadratic in p1=(s + 1.2599)(s^2 - 1.2599s +
                      1.5874)
```

We can combine this procedure with colon notation to select out a subset of a vector:

```
>p8=linspace(-5,5,11);   % create the vector -5 -4 -3 -2 -1 0 1 2 3 4 5
>p9=p8(6:11)             % selects out elements 6 thru 11 of p8
>p10=p8(1:3)             % selects out the first 3 elements of p8
```

What do you think the command **polyder** does? How would you find out? Do it!

zplane

```
>zplane(p1,p2)  % plots the pole-zero diagram of F(s) = p1/p2
```

MATLAB wakes up in a *short fixed-number display* format. The systems we have been working with and will continue to work with are normalized and have nice, simple coefficients. We will learn later how to denormalize them. Once we do, the fixed format display will not be acceptable. Keep in mind that we are discussing how the results are *displayed* by MATLAB, not the number of digits used in calculations.

format

```
>p8=[1 1000]          % creates the polynomial s + 1000
>p8=conv(p8,p8);      % creates the polynomial (s + 1000)^2
```

```
>(up arrow, return)  % hitting the up-arrow key brings the last command
                       back up, hitting return executes it again. Now p8
                       is (s+ 1000)^4
>p8                  % see the result in fixed format
>format short e      % change to short floating point
>p8                  % see the actual value of p8
>format short        % change back
>help format         % will show all the options
```

EXAMPLE 2.9

Find the complete response of the linear differential equation

$$\frac{d^4y}{dt^4} + 2\frac{d^3y}{dt^3} + 8\frac{d^2y}{dt^2} + 18\frac{dy}{dt} + 5y = 3\frac{d^2x}{dt^2} - 12\frac{dx}{dt}$$

if the forcing function is

$$x = -6 + 2e^{-t} + 2\cos(4t + 20°)$$

Solution

The transfer function polynomials are

```
>num=[3 -12 0];   den=[1 2 8 18 5];
```

To find the natural response terms we use the *roots* function

```
>r=roots(den)

r = 0.1364 + 2.8242i
    0.1364 - 2.8242i
    -1.9525
    -0.3203
```

In subsequent presentations of MATLAB results, column vectors will be displayed as if they were row vectors to conserve space.

These roots show the natural response will be

$$y_n(t) = K_1e^{-0.3203t} + K_2e^{-1.9525t} + K_3e^{0.1364t}\cos(2.8242t + \theta)$$

Since the conjugate roots are in the RHP, the system is unstable and the forced response is meaningless. We will find it for the practice.

```
>s=[0 -1 4j];                          % List the phasor frequencies of x
>T=polyval(num,s)./polyval(den,s)      % Evaluate the transfer function at
                                         these frequencies
```

```
T = 0    -2.5000    -0.1775 -0.4356i

>X=[-6 2 2*exp(j*20*pi/180)];        % Create the phasors of x

>Yforced=X.*T
Yforced = 0    -5.0000    -0.0356- 0.9401i

>mag=abs(Yforced(3))                 % find Yforced(3) in polar form
mag = 0.9408
>theta=180*angle(Yforced(3))/pi
theta = -92.1663
```

The forced response is consequently

$$y_f(t) = 0 - 5e^{-t} + 0.9408\cos(4t - 92.2°)$$

and the complete response is $y(t) = y_n(t) + y_f(t)$.

CHAPTER SUMMARY

We have seen how mechanical, electrical, and electromechanical systems obey physical laws that relate system variables through differential equations. If the equations are linear, they are easily solvable for an exponential forcing function, e^{st}. We were also able to handle sinusoidal forcing functions, because linear equations obey superposition. Euler's identity was used to express the sinusoid as two conjugate exponentials and to apply them individually. The multiplier of these exponential functions was, in general, a complex number called a phasor.

With input and output signals represented by phasors of e^{st} signals, linear differential equations took an algebraic form called a transfer function. The transfer function provided a fixed ratio between the input and output phasors at any given s. We were able to relate the output phasor back to a sinusoid formally using Euler phasors or informally using a more convenient phasor transformation.

The poles of the transfer function were shown to be the natural frequencies of the system, and indicate the general form of the system's natural response to any disturbance. If any of the system poles are in the RHP, the natural response grows exponentially with time and the system is unstable and unusable.

The realm of the differential equation is called the time domain. The realm of the transfer function is called the frequency domain or the s domain or the phasor domain.

In future continuous-time systems, we will work almost exclusively with transfer functions and, consequently, with polynomials in s. We have seen that MATLAB has a simplified method for describing polynomials in an array and a variety of functions that interpret an array as a polynomial. Many more such functions will be introduced later.

PROBLEMS

Section 2.1

1. Write the differential equation for $i_1(t)$ for $t > 0$ in the circuit of Figure P2.1.

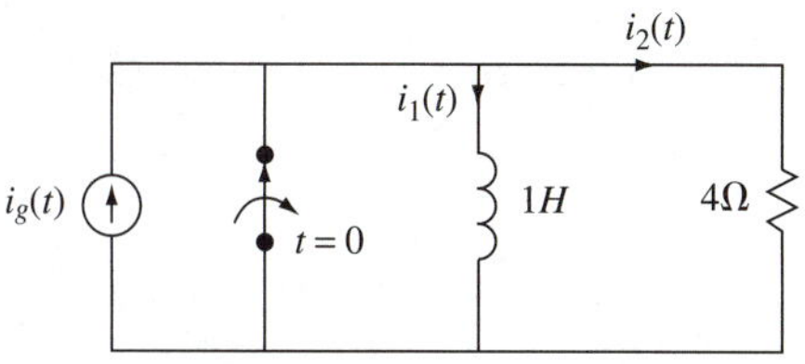

Figure P2.1

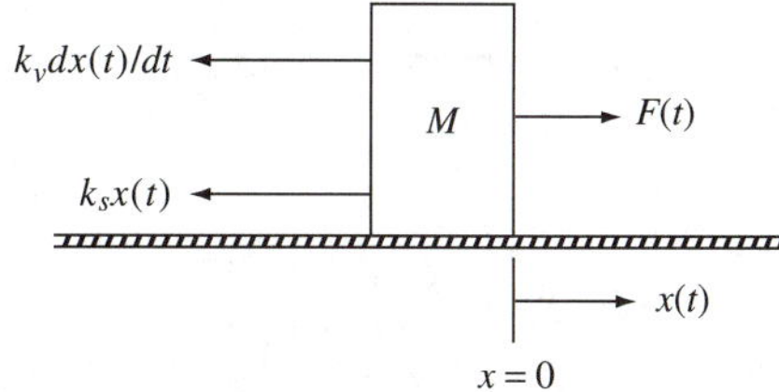

Figure P2.2

2. In Figure P2.2, a mass slides on a frictionless surface due the applied force F. Write the differential equation for the position of the mass M if the force is applied at $t = 0$ and the motion is opposed by a spring and dashpot whose forces are as shown. (A *dashpot* is a piston/cylinder assembly that produces a viscous friction similar to wind resistance.)
3. Write the differential equation for $i(t)$ in the circuit of Figure P2.3 given that the switch closes at $t = 0$. Assume the capacitor is initially uncharged.

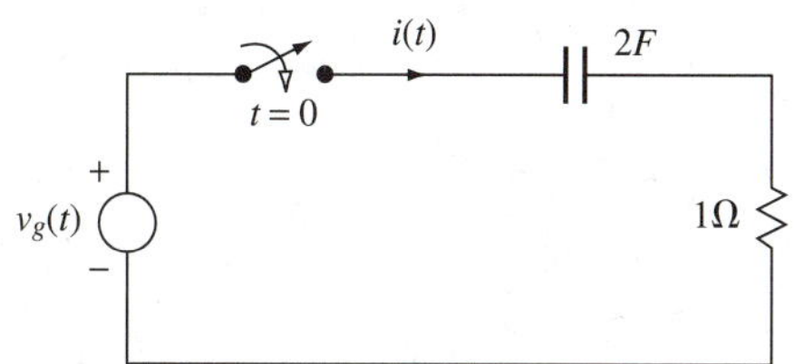

Figure P2.3

pivot

$J = Ml^2$

l

θ

$\theta = 0$

M

$F = Mg$

θ_0

Figure P2.4

4. The pendulum shown in Figure P2.4 is displaced to an initial angle θ_0 and released at $t = 0$.
   a. Ignoring friction and wind resistance, determine the differential equation describing $\theta(t)$. Is it linear?
   b. State an approximation that will make the pendulum equation linear for small θ?

Section 2.2

5. For the following conditions, find the forced response of the differential equation

$$\frac{d^2y}{dt^2} + 4\frac{dy}{dt} + 2y = x(t)$$

a. $x(t) = e^{-t}$ b. $x(t) = 3e^{2t}$

6. Find the forced response of the given differential equation to the $x(t)$ specified,

$$\frac{d^2y}{dt^2} - 2\frac{dy}{dt} = x(t)$$

a. $x(t) = 2 \cos t$ b. $x(t) = e^{-t} \cos t$

7. Identify the Euler phasor and s value of each forcing function.
   a. $x(t) = 3 \cos(2t + 12°)$ b. $x(t) = e^{-3t} \cos t$
   c. $x(t) = 12 \sin(2t + 20°)$ d. $x(t) = 2e^{t} \cos 3t$

Section 2.3

8. Select an appropriate phasor transformation and use it to find the forced response given that $x(t) = 10 \cos(2t + 35°)$ and the system equation is

$$\frac{d^2y}{dt^2} + 3\frac{dy}{dt} + 2y = 2\frac{dx}{dt} + x(t)$$

9. Find the output phasor given that $x(t) = 3e^{-2t} \sin t$ and the system is described by

$$\frac{d^2y}{dt^2} + 3\frac{dy}{dt} + y = \frac{dx}{dt} + x$$

State the phasor transformation used, and identify $\vec{X}$ and s.

Section 2.4

10. Determine the transfer functions of the following systems.

a. $$\frac{d^3y}{dt^3} + 4\frac{d^2y}{dt^2} + 16\frac{dy}{dt} + 8y = 2\frac{d^2x}{dt^2} + 3x$$

b. $$\frac{d^2y}{dt^2} + 3\frac{dy}{dt} = \frac{d^2x}{dt^2} - 9x$$

11. Determine the transfer function of the systems indicated.

a. $$\frac{d^3y}{dt^3} + 3\frac{d^2y}{dt^2} + 3\frac{dy}{dt} + y = 4\frac{dx}{dt} + 4x$$

b. $$\frac{d}{dt}\left[\frac{dy}{dt} + 4y\right] = \frac{d^2x}{dt^2}$$

12. Sketch pole-zero diagrams for the systems of Problem 2.10.

13. Sketch pole-zero diagrams for the following systems.

a. $\frac{\vec{Y}}{\vec{X}} = 2\frac{s+1}{s+2} - \frac{s+2}{s+4}$ b. $\frac{\vec{Y}}{\vec{X}} = \frac{s^2-1}{s^3+2s^2+2s+1}$

Section 2.5

14. Determine the complete response of the systems for $x(t) = 2\cos 2t$.

a. $\frac{\vec{Y}}{\vec{X}} = \frac{1}{s^3+2s^2+2s+1}$ b. $\frac{\vec{Y}}{\vec{X}} = \frac{s-2}{s^2+6s+8}$

15. Determine the natural response of the systems. Classify each as stable, unstable, or marginally stable.

a. $\frac{\vec{Y}}{\vec{X}} = \frac{s^3-2s}{s^2+4s+8}$ b. $\frac{\vec{Y}}{\vec{X}} = \frac{s^2+16}{s^3+4s^2+8s+32}$

c. $\frac{\vec{Y}}{\vec{X}} = \frac{s+24}{(s+8)(s^2-s+4)}$ d. $\frac{\vec{Y}}{\vec{X}} = \frac{2s(s+2)^2}{s^3+2s^2+3s+2}$

Section 2.6

16. Show the MATLAB programming needed to multiply together the polynomials $x^2 - 3x + 4$ and $x^5 + 2x^3 + 3x^2 + 4x + 2$.

17. Show the MATLAB programming that would be used to evaluate the polynomial $y = x^5 + 2x^2 + x - 4$ at the points $x = -2, 3j$, and 4.

18. Show the commands that are needed to correct the z vector after the following command sequence:

```
>x = -2:8;  y = x.^2 - 4;  z = sin(y)./y
```

19. Show the command needed to correct the y vector after the following command sequence:

```
>x =0:20;  num=[1 -2 1];  den = [1 -1];  y=polyval(num,x)./polyval(den,x)
```

Advanced Problems

20. Verify that $y = K_1e^{-t} + K_2te^{-t}$ is a solution of the differential equation $dy/dt + y = e^{-t}$, where $K_2 = 1$ and K_1 is arbitrary.

21. Given $x(t) = 2 + 4e^{-3t} + 6\cos 3t$, use MATLAB to assist you in finding the complete solution of the differential equation

$$\frac{d^4y}{dt^4} + 2\frac{d^3y}{dt^3} + 16\frac{d^2y}{dt^2} + 18\frac{dy}{dt} + 2y = \frac{d^2x}{dt^2} - 2\frac{dx}{dt} + x$$

3

ANALYSIS TECHNIQUES

OUTLINE

OBJECTIVES

1. Define impedance.
2. Use impedances and phasors to analyze circuits.
3. Introduce feedback systems using a block diagram.
4. Simplify block diagrams to obtain the system transfer function.
5. Explain ways to test a system for stability.
6. Use MATLAB for block diagram reduction and system simulation.

INTRODUCTION

Although physical systems are fundamentally described with differential equations, they can also be described by algebraic equations for exponential forcing functions. In circuit theory, this process is expedited by converting the calculus-based laws of the circuit components immediately into phasor ratios. This leads to the definition of *impedance,* and allows circuit properties to be derived directly in the *s* domain.

Most systems are not purely electrical or mechanical, but rather a mix of amplifier circuits, actuators such as motors and hydraulic pistons, and sensors such as tachometers and anemometers. Each of these elements individually may be described with a transfer function and represented with an input/output block. The blocks are then interconnected to form a system, usually in a feedback arrangement where the desired output is compared to the actual output. Feedback has the potential to make a system's forced response very precise, but it simultaneously may make the system's natural response unacceptable or even unstable. The design of elementary feedback systems must consider both responses.

3.1 DRIVING POINT IMPEDANCE

For specialists in electrical systems, it is not necessary to start each problem with a differential equation. By investigating the properties of phasors you will be able to:

- Define impedance.
- Convert circuits directly to the phasor or s domain.
- Combine impedances to simplify circuits.

Suppose we have the three currents i_1, i_2, and i_3 meeting at a junction in a circuit, where

$$i_1(t) = I_1 e^{\sigma t} \cos(\omega t + \theta_1)$$
$$i_2(t) = I_2 e^{\sigma t} \cos(\omega t + \theta_2)$$
$$i_3(t) = I_3 e^{\sigma t} \cos(\omega t + \theta_3)$$

As shown previously, each of these currents can be written in terms of an Euler phasor, and e^{st} signal, and their conjugates as

$$i(t) = \vec{I} e^{st} + \vec{I}^{*} e^{s^{*}t}$$

Kirchhoff's law requires that at every instant $i_1(t) + i_2(t) + i_3(t) = 0$, which means

$$\underbrace{\vec{I}_1 e^{st} + \vec{I}_1^* e^{s^*t}}_{i_1(t)} + \underbrace{\vec{I}_2 e^{st} + \vec{I}_2^* e^{s^*t}}_{i_2(t)} + \underbrace{\vec{I}_3 e^{st} + \vec{I}_3^* e^{s^*t}}_{i_3(t)} = 0$$

$$\underbrace{(\vec{I}_1 + \vec{I}_2 + \vec{I}_3)}_{0} e^{st} + \underbrace{(\vec{I}_1^* + \vec{I}_2^* + \vec{I}_3^*)}_{0} e^{s^*t} = 0$$

Because neither e^{st} nor e^{s^*t} is zero for all time, it follows that the current phasors (and their conjugates) must sum to zero. This establishes a very important fact: *If a set of signals obey Kirchhoff's laws, then the phasors representing those signals also obey Kirchhoff's laws.*

The voltage across an inductor or capacitor is related to the current through that device by a calculus expression. What is the corresponding relationship between the device's phasors? Taking the inductor as an example, if $i(t) = \vec{I}e^{st}$, then

$$v(t) = L\frac{d}{dt}\vec{I}e^{st} = L\vec{I}\,se^{st} = \vec{V}e^{st}$$

and

$$\frac{v(t)}{i(t)} = \frac{\vec{V}e^{st}}{\vec{I}e^{st}} = sL \qquad \textbf{(3.1)}$$

The notation in Equation 3.1 is deceptive and is considered poor practice, because it is only true for the exponential signal assumed initially. Compare it to Ohm's law for a resistor, for instance, where $v(t) = i(t)R$ is true for *any* waveform. The presence of either a phasor or the s variable automatically means that the signal is e^{st}. The $v(t)$ and $i(t)$ notation implies a generality that does not exist, and it is best to avoid using it.

To write Equation 3.1 in a proper form, we retain just the second half of the equation, cancel the e^{st} terms, and define the ratio of the inductor's voltage and current phasors to be the inductive impedance, Z_L:

$$Z_L = \frac{\vec{V}}{\vec{I}} = sL \qquad \textbf{(3.2)}$$

This result is often thought of as an Ohm's law for phasors. The impedances for all of the passive circuit elements are shown in Figure 3.1. Impedance is defined only for e^{st} signals, not for sinusoids, although sinusoids may be represented by an e^{st} signal through a phasor transformation.

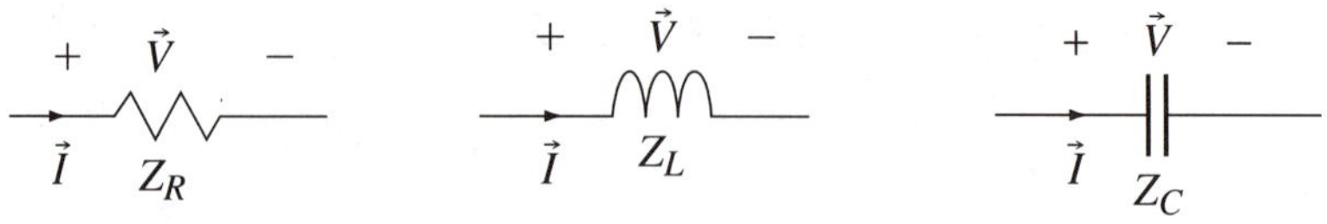

$$\frac{\vec{V}}{\vec{I}} = Z_R = R \qquad \frac{\vec{V}}{\vec{I}} = Z_L = sL \qquad \frac{\vec{V}}{\vec{I}} = Z_C = \frac{1}{sC}$$

Figure 3.1 An "Ohm's law" exists for voltage and current phasors. The ratio of the voltage and current phasor for each circuit element is defined as the element's impedance and is designated by the symbol Z.

All information may be given in the phasor or s domain, and phasor transformations may remain unspecified until a link to the time domain needs to be established. Frequently this link is never stated, because all the important information is contained in the phasors. Since ratios of variables are involved, it makes absolutely no difference if the amplitudes are interpreted as peak, peak to peak, or rms.

It is easy to show (Figs. 3.2a–c) from Kirchhoff's laws and the relationships for phasors that impedances in series add. Admittances, which are the inverse of impedances, add for parallel combinations. The voltage and current divider relationships follow immediately, and are often sufficient for solving simple circuit problems.

$$\vec{V}_T = \vec{V}_1 + \vec{V}_2 + \vec{V}_3 \qquad \text{(KVL)}$$

$$Z_{eq} = \frac{\vec{V}_T}{\vec{I}} = \frac{\vec{V}_1}{\vec{I}} + \frac{\vec{V}_2}{\vec{I}} + \frac{\vec{V}_3}{\vec{I}} = Z_1 + Z_2 + Z_3$$

$$\frac{\vec{V}_3}{\vec{V}_T} = \frac{\vec{I}Z_3}{\vec{I}Z_{eq}} = \frac{Z_3}{Z_1 + Z_2 + Z_3} \qquad \text{(Voltage Divider)}$$

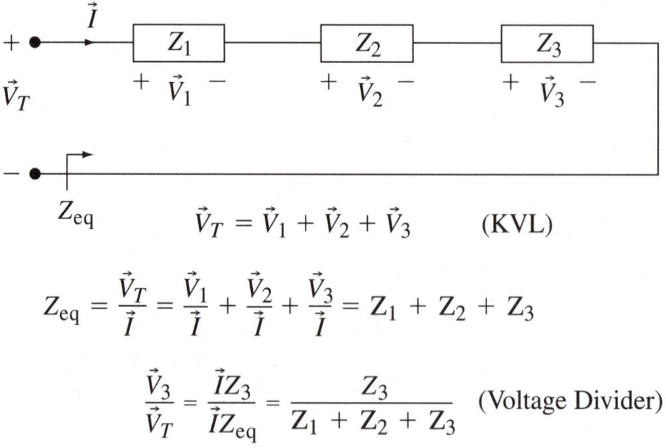

Figure 3.2a Phasor relationships in a series circuit.

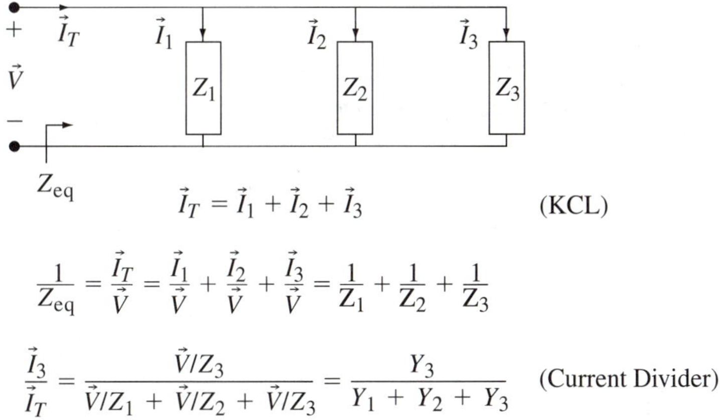

$$\vec{I}_T = \vec{I}_1 + \vec{I}_2 + \vec{I}_3 \qquad \text{(KCL)}$$

$$\frac{1}{Z_{eq}} = \frac{\vec{I}_T}{\vec{V}} = \frac{\vec{I}_1}{\vec{V}} + \frac{\vec{I}_2}{\vec{V}} + \frac{\vec{I}_3}{\vec{V}} = \frac{1}{Z_1} + \frac{1}{Z_2} + \frac{1}{Z_3}$$

$$\frac{\vec{I}_3}{\vec{I}_T} = \frac{\vec{V}/Z_3}{\vec{V}/Z_1 + \vec{V}/Z_2 + \vec{V}/Z_3} = \frac{Y_3}{Y_1 + Y_2 + Y_3} \qquad \text{(Current Divider)}$$

Figure 3.2b Phasor relationships in a parallel circuit: $Y = 1/Z$ = admittance.

The types of circuits of primary interest in this text have a single input signal. We can characterize the effects of the input signal in terms of a transfer function, which relates the output signal at one set of terminals to the input signal at another set of terminals. We might, for instance, determine the output voltage caused by an input current. The result would be a ratio of an output voltage phasor to an input current phasor. Such a transfer function would have the units of impedance and would be called a *transimpedance,* to emphasize that the voltage and current involved are at different points in the circuit. Generally the term *impedance* is understood to apply to a ratio of phasors at the same set of terminals. If there is danger of confusion, the term *driving point impedance* can be used to emphasize that only one set of terminals is involved.

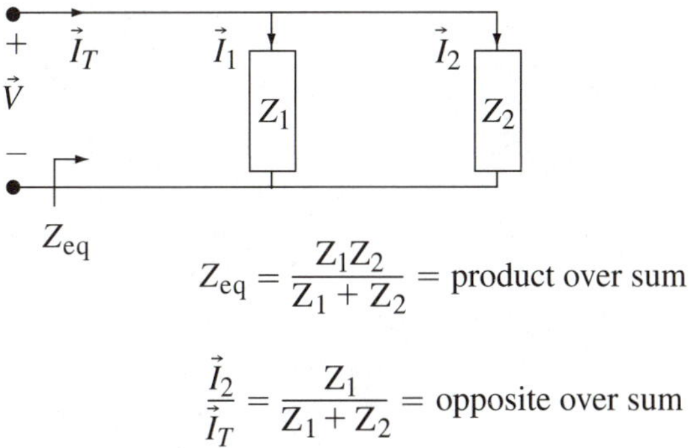

$$Z_{eq} = \frac{Z_1 Z_2}{Z_1 + Z_2} = \text{product over sum}$$

$$\frac{\vec{I}_2}{\vec{I}_T} = \frac{Z_1}{Z_1 + Z_2} = \text{opposite over sum}$$

Figure 3.2c Special case of two impedances in parallel.

EXAMPLE 3.1

Find the driving point impedance of the circuit in Figure 3.3a.

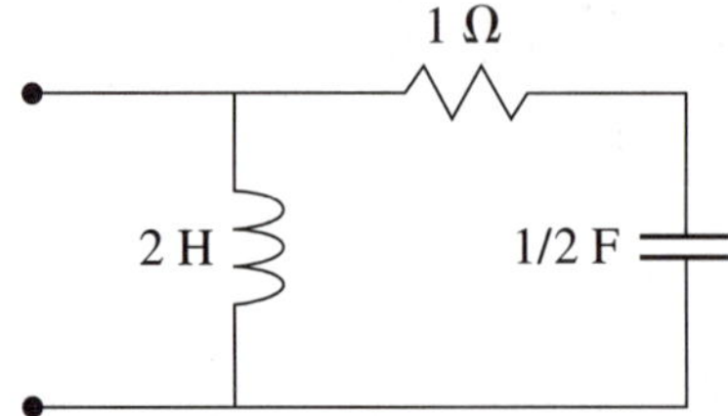

Figure 3.3a

Solution

The circuit components must be converted to impedances or admittances, as in Figure 3.3b. Next the series combination can be combined, as in Figure 3.3c.

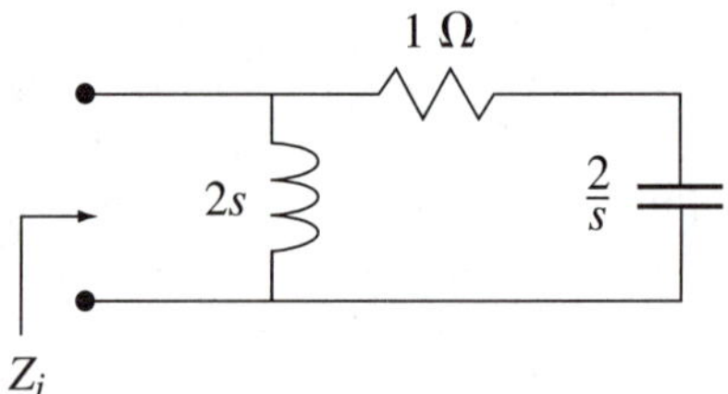

Figure 3.3b

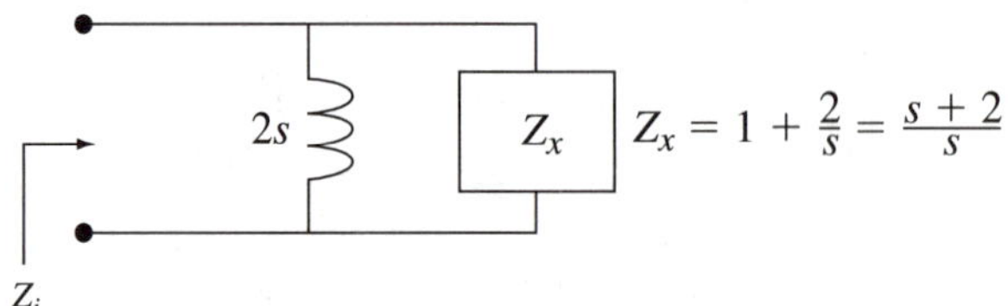

Figure 3.3c

The equivalent admittance is then found by adding the admittances of the parallel paths:

$$Y_i = \frac{1}{2s} + \frac{s}{s+2} = \frac{2s^2 + s + 2}{2s(s+2)} \quad \text{or} \quad Z_i = \frac{2s(s+2)}{2s^2 + s + 2} = \frac{s(s+2)}{s^2 + s/2 + 1}$$

Some algebra is involved in this process, so it is always a good idea to see if the result makes sense. At d-c ($s = 0$) the inductor makes the circuit a short. At high frequencies ($s = \infty$) the inductor is open, the capacitor is a short, and the circuit looks like a 1-Ω resistor. These properties are confirmed by the expression for Z_i.

3.2 CIRCUIT ANALYSIS IN THE *s* DOMAIN

With phasors representing voltages and currents, and impedance representing the passive circuit elements, the solution for circuits in the s domain is superficially no different than for purely resistive circuits. All of the analysis techniques for resistive circuits still apply. Often a transfer function for a circuit can be deduced using simple voltage or current divider relationships. In more complicated circuits, mesh or node equations may be needed. After completing this section you will be able to:

- Combine impedances to find transfer functions of simple circuits.
- Apply voltage and current dividers.
- Select among various analysis procedures.

For a simple series circuit, the voltage divider may be applied directly to obtain a voltage transfer function.

EXAMPLE 3.2

Find $\vec{V}_o/\vec{V}_g$ for the circuit of Figure 3.4.

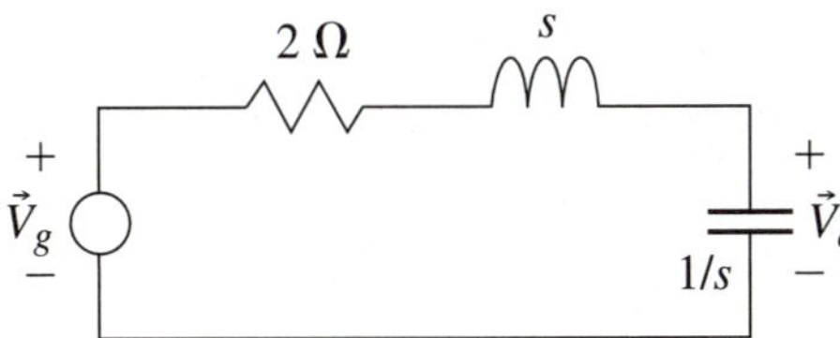

Figure 3.4

Solution

In this case the circuit has already been converted to the phasor domain, and the circuit elements are given as impedances. The result is obtained by inspection from the voltage divider (Fig. 3.2a).

$$\frac{\vec{V}_o}{\vec{V}_g} = \frac{1/s}{2 + s + 1/s} = \frac{1}{s^2 + 2s + 1} = \frac{1}{(s + 1)^2}$$

In some problems it may be necessary to combine some impedances before trying to apply the voltage divider.

EXAMPLE 3.3

Find $\vec{V}_o/\vec{V}_g$ for the circuit of Figure 3.5a.

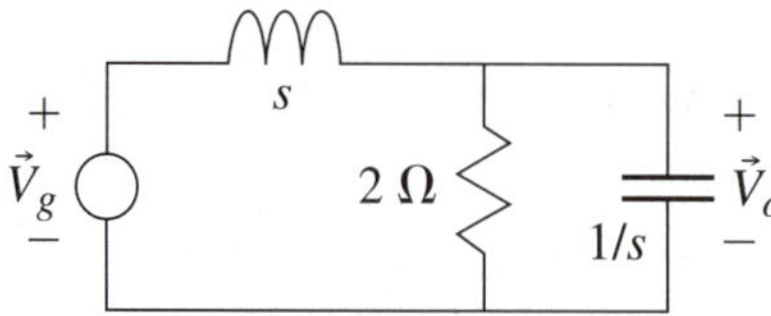

Figure 3.5a

Solution

Combining the parallel combination of 2 Ω and 1/*s* Ω into an equivalent impedance reduces the circuit to a series equivalent, and the voltage divider may be applied, as in Figure 3.5b:

$$Z_{\text{eq}} = \frac{2/s}{2 + 1/s} = \frac{2}{2s + 1} = \frac{1}{s + 1/2}$$

Figure 3.5b

$$\frac{\vec{V}_o}{\vec{V}_g} = \frac{Z_{eq}}{s + Z_{eq}} = \frac{1}{s(s + 1/2) + 1} = \frac{1}{s^2 + s/2 + 1}$$

The current divider is also often useful. A typical procedure is to find the driving point impedance of the circuit to find the total current delivered to the circuit. Then the current divider is used to determine how the total current divides up among several paths.

EXAMPLE 3.4

Find $\vec{I}_o/\vec{V}_g$ in the circuit of Figure 3.6a.

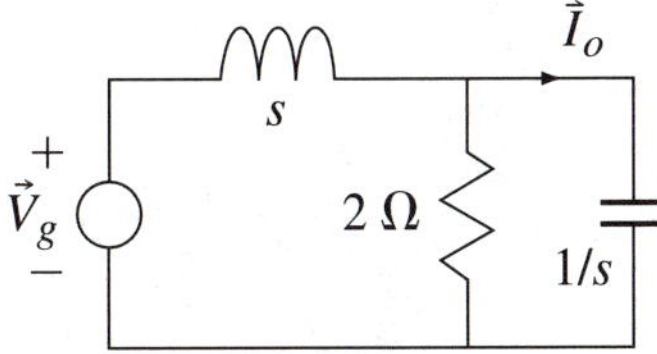

Figure 3.6a

Solution

The driving point impedance seen by $\vec{V}_g$ is

$$Z_T = s + \frac{2/s}{2 + 1/s} = s + \frac{2}{2s + 1} = \frac{2s^2 + s + 2}{2s + 1}$$

The total current into the circuit consequently is as shown in Figure 3.6b.

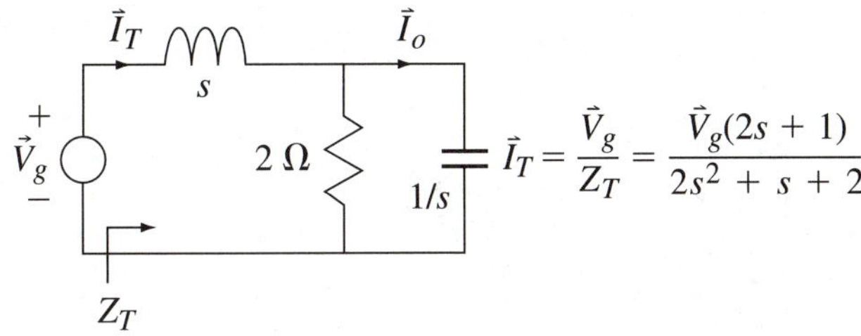

Figure 3.6b

Using the special case of the 2-branch current divider (Fig. 3.2c),

$$\vec{I}_o = \frac{2}{2 + 1/s}\vec{I}_T = \frac{2s}{2s + 1}\left(\frac{\vec{V}_g(2s + 1)}{2s^2 + s + 2}\right) = \frac{s\vec{V}_g}{s^2 + s/2 + 1}$$

This result could also have been obtained from taking Example 3.3 one step further, since the voltage across the capacitor was found in that example.

All of the reduction techniques normally taught in circuit theory can be used to find transfer functions in the s domain. For complicated electrical systems, the more general procedures of writing mesh or node equations may provide easier and more direct routes to the desired result than simple tools like voltage and current dividers. The inclusion of dependent sources requires such an approach. There are also circuit topologies for which the divider equations can not be applied (see Fig. 3.7).

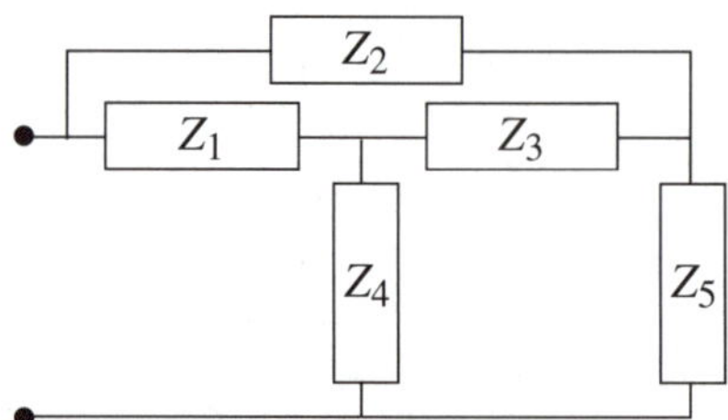

Figure 3.7 A circuit topology that does not lend itself to voltage or current division techniques because there are no series or parallel connections.

EXAMPLE 3.5

Find $\vec{V}_o/\vec{V}_g$ for the circuit of Figure 3.8a.

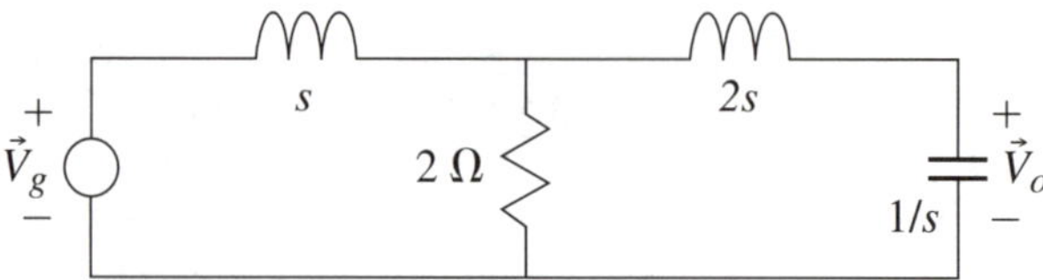

Figure 3.8a

Solution

This circuit could be solved by combining impedances and using divider equations, but it may be easier to assume mesh currents and write simultaneous equations. In this case, the mesh equations are combined to eliminate $\vec{I}_1$ (Fig. 3.8b):

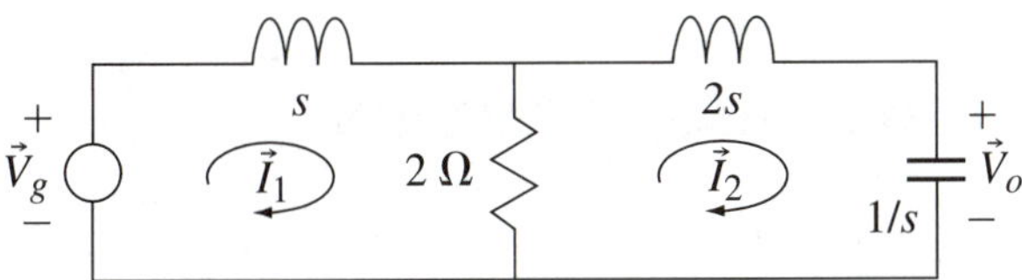

Figure 3.8b

$$\left.\begin{aligned} \vec{V}_g &= (s+2)\vec{I}_1 - 2\vec{I}_2 \\ 0 &= -2\vec{I}_1 + (2+2s+1/s)\vec{I}_2 \end{aligned}\right\} \vec{V}_g = (s+2)(1+s+1/2s)\vec{I}_2 - 2\vec{I}_2$$

$$\vec{I}_2 = \frac{\vec{V}_g}{(s+2)(1+s+1/2s)-2} = \frac{\vec{V}_g}{s+s^2+1/2+2s+1/s}$$

$$\vec{I}_2 = \frac{s\vec{V}_g}{s^3+3s^2+s/2+1}$$

$$\therefore \frac{\vec{V}_o}{\vec{V}_g} = \frac{1}{s^3+3s^2+s/2+1}$$

It is always wise to check the result at $s = 0$ and $s = \infty$ to identify potential algebraic errors. The circuit under these conditions is as sketched in Figures 3.8c and 3.8d; the sketches support the transfer function equation at these s values.

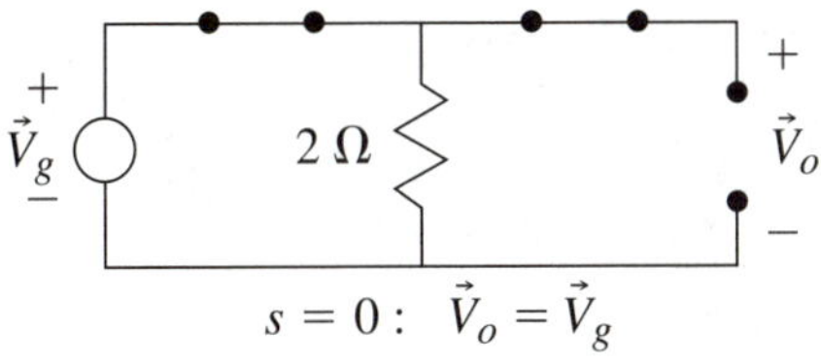

Figure 3.8c

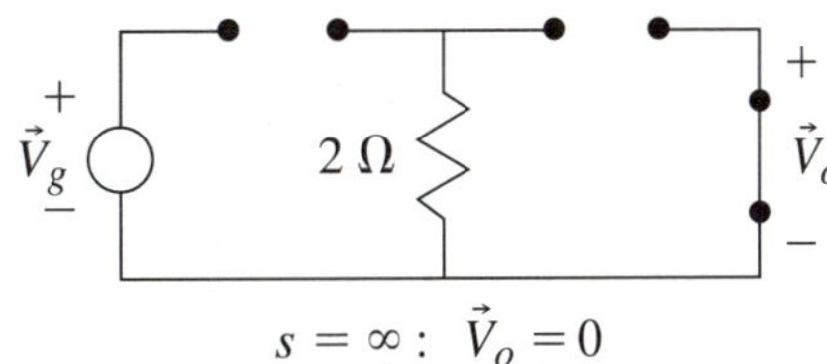

Figure 3.8d

Whatever analysis technique is used to achieve it, the characteristics of a particular circuit can be reduced to those of a transfer function. It may then be simply represented as a mathematical input/output block. In this way a complex system may be created and analyzed without cluttering up the system diagram with circuit details. The same process may be used to provide blocks representing mechanical, electromechanical, hydraulic, and other devices.

3.3 FEEDBACK DIAGRAM ELEMENTS

A block diagram is used to show signal flow in a major system where several types of components interact. The components are represented by their *s* domain transfer functions. In most such systems, a comparison is made between the desired output and the actual output. This comparison is then used to automatically improve the system performance. After completing this section you will be able to:

- Represent feedback systems with transfer function building blocks.
- Describe the function of block diagram elements.
- Explain the main advantage provided by feedback.
- Explain the main danger introduced by feedback.

The basic elements of a negative feedback system are shown in Figure 3.9. The notation used for signals is generic: A reference input $\vec{R}$ is compared to $\vec{F}$, a fed-back sample of the controlled output $\vec{C}$. The difference signal, $\vec{E}$ for "error," adjusts the output. (The difference signal is truly an error only if $H = 1$.) Forward transfer functions are denoted by G, while H is used for reverse transfer functions.

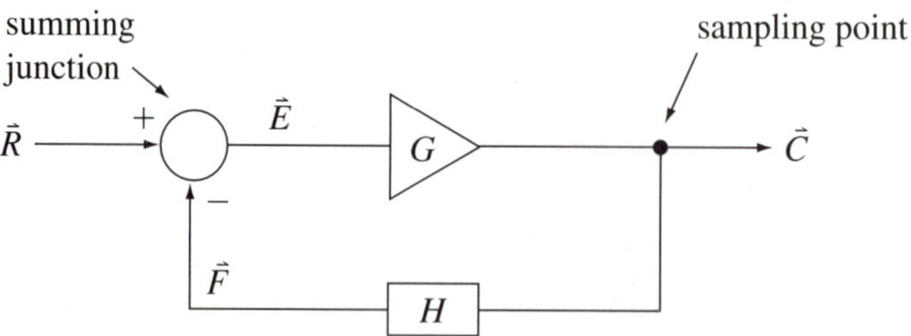

Figure 3.9 The block diagram of an elemental negative feedback loop. Signals are represented by phasors, while *G* and *H* are transfer functions.

$$G = \frac{\vec{C}}{\vec{E}}, \qquad H = \frac{\vec{F}}{\vec{C}}, \qquad \text{and} \qquad \vec{E} = \vec{R} - \vec{F}$$

$$\vec{C} = G\vec{E} = G(\vec{R} - H\vec{C})$$

$$\frac{\vec{C}}{\vec{R}} = \frac{G}{1 + GH} \Rightarrow \frac{1}{H} \qquad \text{if } GH \gg 1 \tag{3.3}$$

Frequently the signals $\vec{R}$ and $\vec{F}$ are electrical, in which case the summing junction can be made with either an op amp differential amplifier or, if the signals are of opposite polarity, an op amp summer (Fig. 3.10).

The sampling point is simply a connection to an input of a transfer function block. The transfer functions are assumed not to load each other down. Either G needs to be derived with the loading of H in place, or H needs to be designed not to load G down.

There are two aspects of the negative feedback loop that are evident in Equation 3.3. The first is the ability to make the overall transfer function virtually independent of the forward gain G. This gain usually involves amplifiers, and amplifiers have gains that depend on transistor parameters known to be widely variable and temperature dependent. As long as the loop gain, GH, is large, the transfer function approaches the ratio of $1/H$, where H consists of carefully chosen, precision components. In other words, the feedback loop allows us to trade a large but highly variable gain for a lower but precisely determined gain.

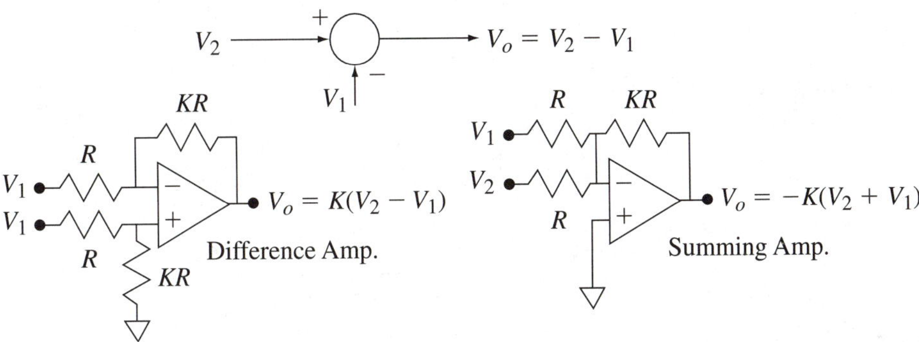

Figure 3.10 Summing junctions are implemented electronically with op amp summing or differencing amplifiers. The input voltages must be of opposite polarity to achieve a difference voltage with the summing amplifier.

EXAMPLE 3.6

Discuss the properties of the noninverting op amp circuit of Figure 3.11a in terms of the feedback model.

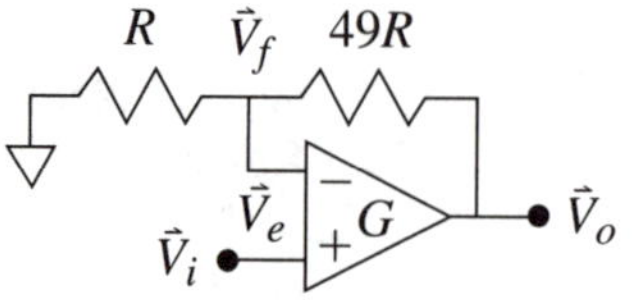

Figure 3.11a

Solution

Assuming the op amp input terminals draw negligible current, the feedback voltage is found from the voltage divider as

$$\frac{\vec{V}_f}{\vec{V}_o} = \frac{R}{R + 49R} = H = \frac{1}{50}$$

Kirchhoff's voltage law provides the equivalent of a summing junction since

$$\vec{V}_e = \vec{V}_i - \vec{V}_f$$

These relationships are summarized in the block diagram of Figure 3.11b, which predicts a closed loop d-c gain of

$$\frac{\vec{V}_o}{\vec{V}_i} = \frac{G}{1 + G/50}$$

This gain approaches 50 if $G \gg 50$.

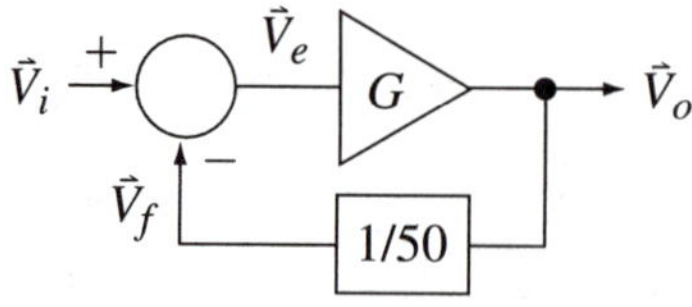

Figure 3.11b

A general purpose op amp has a d-c gain range that may easily vary by a factor of 10 between the minimum and the typical values. Data sheet information on a μA741, for instance, indicates

$$\frac{\vec{V}_o}{\vec{V}_e} = G \qquad G_{\text{min}} = 20V/mV \qquad G_{\text{typ}} = 200V/mV$$

Using these values, the closed loop d-c gain would only vary from

$$\frac{\vec{V}_o}{\vec{V}_i} = \frac{20{,}000}{1 + 400} = 49.88 \qquad \text{to} \qquad \frac{\vec{V}_o}{\vec{V}_i} = \frac{200{,}000}{1 + 4000} = 49.99$$

due to the factor of 10 variation possible in G.

The second important aspect of Equation 3.3 is that G appears in the denominator of the transfer function and will affect the system poles. That, in turn, means it affects the system's natural response and stability. If G remained a positive, real constant, as in Example 3.6, it would not be a problem. Unfortunately that is not the case.

An op amp is a multistage amplifier, and at high frequencies the various stages cut out as capacitive effects start bypassing the signal around the transistors or shorting it to ground. We can model each stage with the circuit of Figure 3.12. The poles introduced by this model unfortunately have a habit of moving into the RHP when encorporated in a feedback loop. This is most easily seen by considering a sinusoidal input signal.

A sinusoidal signal ($s = j\omega$) will be given a phase shift approaching $-90°$ at high frequencies for each amplifier stage that is in cutoff. Two such stages shift the phase by up to $-180°$ which makes the GH product negative. Under these conditions it possible for the denominator of Equation 3.3 to approach zero, which would give the system a very large gain. Having more than two stages just aggravates this problem, forcing the system poles into the RHP.

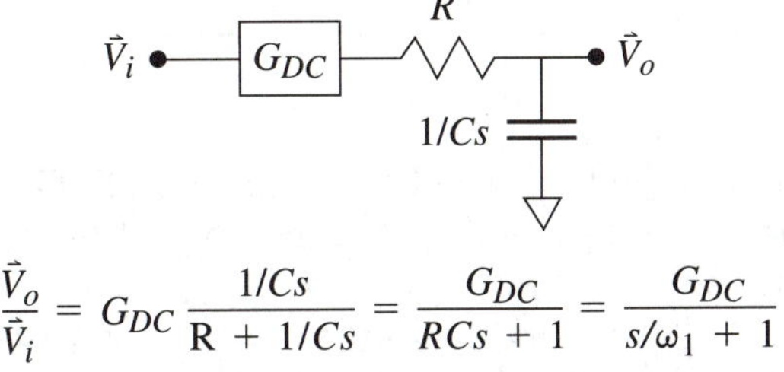

$$\frac{\vec{V}_o}{\vec{V}_i} = G_{DC}\frac{1/Cs}{R + 1/Cs} = \frac{G_{DC}}{RCs + 1} = \frac{G_{DC}}{s/\omega_1 + 1}$$

Figure 3.12 A model for an amplifier stage used to establish the stage's frequency response in terms of its cutoff frequency ω_1.

EXAMPLE 3.7

Is the amplifier of Example 3.6 stable if the op amp consists of three stages that each cut off at 1 rad/s?

Solution

The cutoff frequencies have been chosen to keep the numbers simple, but the choice of cutoff frequency does not affect the conclusion. Assuming the op amp d-c gain is typical, we could represent G as

$$G = \frac{200{,}000}{(s + 1)^3}$$

and since $H = \dfrac{1}{50}$,

$$\frac{\vec{V}_o}{\vec{V}_i} = \frac{G}{1 + GH} = \frac{200{,}000}{(s + 1)^3 + 4000} = \frac{200{,}000}{s^3 + 3s^2 + 3s + 4001}$$

$$\frac{\vec{V}_o}{\vec{V}_i} = \frac{200{,}000}{(s + 16.874)(s - 6.9370 \pm j13.747)}$$

The system has conjugate poles in the RHP and is consequently unstable. This invalidates the results of Example 3.6, since the amplifier will be unusable. Many op amps are internally compensated to prevent this problem. These amplifiers let one stage cause cutoff at such a low frequency that the overall gain is down to unity before the other stages enter cutoff. This makes the op amp frequency response look like that of a single stage and ensures stability for any resistive H.

When the feedback signal is added to $\vec{R}$ instead of subtracted, the result is called positive feedback, and Equation 3.3 becomes $\vec{C}/\vec{R} = G/(1 - GH)$. This is exactly the situation that was created inadvertently in Example 3.7. Intentional positive feedback is occasionally used within an overall negative feedback loop, but otherwise is something to be avoided. It produces poles in the RHP in even the simplest of systems.

EXAMPLE 3.8

Find $\vec{C}/\vec{R}$ for a positive feedback system where

$$G = \frac{100}{s + 1} \qquad \text{and} \qquad H = 0.2$$

Solution

$$\frac{\vec{C}}{\vec{R}} = \frac{G}{1 - GH} = \frac{100}{s + 1 - 20} = \frac{100}{s - 19}$$

The pole at $s = 19$ will cause an ever increasing natural response of Ke^{19t}.

3.4 BLOCK DIAGRAM REDUCTION

Most systems do not fit the form of the elemental feedback block diagram. Multiple summing junctions and sampling points may be present, and the system transfer function is not immediately evident. After completing this section you will be able to:

- Simplify block diagrams to obtain the system transfer function.
- Include systems with multiple inputs.

Generally the more variables sampled in a control system, the more control that can be exercised. Additional summing junctions are then also needed, and the block diagram complexity increases. If the block diagram can be reduced to the basic elements of Figure 3.9, the system transfer function can be found by inspection. Figure 3.13 shows legitimate consolidation operations.

Although we will limit our concern to simplifying given block diagrams, it is instructive to see how the permanent magnet d-c motor whose differential equation was derived in Chapter 2 could be represented by block diagrams. Figures 3.14a and 3.14b show the two main equations of the system and their corresponding block diagram representation. We have added a load torque so that we might investigate how the motor speed reacts to loading.

Combining the electrical and mechanical diagrams gives the overall block diagram (Fig. 3.14c). The torque summing junction has been broken down into two summing junctions. This makes the inner feedback loop more obvious so that it can be immediately replaced by a single block ($G = 1/Js$ and $H = k_v$). The sign of the load torque is also changed, with a corresponding change in the sign at its summing junction (Fig. 3.14d).

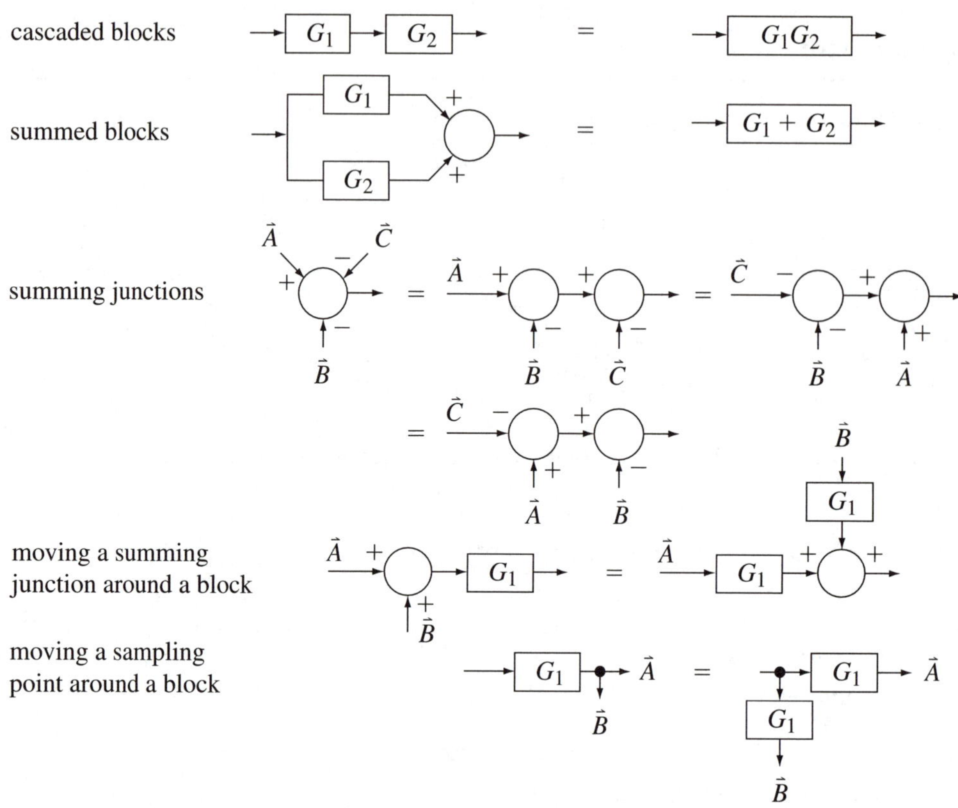

Figure 3.13 Block diagram relationships used to reduce and simplify complicated systems.

$$J\,\frac{d\omega}{dt} = \Sigma T = T_m - k_v\omega - T_l$$

T_m = motor torque

T_l = a variable loading torque

$\omega = \vec{\omega}e^{st}$, $T_m = \vec{T}_M e^{st}$, and $T_l = \vec{T}_L e^{st}$

$\vec{T}_L$, $\vec{T}_M$, $1/Js$, $\vec{\omega}$, k_v

$$Js\vec{\omega} = \vec{T}_M - k_v\vec{\omega} - \vec{T}_L$$

Figure 3.14a Mechanical portion of the block diagram for a permanent magnet d-c motor.

$$T_m = k_m i_a = \frac{k_m}{R_a}(v_i - e_b)$$

$v_i =$ motor input voltage
$e_b = k_m \omega$ induced back *emf*

$$v_i = \vec{V}_i e^{st}$$

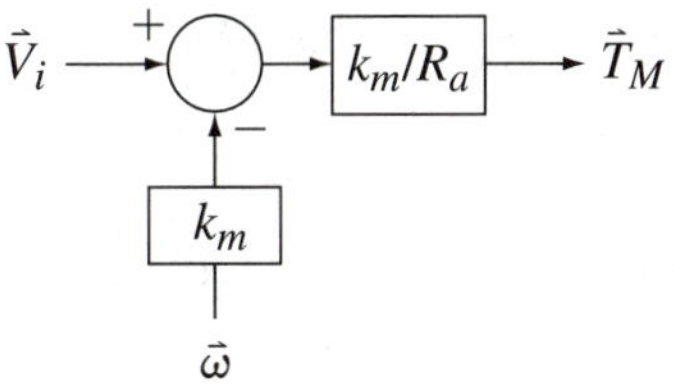

$$\vec{T}_M = \frac{k_m}{R_a}(\vec{V}_i - k_m \vec{\omega})$$

Figure 3.14b Electrical portion of the block diagram for a permanent magnet d-c motor.

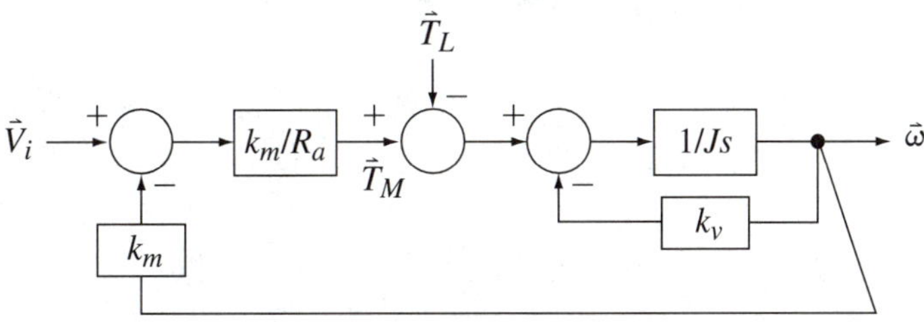

Figure 3.14c Block diagram of a permanent magnet d-c motor.

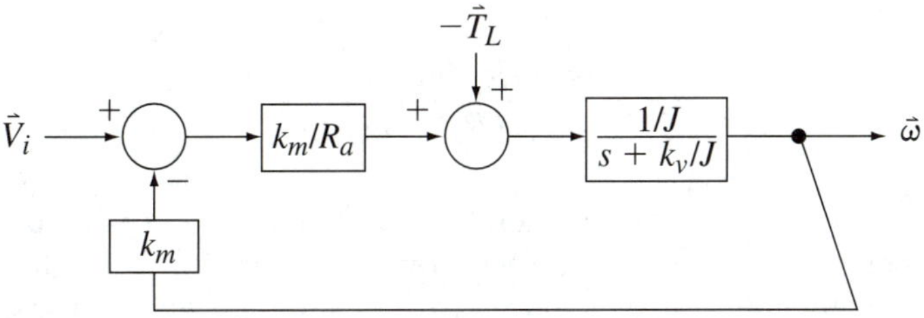

Figure 3.14d Reduced block diagram of a permanent magnet d-c motor.

Since there are two input signals, we will use superposition to determine how each affects the motor speed. We first let $\vec{T}_L$ be zero, which allows the torque summing junction to be removed and the effective forward and reverse transfer functions to be identified as

$$G = \frac{k_m/JR_a}{s + k_v/J} \qquad \text{and} \qquad H = k_m$$

the resulting transfer function is

$$\frac{\vec{\omega}}{\vec{V}_i} = \frac{\left(\dfrac{k_m/JR_a}{s + k_v/J}\right)}{1 + k_m\left(\dfrac{k_m/JR_a}{s + k_v/J}\right)} = \frac{k_m/JR_a}{s + k_v/J + k_m^2/JR_a}$$

Next $\vec{V}_i$ is reduced to zero. The minus sign on the voltage summing junction can be transferred to the torque summing junction. Now with the voltage summing junction removed, the forward and reverse transfer functions can be identified as,

$$G = \frac{1/J}{s + k_v/J} \qquad \text{and} \qquad H = k_m^2/R_a$$

giving

$$\frac{\vec{\omega}}{-\vec{T}_L} = \frac{\dfrac{1/J}{s + k_v/J}}{1 + (k_m^2/R_a)\dfrac{1/J}{s + k_v/J}} = \frac{1/J}{s + k_v/J + k_m^2/JR_a}$$

The superposition of these results gives the motor's speed in reaction to applied voltage and load torque as

$$\vec{\omega} = \frac{(k_m/JR_a)\vec{V}_i - (1/J)\vec{T}_L}{s + k_v/J + k_m^2/JR_a} \tag{3.4}$$

Although the derivation of Equation 3.4 has provided an opportunity to demonstrate some block diagram reduction techniques, the result has further significance. First, notice that the poles of the system are the same for either input. That means that an abrupt change in the input voltage, or in the mechanical loading will induce exactly the same type of natural response. This again confirms that a system's natural response does not depend on the forcing function. Secondly, we note that feedback is inherent in nature and is just a way of looking at the equations of a system. The feedback provided by the motor's back emf term can be duplicated and enhanced by having the motor drive an unloaded generator (tachometer) and adding its signal to the voltage summing junction. This is the basis of many automated speed control systems.

Finally, for constant input voltage and load torque, $s = 0$, and phasors become identical to the variables they represent ($v_i = \vec{V}_i e^{st} = \vec{V}_i$, etc.). Under these conditions Equation 3.3 for a given motor reduces to

$$\omega = k_1 v_i - k_2 T_l$$

which simply says a motor's steady state speed is proportional to the applied d-c voltage, and decreases with loading. This is consistent with our everyday experiences.

Suggestions for reducing block diagrams are about the same as for reducing circuits. Make the major simplifications one at a time to avoid confusion, and fully simplify intermediate transfer functions. Identify each simplification as an operation permitted by Figure 3.13. Notice that there is no simplification that moves a summing junction around a sampling point. Different but legitimate simplification routes always lead to the same final result.

EXAMPLE 3.9

Find the transfer function for the system in Figure 3.15a.

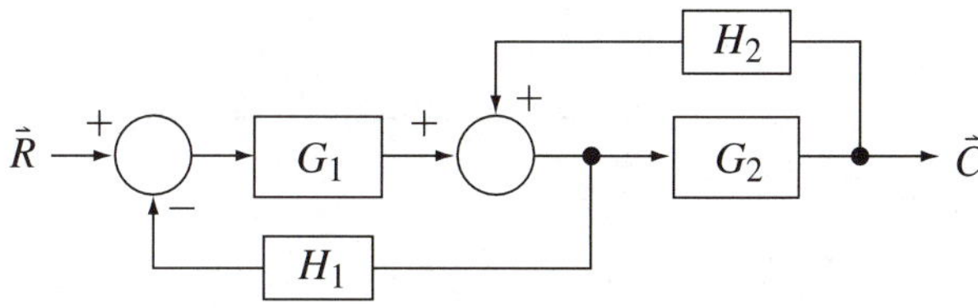

Figure 3.15a

Solution

The sampling point for H_1 and the summing junction next to it prevent us from seeing any elemental feedback loops. Either the summing junction must be moved around G_1 or the sampling point moved around G_2. Taking the latter approach, the signal sampled will be G_2 larger than before, so it must be reduced by the same factor in the feedback path (Fig. 3.15b).

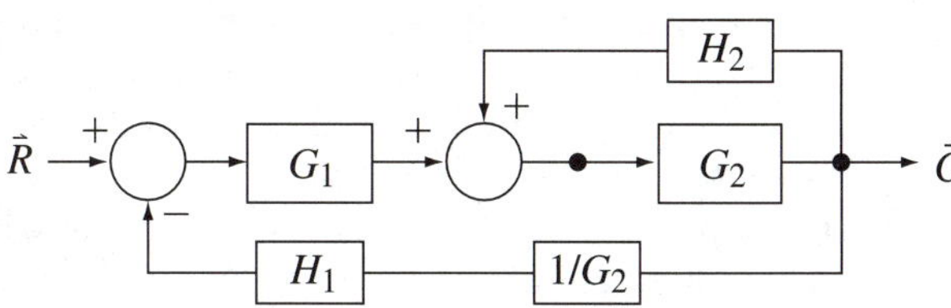

Figure 3.15b

Now the positive feedback loop formed by G_2 and H_2 is obvious, and reduced to a single block (Fig. 3.15c).

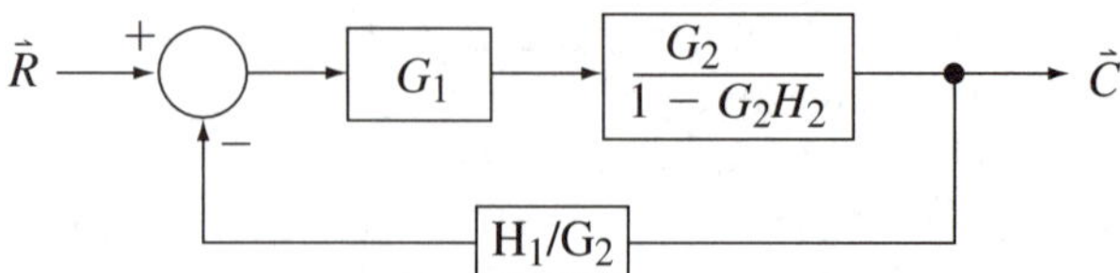

Figure 3.15c

At this point the final transfer function can be obtained as

$$\frac{\vec{C}}{\vec{R}} = \frac{\dfrac{G_1G_2}{1 - G_2H_2}}{1 + \dfrac{G_1G_2}{1 - G_2H_2}\dfrac{H_1}{G_2}} = \frac{G_1G_2}{1 + G_1H_1 - G_2H_2}$$

3.5 STABILITY TESTS

Any system with poles in the RHP is unusable. It would be unfortunate to spend a considerable amount of time investigating the forced response of a system only to find that it has an unstable natural response. The importance of the stability question has led to the development of a number of tests for stability. After completing this section you will be able to:

- Explain ways to test a system for stability.
- Define the unit impulse function.
- Use the Routh–Hurwitz stability test for polynomials.

One of the things that usually cannot be done to test a system's stability is to turn it on and see what happens. Watching a motor snap its shaft in an unstable system is both expensive and potentially dangerous to test personnel. Instead, a great deal of effort has been expended in developing computer simulation tools to evaluate stability.

Given a system, we can test it for stability simply by finding where its poles are located. If the system gain is a parameter, the poles will move as the gain varies. This movement is smooth and gradual, with each pole tracing a path on the s plane. A plot of these paths as a function of the gain is called a *root locus diagram.* This diagram is particularly useful in control theory, because it not only identifies when poles move in or out of the RHP, but also shows how close poles come to the $j\omega$ axis, which measures the system's relative stability. We will explore this technique briefly in the next section.

If a rectangular pulse is applied to a system, its application stimulates the system's natural response along with the forced response. When it turns off, it again stimulates the natural response, but is no longer forcing a response. By reducing the pulse's width to zero while simultaneously increasing its amplitude to retain a finite energy, we obtain an impulse function. The impulse is over so quickly that we only see the results of its turning off, which is the natural response. Using simulation, therefore, we can determine the stability of a system by finding its impulse response and seeing if it dies out or just keeps growing.

Mathematically, the unit impulse has properties that make it very useful in other areas of system or signal analysis. Its definition is given in Figure 3.16.

The *Routh–Hurwitz Criterion* is a stability test that does not require use of a computer. It can, however, become cumbersome for polynomials higher than the fourth degree, so we will limit our use to polynomials of up to the fourth degree. For such a polynomial, i.e., $s^4 + a_3s^3 + a_2s^2 + a_1s + a_0$, to have no roots in the RHP, it must meet two tests:

Test 1: All the lower a_n must be present and have the same sign. This test is *sufficient* for polynomials of the first or second degree. This test simply recognizes that a root in the RHP would have the form $(s - r)$. Its presence will produce a negative term in first- or second-degree polynomials, but its sign may be swamped out by the contributions of LHP roots in larger polynomials.

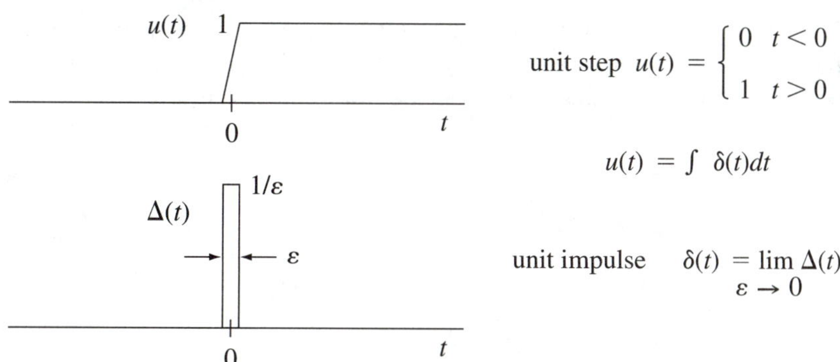

Figure 3.16 Step and impulse functions are frequently used test signals. When they turn on, they activate a circuit's natural response. The step also produces a d-c ($s = 0$) forced response, while the impulse dumps energy into the circuit at $t = 0$ and then lets the circuit's natural response take over. Although the ideal impulse occurs in zero time, any pulse whose duration is short compared to the time constants of the circuit under test will produce the same effect.

Test 2: For polynomials of the third or fourth degree, form the array

$$\begin{matrix} a_1 & a_0 & 0 \\ a_3 & a_2 & a_1 \\ a_5 & a_4 & a_3 \end{matrix} \qquad (a_5 = 0, \text{ but helps show the array pattern})$$

For a third-degree polynomial, evaluate the determinant for the 2 × 2 square array whose upper left-hand corner is a_1 For a fourth-degree polynomial, evaluate the determinant of the 3 × 3 array whose upper left-hand corner is a_1. The value of these determinants must be greater than zero for stability. The determinants needed for these two cases are shown next. (Determinants for arrays of the fifth or higher degree polynomials require use of a technique called the *method of minors.* See an appropriate mathematics text.)

$$D_2 \equiv \begin{vmatrix} a_1 & a_0 \\ a_3 & a_2 \end{vmatrix} = \begin{vmatrix} a_1 & a_0 \\ a_3 & a_2 \end{vmatrix} = a_1a_2 - a_0a_3 > 0$$

The arrows show the terms multiplied and the sign of the product.

$$D_3 \equiv \begin{vmatrix} a_1 & a_0 & 0 \\ a_3 & a_2 & a_1 \\ 0 & a_4 & a_3 \end{vmatrix} = a_1a_2a_3 - a_1^2a_4 - a_0a_3^2$$

The pattern for finding D_3 is most easily remembered by repeating the first two columns and considering only three-term diagonals. Remember, the procedure given here has been specialized to polynomials of up to the fourth degree.

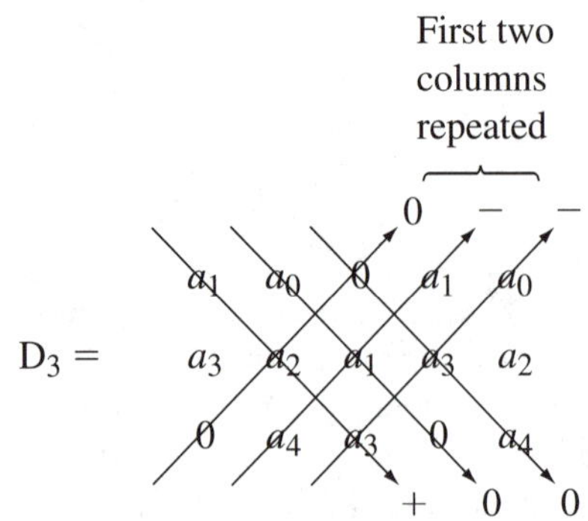

EXAMPLE 3.10

Determine the conditions on K for stability if a system has the characteristic equation

$$s^3 + 2s^2 + Ks + 8$$

Solution

Test 1: The cubic polynomial has all of its lower powers of s terms present, and they are all positive, provided $K > 0$.

Test 2: The array is

$$\begin{matrix} K & 8 & 0 \\ 1 & 2 & K \\ 0 & 0 & 1 \end{matrix}$$

and since the polynomial is of the third degree, we only need to evaluate the 2×2 determinate

$$D_2 = \begin{vmatrix} K & 8 \\ 1 & 2 \end{vmatrix} = 2K - 8$$

$$2K - 8 > 0 \quad \text{for stability}$$

$$\therefore K > 4 \quad \text{for stability}$$

The most restrictive requirement must be met.

Do not gloss over Test 1 just because it is simple. Its requirements are essential if incorrect conclusions are to be avoided. The conditions set by Test 2 can become a little trickier to deduce in a fourth-degree polynomial.

EXAMPLE 3.11

A particular system is tested for stability using Routh–Hurwitz. Test 1 requires $K > 0$ for stability. Test 2 results in the requirement that

$$-K^2 + 10K + 11 > 0$$

Provide an unambiguous statement on the K values for which the system is stable.

Solution

The Test 2 requirement is converted to

$$K^2 - 10K - 11 < 0$$

and factored, temporarily assuming an equality. The result is

$$(K + 1)(K - 11) < 0$$

The inequality says that the product of the two factors must be negative for stability. That, in turn, requires the factors to be opposite in sign, which occurs if $-1 < K < +11$. However, Test 1 requires $K > 0$, so the system is stable only as long as $0 < K < 11$. If $K = 0$ or 11, at least one pole is on the $j\omega$ axis.

3.6 MATLAB LESSON 3

This lesson will demonstrate some tools for simplifying block diagrams, investigating stability, and expanding our graphing options. In addition, conditional branching options will be demonstrated. After completing this section you will be able to:

- Use MATLAB for block diagram reduction and system simulation.
- Select a multiple-graph option.
- Create a root locus plot.
- Discuss conditional branching options.

As with previous MATLAB lessons, you need to perform the command statements indicated to benefit from the lesson.

MATLAB EXAMPLES

MATLAB Toolboxes are available for specialized subject areas. The Controls toolbox contains functions that are helpful for some of the tasks encountered in this chapter. If you are in an unfamiliar computer lab, you may find that the MATLAB commands you are used to using are not available. This indicates that the corresponding toolbox has not been installed.

series feedback cloop

```
>gnum=10; gden=[1 2];                         % defines the transfer function
                                                G = 10/(s + 2)
>hnum=1; hden=[1 0];                          % defines the transfer function
                                                H = 1/s
>[nu,de]=series(gnum,gden,hnum,hden)          % finds the transfer function
                                                GH
>[num,den]=feedback(gnum,gden,hnum,hden)      % finds the transfer function
                                                G/(1 + GH)
>[num,den]=feedback(gnum,gden,hnum,hden,1)    % finds the transfer function
                                                G/(1 - GH)
```

The fifth argument of the **feedback** function describes the sign used for the feedback connection at the summing junction. A -1 defines negative feedback and is the default value. In control theory, the H transfer function often is unity. The **cloop** command is this special case of **feedback.**

```
>[num,den]=feedback(gnum,gden,1,1,-1)   % finds the function G/(1 + G)
>[num,den]=feedback(gnum,gden,1,1)      % does the same thing
>[num,den]=cloop(gnum,gden)             % so does this
```

impulse step

The response of a system to a step or impulse input may be obtained by specifying the transfer function numerator and denominator and the time range of interest. The impulse response is a particularly good way to check a system for stability.

```
>num=[1 0]; den=[1 1 4];   % a slightly resonant transfer function
>impulse(num,den)          % or step(num,den)—lets program determine range
                             of t.
>t=linspace(0,4);          % if you want to see details of first two peaks
>impulse(num,den,t)
>den=[1 -1 4];             % set up an unstable system
>impulse(num,den)          % instability is obvious
```

Although it is rarely of interest, we can fool the program into giving the forced response to a ramp or parabolic input by multiplying the denominator by s or s^2, respectively, and using the step command.

```
>num=4            % get a low-pass filter
>den=[1 1 4];     % regain a stable system
>step(num,den)    % gives the response to a unit step input
>den=[den 0];     % multiply denominator by s
>step(num,den)    % this is the response to a unit ramp forcing function
>den=[den 0];     % multiply denominator by s again
>step(num,den)    % this is the response to a 0.5* unit parabola forcing
                    function
```

subplot

The Figure Window may be subdivided into multiple plots. Each plot may then be addressed separately, and all of the graph related commands are available for modifying and labeling each plot. **subplot(a,b,c)** specifies that the Figure Window be divided into **a** rows and **b** columns of plots, and selects plot **c** as the active plot. The plots are numbered sequentially from left to right across the first row, followed by the second row, etc. **subplot(3,3,5)** would specify nine graphs, arranged in 3 rows and 3 columns, and selects the plot in the center of the array as the active plot.

```
>x=linspace(-2,2);          % define an independent variable
>y=2*x;                     % define a dependent variable
>z=y.^2;                    % define another dependent variable
>subplot(1,2,1);            % sets up two side-by-side plots, and makes the
                              left plot active
>plot(x,y); title('y=2x')   % all of the graphical commands apply only to
                              the active plot
                            % check the next command carefully before
                              executing!
>subplot(1,2,2);            % the right-hand graph is now the active one
>plot(x,z); ylabel('z')     % all of the graphical commands apply only to
                              the active plot
```

If we change the first two arguments of the subplot command, even unintentionally, it forces a change in the graph format and may destroy any existing graphs.

```
>subplot(2,1,1)   % wanted to make labeling changes on the left graph, but
                    goofed
```

When dealing with complex numbers, we frequently want to graph both magnitude and phase (or real and imaginary parts) on the same sheet of paper. The subplot feature allows us to do this. Many MATLAB functions also use the subplot feature, and we need to know the subplot command if we want to modify the axis scale or labeling on a graph it has created. After printing or saving the page, you need to restore the original single-graph format.

```
>subplot(1,1,1)   % returns the Figure Window to single-graph format
```

for-end while-end if-elseif-else-end

MATLAB permits the same types of conditional programming loops found in general computer languages. When they are used within the command window, the loop execution is suspended until the **end** command is entered. Errors will consequently not be detected until then. The MATLAB cursor is also turned off within the loop.

For loops evaluate a set of commands a fixed number of times.

```
>for i=0:306
     x(i+1)=sin(i*pi/307);  % lowest allowed element number is 1 (the
                              indenting is optional)
end
>x
```

For loops can be nested.

```
>for i=1:3
     for j=1:3
          x(i,j)=i*j;
     end
end
>x
>clear
```

While loops evaluate a group of expressions until a test condition is met.

```
>t=0; x=0; y=10;  % make sure loop starts
while x<y
     t=t+.0001;   % Solve to find t, where 1.5t=sin πt, t ≠ 0
     x=1.5*t;
     y=sin(pi*t);
end
>t
>clear
```

If-else constructs provide alternative paths based on the test condition. The "=" sign is used as an assignment operator in MATLAB. It assigns the value of the expression on its right to the variable named on its left. To test if an integer variable k "equals" zero, we need to use "==".

```
for t=1:300
     if t==100
          y(t)=pi/10;  % correct the indeterminate value of y(100)
     elseif t==200
          y(t)=pi/10;  % correct the indeterminate value of y(200)
     else y(t)=1000*sin(pi*t/100)/((t-100)*(t-200));
     end
end
>plot(y,'o')
```

The following commands will plot the location of the roots of a polynomial. We want them all on one line, so we can repeat the entire line of commands.

```
>k=0; p=[1 3 10 k+1 k]; r=roots(p); plot(r,'*')  % the polynomial is of
                                                    fourth degree
>hold
```

Now use the up arrow to bring the commands back, modify the value of k, and execute the line.

```
>k=1; p=[1 3 10 k+1 k]; r=roots(p); plot(r,'*')
>k=2; p=[1 3 10 k+1 k]; r=roots(p); plot(r,'*')   % etc. continue as long
                                                    as you wish
>hold
```

The result is an example of a root locus plot. This is also an obvious application for a **for** loop. We may as well add a few embellishments.

```
>k = 0; p=[1 3 10 k+1 k]; r=roots(p); plot(r,'o')  % mark the k = 0 points
                                                     with a 'o'
>hold
>for k = 1:100
p=[1 3 10 k+1 k]; r=roots(p); plot(r,'.')          % use '.' to draw the
                                                     locus
end
p=[1 3 10 k+1 k]; r=roots(p); plot(r,'*')          % mark the k = max
                                                     ending points with
                                                     '*'
>hold
```

CHAPTER SUMMARY

Specializing the voltage–current relationships of the three passive circuit elements for the e^{st} signal allowed their characteristics to be expressed in the form of a ratio of voltage and current phasors called an impedance. It was shown that if a set of signals obey Kirchhoff's laws, then the phasors representing those signals also obey them. With signals represented by phasors and circuit elements represented by impedance, all of the circuit simplification techniques typically introduced in a d-c circuits course can be used. Transfer functions can be derived entirely in the s domain, without reverting to differential equations.

Systems are formed by interconnecting various kinds of devices. The systems may be described using block diagrams, where each block states the transfer function of the device it represents. These interconnections usually take the form of a negative feedback system, in which the actual and desired outputs are compared so that the system can automatically correct itself. Adding gain to such a system can set the system's transfer function at a reliable value, but may also make its natural response unstable. Block diagrams may be systematically simplified and reduced to a final overall transfer function for the system.

If a system is unstable, it is unusable. The Routh–Hurwitz test provides one method of predicting values of a parameter, usually gain, that can cause the system to become unstable. A unit impulse is a good simulation input to use when testing for stability.

MATLAB provides tools that can assist us in simplifying feedback systems. It also provides a means of testing for stability through simulation or through pole locations in the s plane.

PROBLEMS

Section 3.1

1. Determine the driving point impedance of the circuit in Figure P3.1.

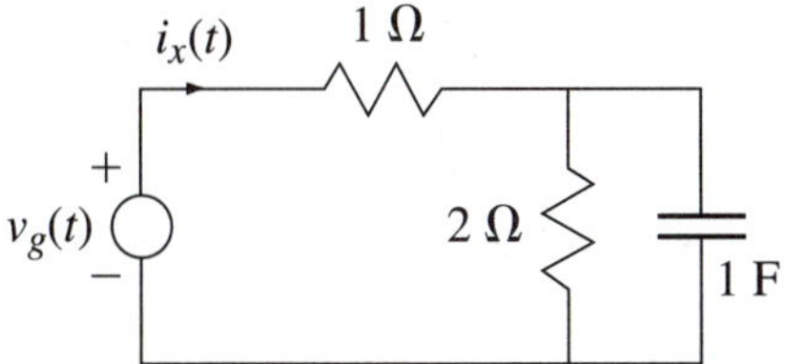

Figure P3.1

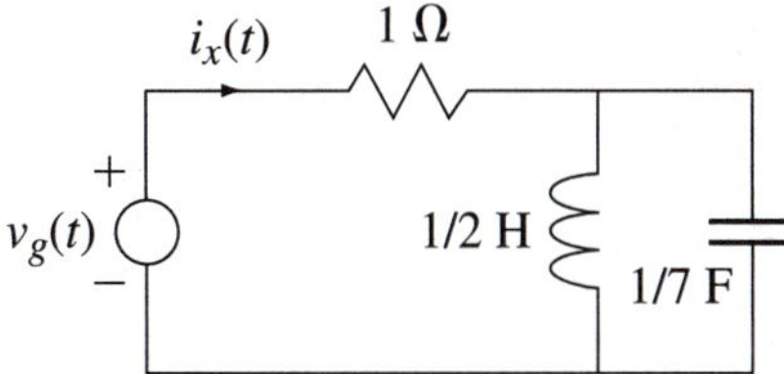

Figure P3.2

2. Determine the driving point impedance of the circuit in Figure P3.2.
3. Determine the driving point admittance of the circuit of Figure P3.3.

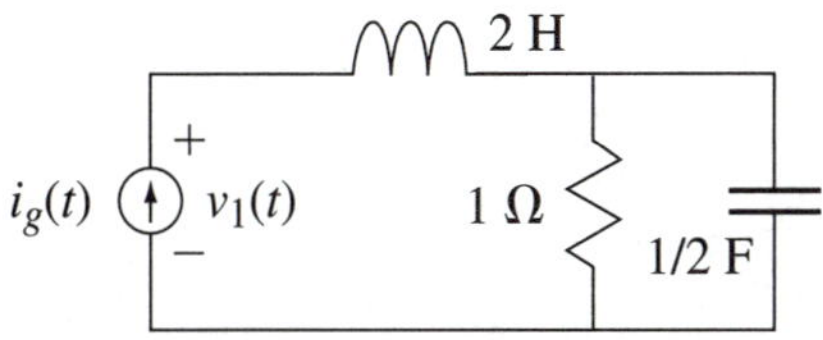

Figure P3.3

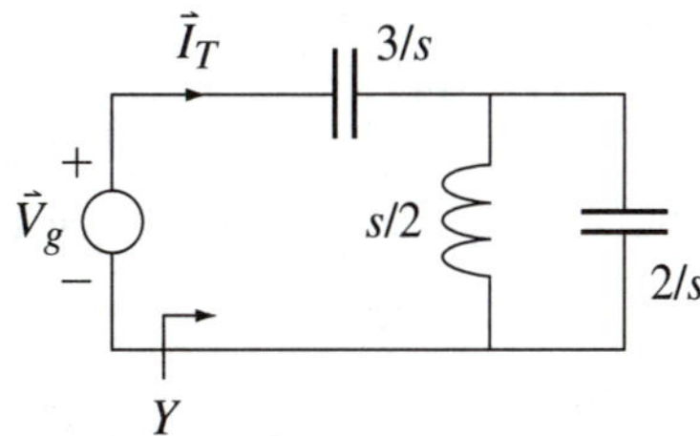

Figure P3.4

4. Determine the driving point admittance of the circuit of Figure P3.4.

Section 3.2

5. For the circuit of Figure P3.5, find and sketch the pole-zero diagram for:
   a. the transimpedance $\vec{V}_o/\vec{I}_g$.
   b. the driving point impedance Z.

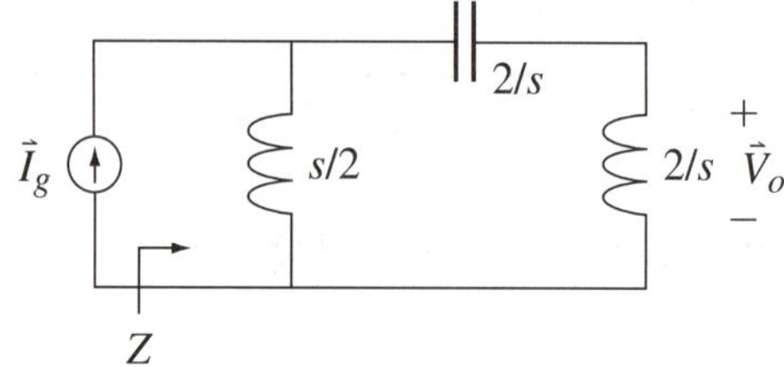

Figure P3.5

6. Find the transfer function $\vec{I}_x/\vec{V}_g$ in the circuit of Figure P3.6.

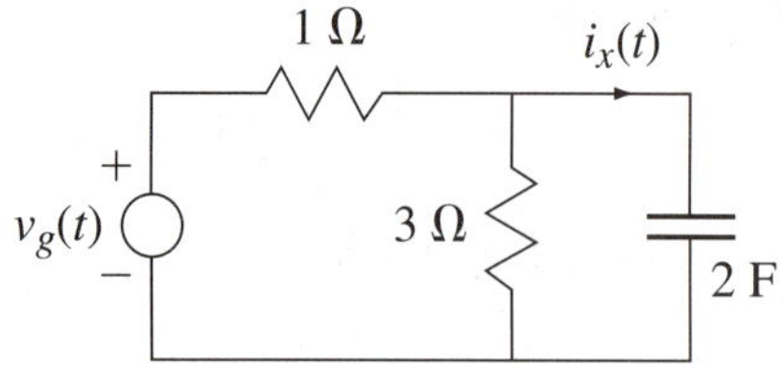

Figure P3.6

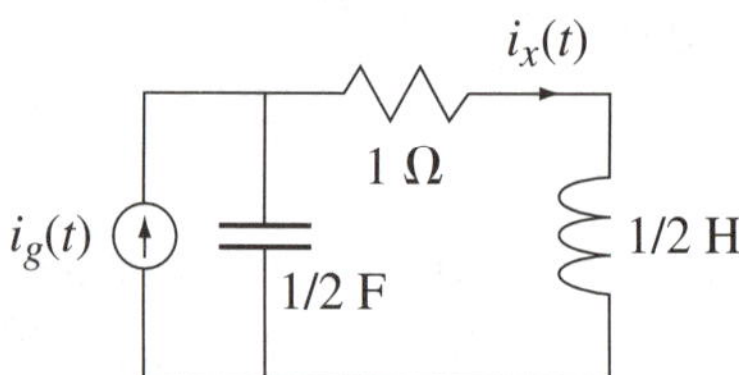

Figure P3.7

7. Find the transfer function $\vec{I}_x/\vec{I}_g$ in the circuit of Figure P3.7.
8. Find the transimpedance $\vec{V}_o/\vec{I}_g$ in the circuit of Figure P3.8.

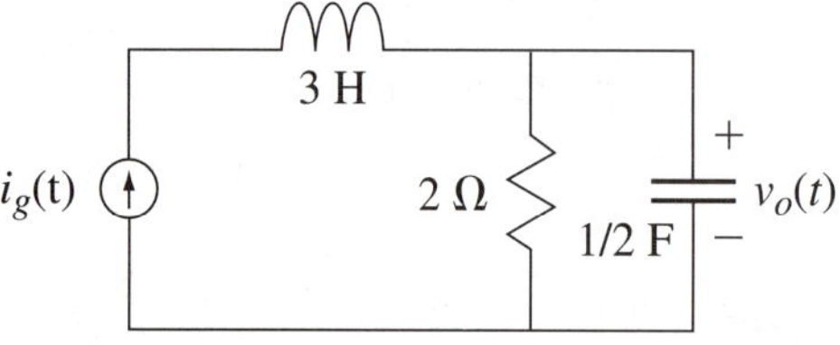

Figure P3.8

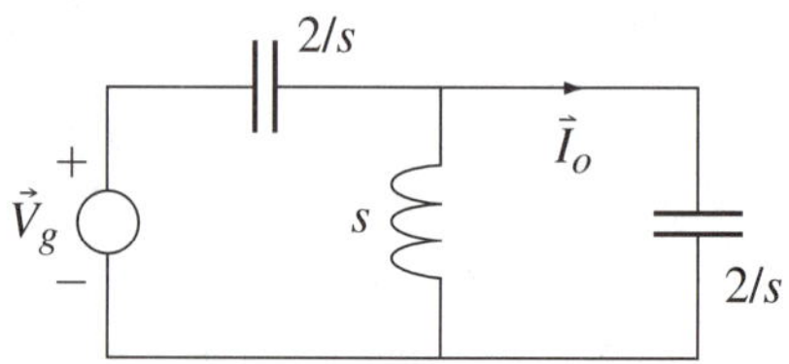

Figure P3.9

9. Find the transadmittance $\vec{I}_o/\vec{V}_g$ in the circuit of Figure P3.9.
10. Find the transfer function for the Wien bridge notch filter of Figure P3.10.

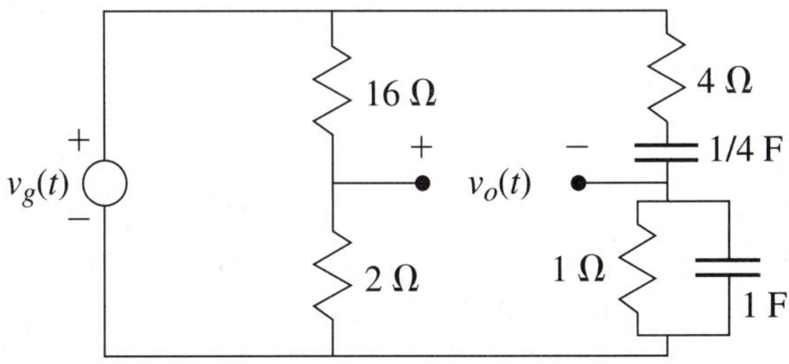

Figure 3.10

11. The circuit of Figure P3.11 is to be used to implement the transfer function indicated. Find expressions for the values of L_1, L_2, and C in terms of the a coefficients.

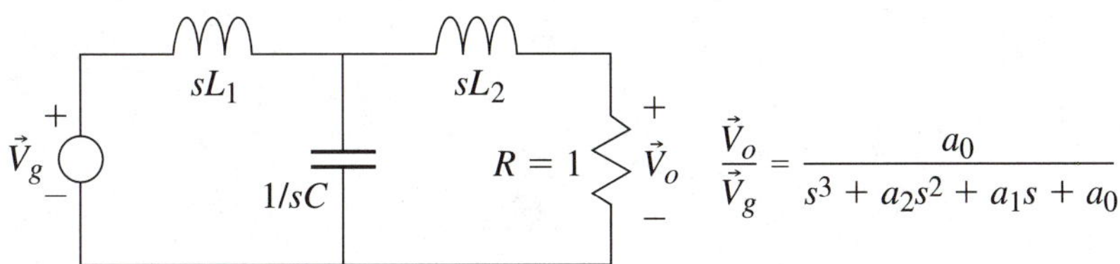

Figure P3.11

12. Find $\vec{V}_o(s)/\vec{V}_g(s)$ for the low-pass filter of Figure P3.12.

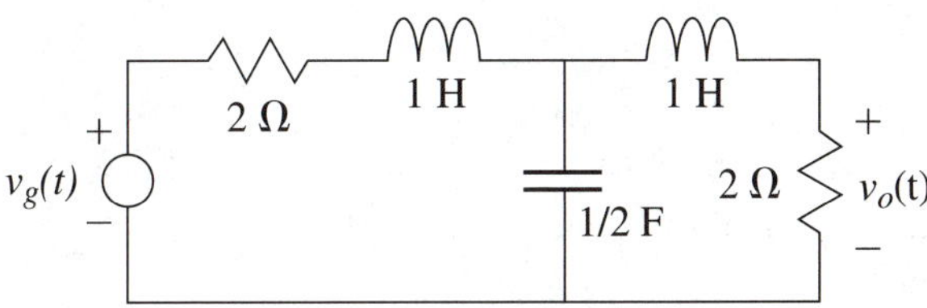

Figure P3.12

13. The circuit of Figure P3.13 is a Colpitts oscillator and contains a dependent current source. Determine the conditions necessary for oscillation. (*Hint:* $s = j\omega$ for an oscillator.)

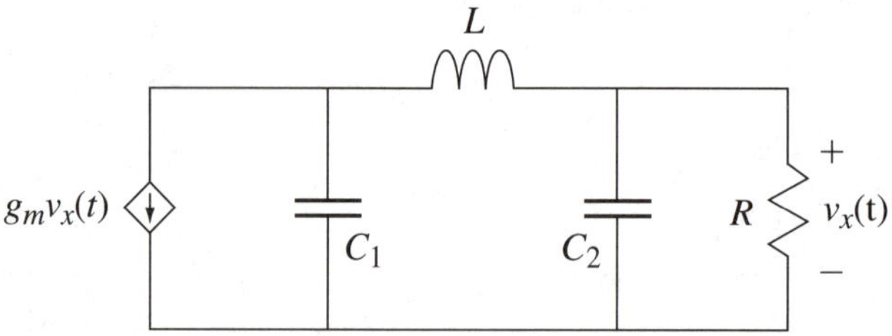

Figure P3.13

Section 3.3

14. The system of Figure P3.14 has $100 < G < 10{,}000$. Determine the range in $\vec{C}/\vec{R}$ if $H = 0.10$.

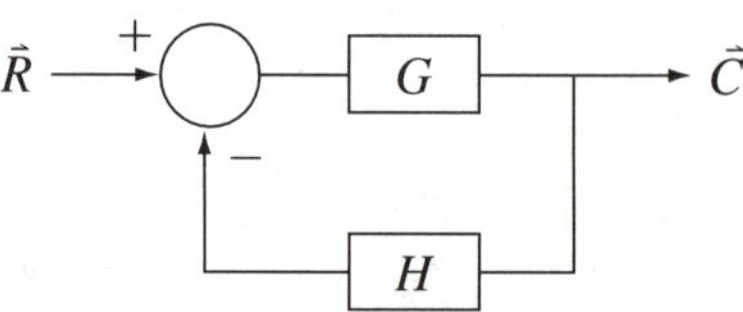

Figure P3.14

15. In the system of Figure P3.14, what value of G is needed if $\vec{C}/\vec{R}$ is to be within 1% of $1/H$?

16. In the system of Figure P3.14,

$$G = \frac{10^3}{s^3 + 3s^2 + 8s + 5}$$

Determine the d-c gain and system stability if:
a $H = 0.10$ b. $H = 0.002$

17. The speed control system of Figure P3.17 is tested to obtain the calibration chart on the left at unit loading. The motor's back-emf may be assumed to equal the tach voltage magnitude. The motor armature resistance is 0.01 Ω, and the power amp has unity gain. The tach is connected such that $\vec{V}_{\text{tach}}$ and $\vec{V}_i$ are always of opposite polarity.
    a. With the tach disconnected and $\vec{V}_x$ set to 12.0 V, the motor runs at 1100 RPM. If the loading now doubles so that the armature current doubles, determine the new motor speed.

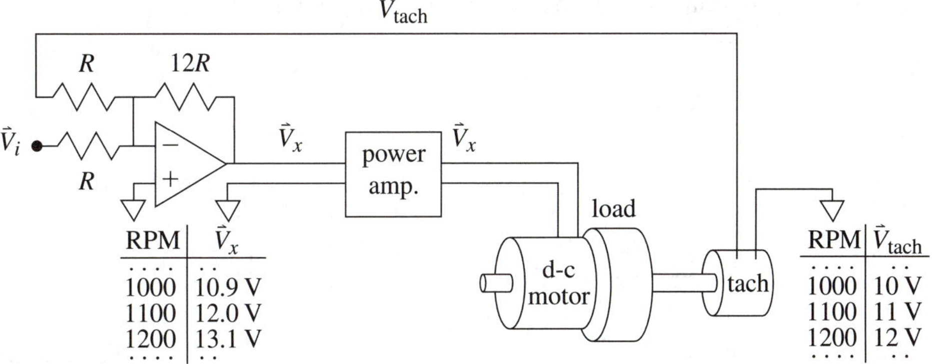

Figure P3.17

b. With the tach connected, the system is readjusted for 1100 RPM at unit load. Determine the $\vec{V}_i$ required.
c. Determine the new motor speed if the loading doubles in the closed loop system.

Section 3.4

18. Find an expression for $\vec{C}/\vec{R}$ in the system of Figure P3.18.

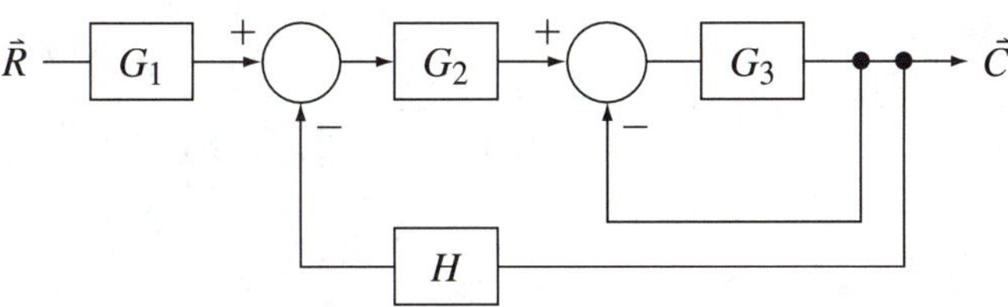

Figure P3.18

19. Find an expression for $\vec{C}/\vec{R}$ in the system of Figure P3.19.

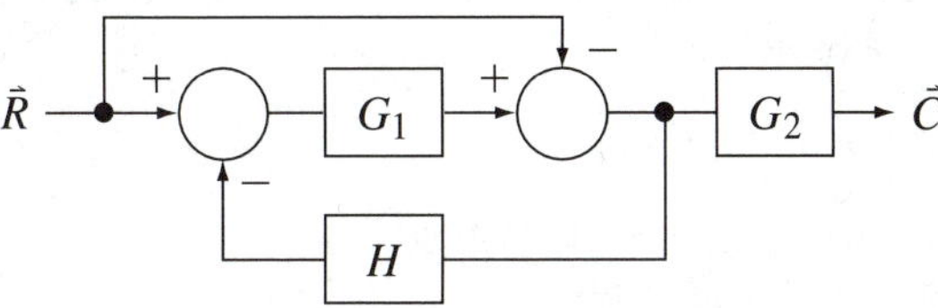

Figure P3.19

20. Find an expression for $\vec{C}$ in the system of Figure P3.20.

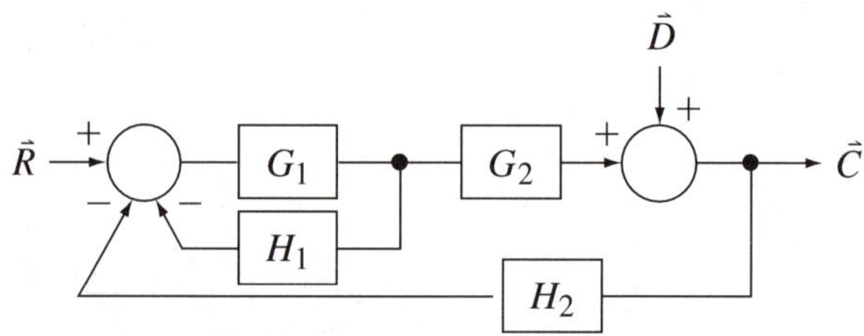

Figure P3.20

21. The transfer functions in the motor speed control system of Figure P3.21 are based on speed in radians/second, torque in newton-meters, and voltage in volts. K is provided by a variable-gain amplifier.
   a. Given $K = 5$, determine the motor's no-load speed for a d-c input of 6 V.
   b. Given the full-load torque is 20 N-m determine the full-load speed.
   c. If K is increased to 50, what new input voltage would be required to achieve the same no-load speed, and what would be the new full-load speed?

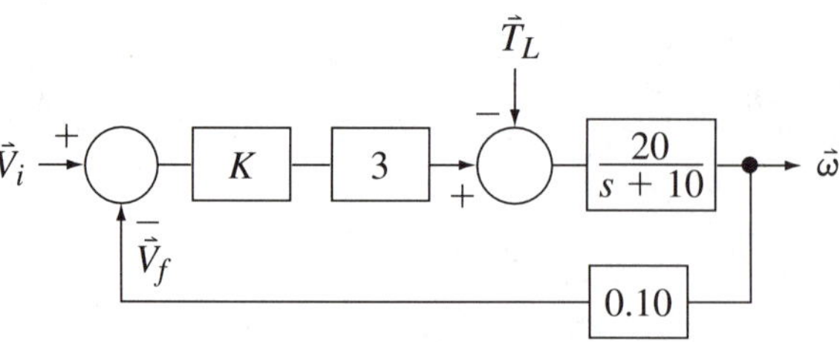

Figure P3.21

Section 3.5

22. Use the Routh–Hurwitz test to determine the range of K for which the following negative feedback systems are stable.

   a. $G = \dfrac{K}{s(s^2 + 2s + 6)} \qquad H = 1$

   b. $G = \dfrac{K}{s(s^2 + 2s + 6)} \qquad H = \dfrac{1}{s + 1}$

23. Use the Routh–Hurwitz test to find the range of K for which the following polynomials will have no roots in the RHP.
    a. $s^3 + 2Ks^2 + Ks + 24$
    b. $s^3 + 10s^2 + Ks + 24$
    c. $s^4 + Ks^3 + 20s^2 + 8s + 16$
    d. $s^4 + 8s^3 + 20s^2 + 8s + K$
    e. $s^4 + 8s^3 + 20s^2 + Ks + 24$

Section 3.6

24. Provide a set of MATLAB instructions that will produce the graph shown in Figure P3.24 using a 900-element vector for $0 \leq x \leq 3.9$.

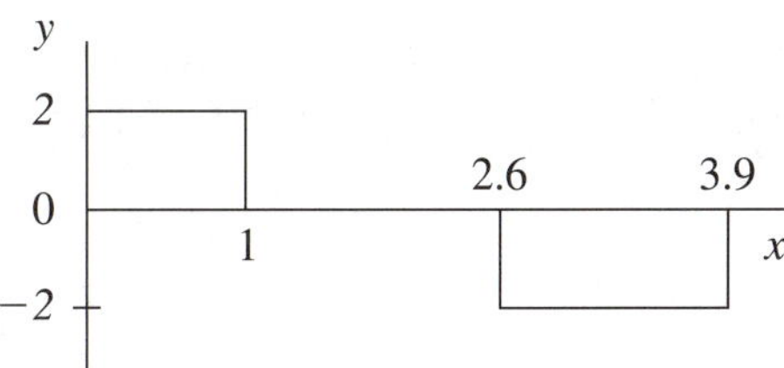

Figure P3.24

25. A transfer function $H(s)$ has been evaluated along the path $s = j\omega$ where ω runs from 0 to 20. Give a MATLAB command sequence that will plot $|H|$ and $\angle H$ vs. ω in two properly titled graphs arranged in a column.
26. Given numG, denG, numH, and denH for a negative feedback system, give a MATLAB two-command sequence that will determine if the system is stable.
27. Explain what is produced by the indicated command sequence.
    a.
    ```
    >x=[];
    >for i=1:10
         x=[x i 0];
    end
    ```
    b.
    ```
    >x=[]; i=3;
    >while size(x) < 5
    i=i-1;
    x=[x i];
    end
    ```

4

FREQUENCY RESPONSE

OUTLINE

OBJECTIVES

1. Predict the features of a circuit near resonance.
2. Estimate a frequency response using Bode straight-line approximations.
3. Define the decibel, and discuss its uses.
4. Apply the Bode SLA technique to systems with mild resonances.
5. Use the MATLAB frequency response tools.

INTRODUCTION

By retaining the s variable in our transfer functions, we are poised to handle questions regarding any of the possible e^{st} signals. In point of fact, however, exponentially increasing forcing functions would eventually destroy real systems, and exponentially decreasing forcing functions might easily be obscured by even transient natural responses. The signal of greatest interest, by far, is the pure sinusoid, and the frequency response of a system is understood to mean evaluating the transfer or driving point functions along the path $s = +j\omega$.

Techniques have been developed over the years for estimating the frequency response of a system while making a minimum number of calculations. These techniques often provide a point of view that can also be helpful in the design process.

Today, computers are routinely used to plot the exact frequency response of driving point or transfer functions. The frequency response of a system is measurable and is likely to be a criterion in its selection among competing products. The abilities to estimate frequency response and to use the tools that graph or measure frequency response are essential skills in engineering and technology.

4.1 RESONANT SYSTEMS: *L-C* CIRCUITS

Systems with poles or zeros very close to the $j\omega$ axis are resonant systems. They have very vigorous responses over small frequency ranges, and their behavior at frequencies far from resonance is generally considered unimportant. A resonant circuit is dominated by inductive and capacitive circuit elements. After completing this section you will be able to:

- Predict the features of a circuit near resonance.
- Estimate the resonant frequency and bandwidth of the circuit.
- Estimate the maximum or minimum value of the function.
- Use a pole-zero diagram to graphically calculate function values.

When evaluated for $s = j\omega$, the impedances of inductors and capacitors are purely imaginary and of opposite sign. As a result, although each separately may help limit current, a combination of the two may cause one to cancel the effect of the other, leading to a striking circuit behavior known as *resonance.*

$$s = j\omega$$

$$Z_L = sL = j\omega L \qquad Z_C = \frac{1}{sC} = \frac{1}{j\omega C} = \frac{-j}{\omega C} \tag{4.1}$$

An R-L-C circuit is said to be resonant when its driving point impedance is purely real (resistive). If an inductor and capacitor are in series, the circuit *resonates* when

$$Z(s) = Ls + \frac{1}{Cs} = \frac{L(s^2 + 1/LC)}{s}$$

$$Z(j\omega_o) = 0 \text{ where } \omega_o = \frac{1}{\sqrt{LC}}$$

$$Z(s) = Ls // \frac{1}{Cs} = \frac{s}{C(s^2 + 1/LC)}$$

$$Z(j\omega_o) = \infty \text{ where } \omega_o = \frac{1}{\sqrt{LC}}$$

Figure 4.1 Ideal series and parallel resonant circuits.

the magnitudes of the inductive and capacitive impedances are equal, and the resulting circuit impedance looks like a short circuit. Parallel resonance, sometimes called *antiresonance,* occurs under the same conditions, but the circuit impedance looks like an open circuit. Figure 4.1 summarizes these facts.

EXAMPLE 4.1

Determine the resonant frequencies in the circuit of Figure 4.2a.

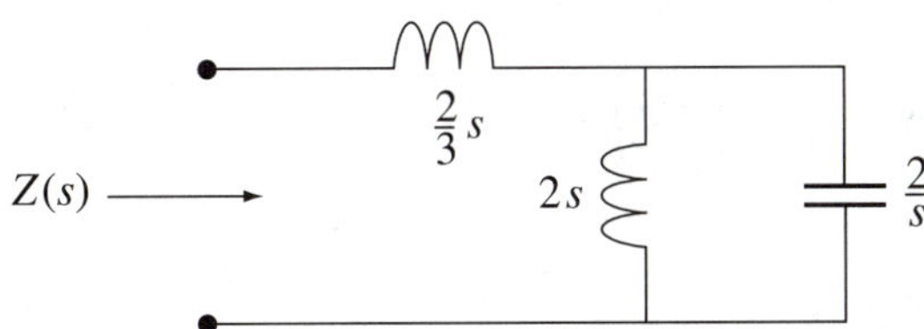

Figure 4.2a

Solution

The 2 H–1/2 F combination will produce a parallel resonance at $\omega_o = 1$ rad/s. Above this frequency the parallel combination looks capacitive and can produce series resonance with the (2/3)-H inductor. To identify where that occurs, an expression for $Z(s)$ is found by combining impedances, and a common denominator is obtained:

$$Z(s) = \frac{2}{3}s + \frac{(2s)(2/s)}{2s + 2/s} = \frac{\frac{4}{3}s^2 + \frac{4}{3} + 4}{2s + 2/s} = \frac{2}{3}\,\frac{s(s^2 + 4)}{s^2 + 1}$$

The usual checks should be made at $s = 0$ and at $s = \infty$ to show any discrepancies between the equation and the circuit (Figure 4.2b). Then we factor each polynomial to find the pole-zero form for $Z(s)$. In this case it can be done by inspection, with the result as shown in Figure 4.2c. The impedance for sinusoids is shown in Figure 4.2d. It is the impedance along the $+j\omega$ axis.

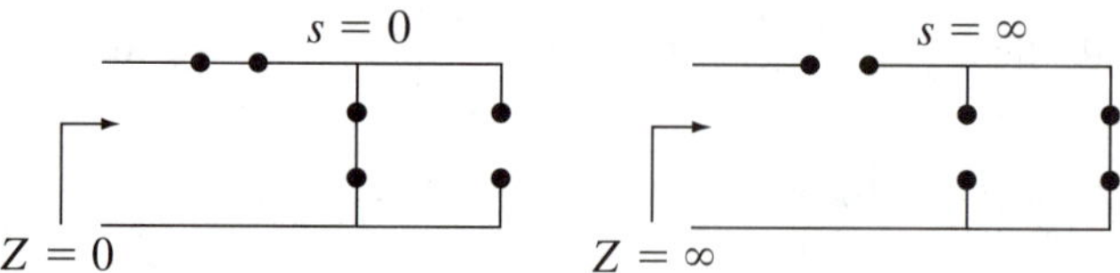

Figure 4.2b Sketches used to check the impedance expression at extreme frequencies.

$$Z(s) = \frac{2}{3}\,\frac{s(s+j2)(s-j2)}{(s+j)(s-j)}.$$

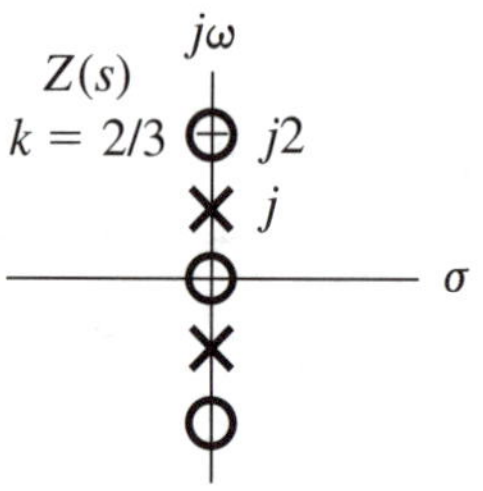

Figure 4.2c The pole-zero diagram of the circuit of Figure 4.2a.

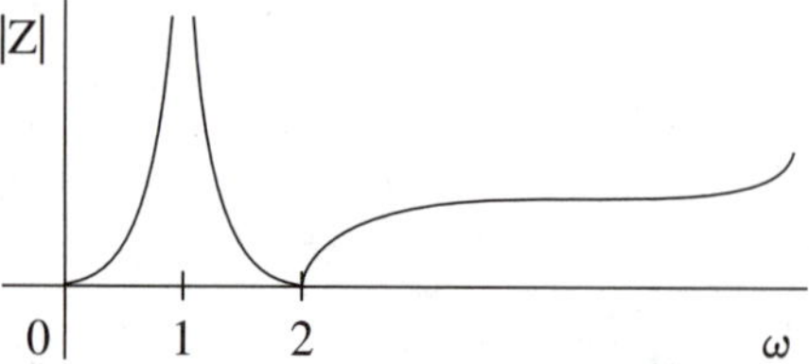

Figure 4.2d A sketch of the impedance's frequency response.

Both the expression for $Z(s)$ and its graphical representation in the form of the pole-zero plot show that the series resonant effect will occur at $\omega = 2$ rad/s.

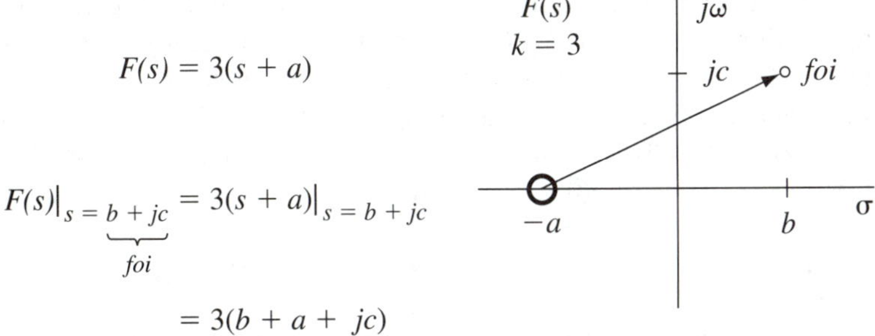

Figure 4.3 *F*(*s*) has a single zero and a multiplier of 3. Evaluating *F* at any arbitrary value of *s*, called the *frequency of interest (foi)*, is accomplished by substituting that *s* into *F*. Equivalently, it may be accomplished by finding the magnitude and angle of the vector from the zero to the frequency of interest on a properly scaled pole-zero plot and then multiplying by *k*.

Notice that a circuit containing only ideal inductors and capacitors has poles and zeros along the $j\omega$ axis. This makes them easy to find, and we may intentionally idealize the components in order to get a first estimate of the circuit's resonant frequencies. The transfer functions of such circuits would be classified as marginally stable. Any initial disturbance would cause a sinusoidal natural response that continues forever.

All actual inductors and capacitors have some inherent resistance. Inductors are easily the least ideal, so a more realistic look at resonant effects would require including at least the resistance of inductors. Doing so will make the impedance expressions more complicated. The poles and zeros will be shifted off of, but still be close to, the $j\omega$ axis. Our visualization of a pole-zero plot as a topographical map suggests that the path $s = j\omega$ will take us up the mountain sides, but no longer to the infinite peaks or to zero elevation in the valleys. We would like to find a simple way to estimate how high or low we actually get and how steep the terrain is.

Consider the simple $F(s)$ and its corresponding pole-zero plot shown in Figure 4.3. With the multiplier indicated on the pole-zero plot, the plot contains all of the information available in the expression for $F(s)$. Given the plot, we can find F; given F, we can make the plot. If we want to evaluate F at a particular *frequency of interest (foi),* we can see that the value contributed by any root is exactly equal to the magnitude and angle of a vector drawn from that root to the frequency of interest. Generalizing this observation, the value of an $F(s)$ at any arbitrary frequency of interest may be found from its pole-zero diagram as

$$F(s_o) = k\frac{\text{product of vectors from the finite zeros to } s_o}{\text{product of vectors from the finite poles to } s_o} \tag{4.2}$$

By measuring magnitudes with a ruler and measuring angles with a protractor, we could rather quickly evaluate the impedance or transfer function of a circuit using this observation. Now, we are not going to seriously advocate this graphical calculation method in the computer age, but suppose all of the poles and zeros are close to the $j\omega$ axis. For a frequency of interest on the $j\omega$ axis, most of the vectors would have angles of nearly $\pm 90°$, and their magnitudes would be essentially their vertical components, which can be read directly on the frequency scale. In this case, no measuring tools are actually needed, and the pole-zero diagram does not need to be drawn to scale.

Consider the pole-zero diagram of the realistic parallel resonant circuit shown in Figure 4.4. It certainly seems clear that a sharp peaking could be expected as we move out the $j\omega$ axis and get close to the pole at $s = -a + jb$. Our first question is how high it gets. Let's assume that the maximum occurs at our closest approach point, $s = jb$. Using this as the *foi,* we mentally draw vectors from the poles and zeros to the $s = jb$ point. The vector from the zero has a much larger vertical component than a horizontal component. We will approximate the length of this vector by its larger component, which is b. This is within 2% of the correct value as long as one component is 5 times as long as the other. Applying the same criterion to the lower pole, it is even more accurate to approximate its magnitude by $2b$. The vector

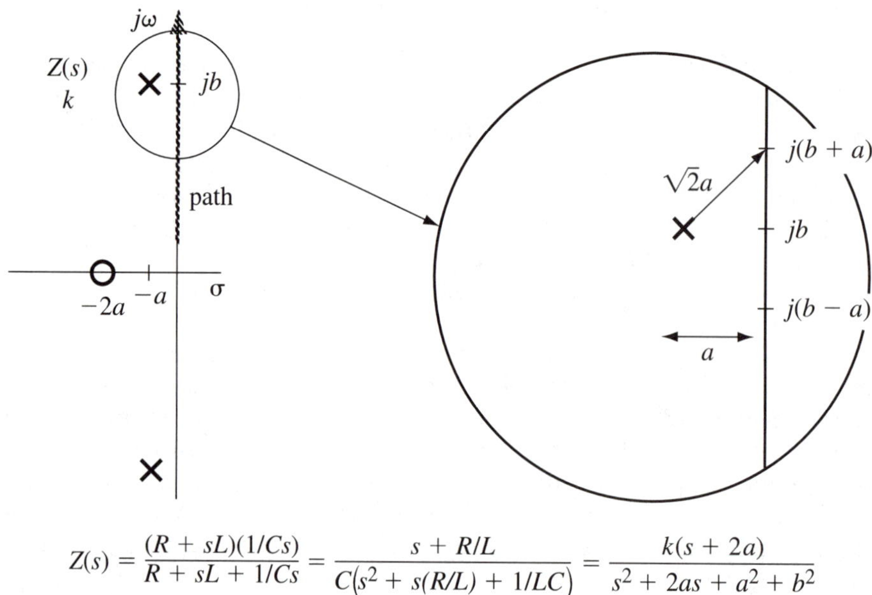

$$Z(s) = \frac{(R + sL)(1/Cs)}{R + sL + 1/Cs} = \frac{s + R/L}{C\left(s^2 + s(R/L) + 1/LC\right)} = \frac{k(s + 2a)}{s^2 + 2as + a^2 + b^2}$$

Figure 4.4 The pole-zero diagram for the impedance of a nonideal parallel resonant circuit. The inset shows the details of the frequency range around resonance.

from the pole adjacent to the *foi* is an exception, in that it has only a horizontal component, and we can estimate the maximum impedance value as

$$Z(jb) = k\frac{b\angle 90^\circ}{a\angle 0^\circ(2b\angle 90^\circ)} = \frac{k}{2a}\angle 0^\circ$$

The next question is how steep the response is. A standard way to answer that question is to give the *bandwidth (B)* of the peak, defined as the frequency range over which the impedance stays at or above 0.7071 of its maximum value. To estimate that range, consider what happens to the vectors as the frequency moves up and down from $s = jb$ an amount equal to the distance of the pole from the $j\omega$ axis. As long as the pole is close to the axis, the percentage change in the length of the vector from the zero is small over that frequency range, and can be ignored for a first approximation. This is even more true for the vector from the lower pole, and since those two vectors change in the same direction, their ratio changes even less. The only vector that experiences a big change, in percentage terms, is the small vector from the pole whose peaking we are trying to evaluate. As the inset of Figure 4.4 shows, its length increases by a factor of $\sqrt{2}$ at the ends of the specified range, which is exactly the factor needed to reduce the impedance to 0.7071 of its maximum value. The conclusion is that the bandwidth is approximately $2a$, where a is the distance of the pole from the $j\omega$ axis:

$$B \approx 2a \tag{4.3}$$

The equation in Figure 4.4 shows that the bandwidth is also the coefficient of the s term in the quadratic causing the resonance.

To generalize these observations, how large a function gets at a frequency adjacent to one of its poles, or how low it is at a frequency adjacent to one of its zeros, can be estimated by evaluating the transfer function at that *foi.* In doing this, vectors having one component much larger than their other component may be approximated by their larger component with little loss in accuracy. The sharpness of the response may be measured by the bandwidth, which may be determined either by inspection of the coefficient of the s term in the quadratic responsible for that pole or zero or from $B = 2a$, where a is the distance of that pole or zero from the $j\omega$ axis. This remains a good approximation if the only vector changing significantly in the vicinity of the pole or zero is the one from the pole or zero itself. For this to be true, other poles and zeros must be at least $10a$ away from the one whose resonance is being evaluated.

An alternate method of stating the steepness of the response is to specify the effective Q of the resonance, where Q is the ratio of resonant frequency to bandwidth. The higher the Q, the steeper the response curve:

$$Q = \frac{\omega_o}{B} \tag{4.4}$$

EXAMPLE 4.2

Estimate the impedance and bandwidth of the resonant conditions in the circuit of Figure 4.5, given that both inductors have a resistance of 0.1 Ω.

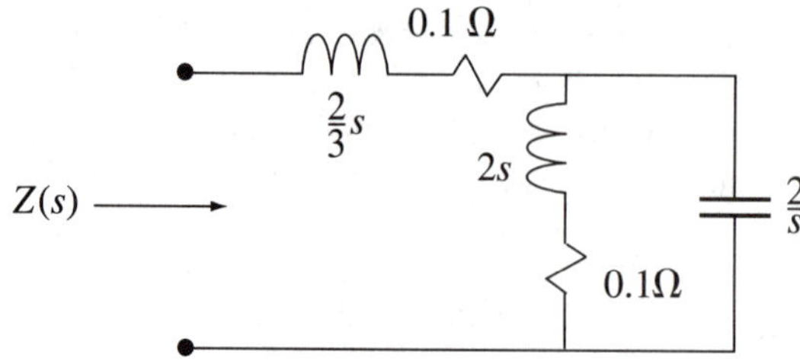

Figure 4.5

Solution

The derivation for the impedance starts off as before, but is more involved due to the presence of the resistive terms:

$$Z(s) = \frac{2}{3}s + 0.1 + \frac{(2s + 0.1)(2/s)}{2s + 0.1 + 2/s} = \frac{2}{3}s + 0.1 + \frac{2s + 0.1}{s^2 + 0.05s + 1}$$

Getting a common denominator and continuing to simplify to standard form gives

$$Z(s) = \frac{(2/3)s^3 + (0.4/3)s^2 + (8/3 + 0.005)s + 0.2}{s^2 + 0.05s + 1} = \frac{2}{3}\frac{(s^3 + 0.2s^2 + 4.008s + 0.3)}{s^2 + 0.05s + 1}$$

Now either we need a root-solving program, or we must find the one real root of the cubic by trial and error, divide that root out of the cubic to get the remaining quadratic, and so on. A sketch of the pole-zero diagram will help in visualizing the vectors, even though the critical values are taken from the result for *Z:*

$$Z(s) = \frac{2}{3}\frac{(s + 0.075)(s^2 + 0.125s + 4)}{s^2 + 0.05s + 1} = \frac{2}{3}\frac{(s + 0.075)(s + .0625 \pm j1.9987)}{s + 0.025 \pm j0.9997}$$

Estimating the $|Z|$ and B of the parallel resonance at $s = j1$,

$$|Z(j1)| \approx \frac{2}{3}\frac{(1)(3)(1)}{(0.025)(2)} = 40\ \Omega$$

$$B_1 \approx 0.05 \text{ rad/s}$$

$$Q_1 = \frac{\omega_o}{B_1} = \frac{1}{0.05} = 20$$

Estimating the $|Z|$ and B of the series resonance at $s = j2$,

$$|Z(j2)| \approx \frac{2}{3}\frac{(2)(4)(0.0625)}{(1)(3)} = 0.111\ \Omega$$

$$B_2 \approx 0.125 \text{ rad/s}$$

$$Q_2 = \frac{\omega_o}{B_2} = \frac{2}{0.125} = 16$$

4.2 NONRESONANT SYSTEMS: *R-C* CIRCUITS

This section also applies to *R-L* circuits, but anything that can be done with an *R-L* circuit can also be done with a cheaper and smaller *R-C* circuit. The poles and zeros for *R-C* circuits all fall along the σ axis in the s plane. How such poles and zeros affect the terrain along the path $s = j\omega$ is much less dramatic and intuitive than for *L-C* circuits, and a different approach is needed. After completing this section you will be able to:

- Estimate a frequency response using Bode straight-line approximations (SLAs).
- Use log-log graph paper for frequency response plots.
- Define and identify break frequencies.
- Explain the significance of low- and high-frequency asymptotes.
- Use symmetry to identify certain frequency response features.
- Explain the difference between wrapped and unwrapped phase.

A driving point or transfer function will take the general form

$$Z(s) = k\, s^n \frac{(s + a_1)(s + a_2)\cdots(s + a_q)}{(s + b_1)(s + b_2)\cdots(s + b_p)} \tag{4.5}$$

where a and b are real and n is a positive or negative integer. Consider the factor $(s + a_1)$, where $s = j\omega$ and a_1 is real. As frequency varies, there will be times when ω is so low that it is negligible compared to a_1. On the other hand, when the frequency is very high, the a_1 value will be negligible compared to ω. *The simplest possible approximation for this type of factor is just to keep the larger number.* At low frequencies the factor is approximated by the constant, at high frequencies it is approximated by the s. We break from one approximation to the other when the two terms are the same size; consequently a_1 is called a *break frequency.* The approximation is least accurate at the break, where we would be approximating the factor by either a_1 or ja_1, when, in fact, its true magnitude and phase is $\sqrt{2}a_1\angle 45°$.

In any frequency range between break frequencies, some of the factors in Equation 4.5 will be approximated by their constants, and some will be approximated by their *s*. As a result, the expression for $Z(s)$ in any frequency range reduces to the form

$$Z(s) = Cs^m \tag{4.6}$$

where all those factors being approximated as constants combine with k to produce the constant C and all those factors being approximated as s combine to produce a net s^m term. Now we need to separate the magnitude and phase information in preparation for plotting:

$$Z(j\omega) = C(j\omega)^m = C(\omega\, e^{j\pi/2})^m = C\omega^m \angle m90°$$

$$\angle Z(j\omega) = m90° \tag{4.7}$$

$$|Z(j\omega)| = C\,\omega^m \tag{4.8}$$

The phase is approximated as a constant over each frequency range between breaks, so plotting it does not present a problem. All that is needed is the power of the *s* term for that particular frequency range. The magnitude is more complicated, but Equation 4.8 will look like a straight line if appropriate graph scales are used. If the log is taken of each side of $|Z(j\omega)|$, the result is

$$\underbrace{\log|Z(j\omega)|}_{y} = \underbrace{\log(C)}_{b} + m\,\underbrace{\log\omega}_{x} \tag{4.9}$$

which is the equation of a straight line with intercept b and slope m in the variables $\log|Z|$ and $\log\,\omega$. Commercial graph paper having this type of scale is called *log-log* paper. Frequency response curves drawn using these techniques are referred to as *Bode straight-line approximations* (SLAs).

For those unfamiliar with a log scale, the following information may be helpful. Refer to Figure 4.6 as necessary.

1. A $\log\omega$ scale has the numerical value of ω indicated on the scale, but its distance from the origin is proportional to $\log\omega$. So, for example, the frequency $\omega = 1$ is located $\log(1) = 0$ units from the origin (i.e., it is the origin), while the value $\omega = 10$ is located $\log 10 = 1$ unit from the origin. The distance between the number 10 and the number 1 sets the *measure* of the scale. Suppose we arbitrarily make that distance 4 inches. Then the number 3 should be placed $4 \times (\log 3) = 1.909$ inches from the number 1.
2. A ratio of 10:1 is called a *decade,* and a ratio of 2:1 is called an *octave.* The distance occupied by any constant ratio is a constant on log scales. In other words, the octave represented by the frequency range from $\omega = 1$ to $\omega = 2$ takes exactly the same number of inches as the octave from 2 to 4 or the octave from 3 to 6. Musically

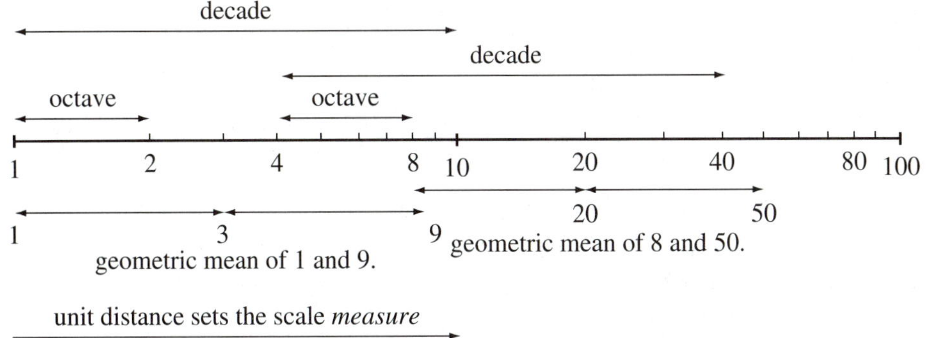

Figure 4.6 Features of a logarithmic graph scale.

speaking, one octave is as important as another, and the log scale gives every octave equal emphasis. For this reason, the log ω scale is preferred for all but resonant types of frequency response curves.

3. If frequency ω_2 is a distance log ω_2 units from the origin, and frequency ω_1 is a distance log ω_1 from the origin, then the frequency ω_o midway between them is a distance equal to the average of these two distances. Thus

$$\log \omega_o = \frac{\log \omega_2 + \log \omega_1}{2} \qquad \text{or} \qquad \omega_o = (\omega_1 \omega_2)^{1/2} = \sqrt{\omega_1 \omega_2} \tag{4.10}$$

The midpoint number,ω_o is the geometric mean of the other two numbers.

4. The slope of our straight line is the integer m. On commercial graph papers, the vertical and horizontal scales usually have the same measure. With computer-generated log-log graphs, that is not necessarily true. A slope of $+1$ means that the line rises one decade (octave) vertically as it increases 1 decade (octave) horizontally.

EXAMPLE 4.3

Sketch the frequency response of the impedance for the circuit in Figure 4.7a.

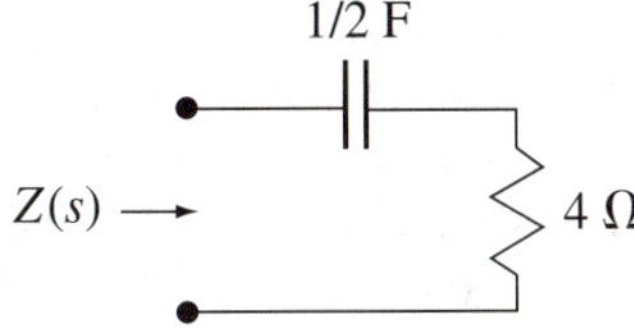

Figure 4.7a

Solution

The expression for the impedance is

$$Z(s) = \frac{2}{s} + 4 = \frac{2 + 4s}{s} = 4\frac{(s + 1/2)}{s}$$

Comparing this to Equation 4.5, we see that $k = 4$, $n = -1$, and the only break frequency is at $\omega = 1/2$ rad/s.

At frequencies below the (1/2)-rad/s break frequency,

$$Z_{LF}(s) = \lim_{s \to 0} Z(s) = 4\frac{1/2}{s} = \frac{2}{s}$$

$$\therefore C = 2 \qquad \text{and} \qquad m = -1$$

Z_{LF} is called the *low-frequency asymptote.* It will be a straight line with a slope of -1, and it will pass through 1 Ω when $\omega = 2$ rad/s, 20 Ω when $\omega = 0.1$ rad/s, and so on (just plug ω values into the equation for $|Z_{LF}|$).

At frequencies above the (1/2)-rad/s break frequency,

$$Z_{HF}(s) = \lim_{s \to \infty} Z(s) = \frac{4s}{s} = 4$$

$$\therefore C = 4 \qquad \text{and} \qquad m = 0$$

Z_{HF} is called the *high-frequency asymptote.* In this case it is a horizontal line at an impedance of 4 Ω. The asymptotes and actual response for $|Z(s)|$ are shown in Figure 4.7b. Notice that the actual impedance at the break is $\sqrt{2}(4) \approx 5.7$.

Since $m = -1$ at low frequencies, the phase is $-1(90°)$, and at high frequencies $m = 0$, so the phase is zero. The approximation for phase is cruder than for the magnitude plot, but it can be significantly improved in the special case where there is only one break frequency and it is caused by a real root. A ramp drawn from the low-frequency phase asymptote starting a decade below the break frequency and running to the high-frequency asymptote a decade above the break frequency provides a superior phase approximation (Figure 4.7c).

Now the results should be compared to what can be expected from the original circuit. At low frequencies the series capacitor blocks signals, producing a high-capacitive impedance. At high frequencies, the capacitor approaches a short circuit, and the circuit looks like the resistor. The changeover from a capacitive to a resistive circuit occurs at the frequency where the magnitudes of their impedances become equal.

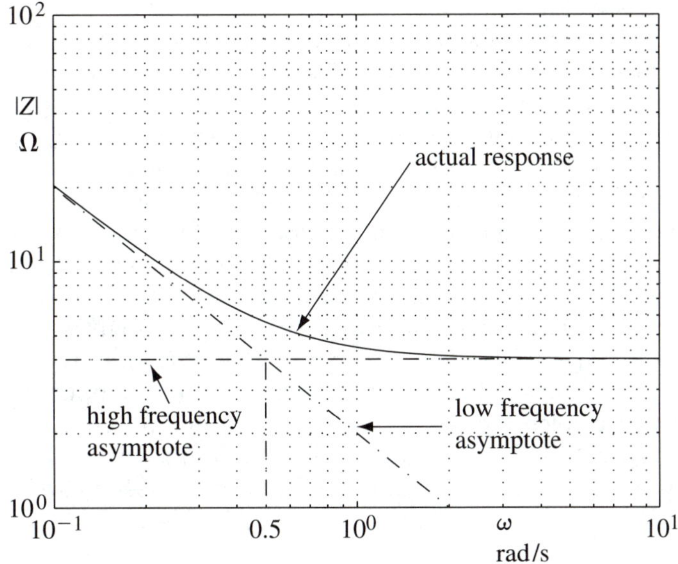

Figure 4.7b Bode SLA and actual response for the magnitude of the impedance of Example 4.3.

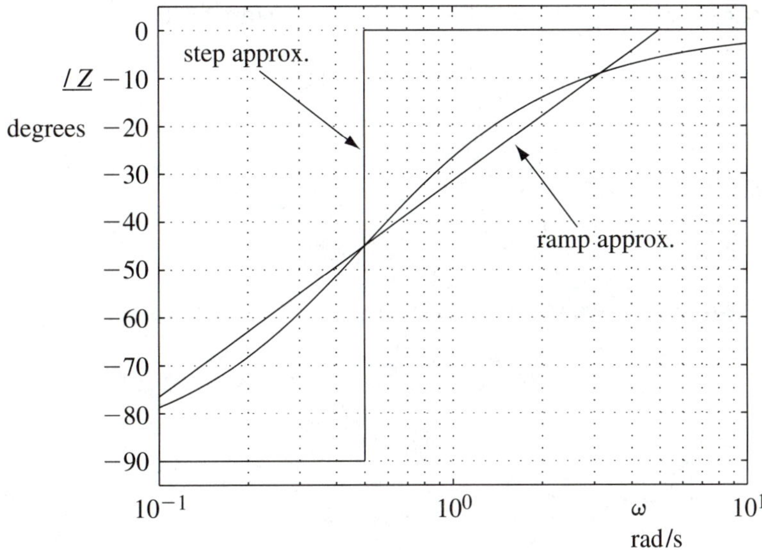

Figure 4.7c Bode step and ramp approximations for the phase of the impedance of Example 4.3.

The high- and low-frequency asymptotes are found from the same approximations used between break frequencies. They are especially important, however, because the actual response must ultimately approach them. The SLAs between break frequencies may be approached but never reached by the actual response. They may also be crossed through. They are essentially asymptotes that become overruled by the effects of subsequent break frequencies.

Before the computer era, the Bode SLA was a primary tool for obtaining frequency response information with a minimum of calculations. A variety of rules of thumb exist for estimating the deviation of the actual plot from the SLA at break frequencies. Using these rules, a few calculated points, and the best smooth curve through the SLAs, a surprisingly accurate response curve can be obtained. The value of this analysis technique today is more in its ability to indicate where, in frequency, special things will happen, so we can tell the computer where we want the actual response calculated. It is also useful in design when we can identify what components set the value of a particular break frequency or when symmetries in the SLA show practical limitations or other features we need to know about. Frequently a sketch of the response is all that is needed.

EXAMPLE 4.4

At what frequency and gain does the circuit whose transfer function is

$$\frac{\vec{V}_o}{\vec{V}_i} = 12\frac{(s + 3/2)}{s(s + 3)}$$

have the least phase shift between its input and output?

Solution

There are two break frequencies. We obtain the following approximations for the SLA:

$$\frac{\vec{V}_o}{\vec{V}_i} = 12\frac{(s + 3/2)}{s(s + 3)} = \begin{cases} 12\frac{3/2}{s(3)} = 6s^{-1} & \omega < 3/2 \\ 12\frac{s}{s(3)} = 4 & 3/2 < \omega < 3 \\ 12\frac{s}{s(s)} = 12s^{-1} & \omega > 3 \end{cases}$$

The low-frequency asymptote has a slope of -1 and passes through 4 at 1.5 rad/s. For $1.5 < \omega < 3$, the slope is 0 and the gain sits at 4. For $\omega > 3$, the slope returns to -1 and passes through 4 at $\omega = 3$. A sketch incorporating these facts is shown in Figure 4.8a. Since it is only a sketch, the axes have been labeled to indicate that a logarithmic measure is being used.

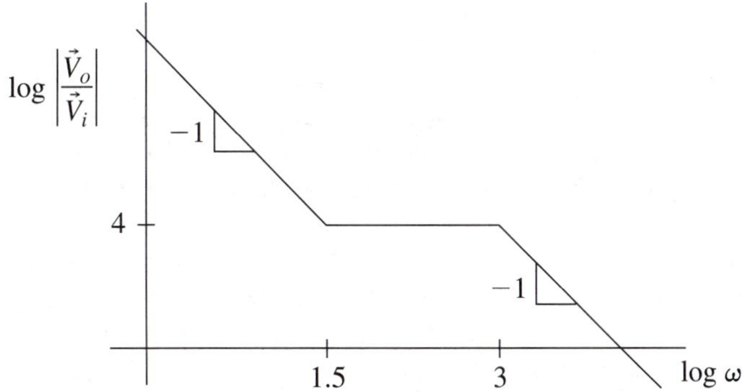

Figure 4.8a Bode SLA magnitude response for the transfer function of Example 4.4.

The asymptotic phase sits at $-90°$ except for the short region between the break frequencies. By symmetry, the phase will make its closest approach to zero at the geometric mean of the break frequencies. (See Figure 4.8b) Symmetry also indicates that the actual gain will pass through 4 at that frequency. A calculation shows

$$\frac{\vec{V}_o}{\vec{V}_i} = 4\angle -70.5° \ @\ \omega = \sqrt{(1.5)3} = 2.12 \text{ rad/s}$$

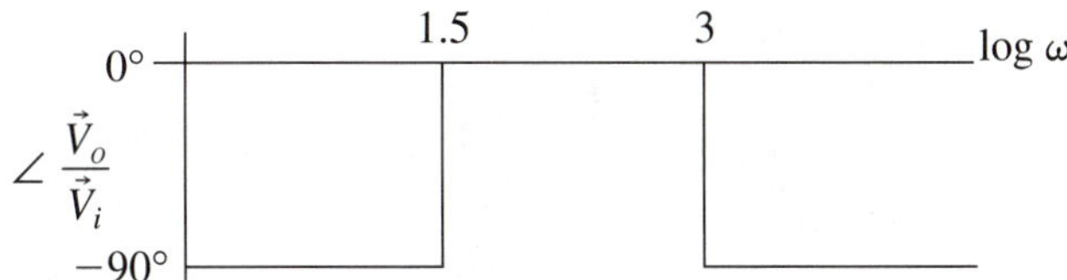

Figure 4.8b Bode SLA phase response for the transfer function of Example 4.4.

Transfer functions may occasionally also have zeros of the form

$$(s - a) = \begin{cases} -a & |s| < a \\ s & |s| > a \end{cases} \qquad \textbf{(4.11)}$$

Such zeros are in the right half of the s plane (RHP), and are approximated in exactly the same fashion as before, but the phase jump at the break requires a little extra thought. Specifically, should the factor be described as having an initial phase of $+180°$ or $-180°$? The issue is important in drawing the correct continuous phase curve. This is resolved by not allowing s to vanish completely. If $s = j\varepsilon$, where ε is a very small value, it becomes clear that the factor $-a + j\varepsilon$ has an angle in the second quadrant and is approaching 180° as a positive angle at low frequencies.

It is common to *wrap* phase information so that all angles are given as values between $+180°$ and $-180°$. This is justified by the fact that, if only a single frequency sinusoid is available, it is impossible to tell the difference between a phase of $+270°$, $-90°$, and $-810°$, etc. But these phase angles do represent different *delay times* for the sinusoid, and this can be measured if frequency can be varied or if more complicated waveforms are used. If the source frequency can be increased steadily, we could use a scope to observe the smooth transition of phase through its progression of values. The SLA phase plots provide an *unwrapped* phase, which properly represents the continuous phase shift produced by real circuits.

4.3 THE DECIBEL (dB)

The *bel* (after Alexander Graham Bell) is a unit that originated in studies on human hearing, but it has become used and abused in numerous applications where its logarithmic measure is the desired feature. After completing this section you will be able to:

- Define the decibel, and discuss its uses.
- Express Bode SLA gains in terms of a decibel scale.

The most common application of the decibel is for describing power ratios. For power gains the decibel is defined as

$$1 \text{ dB} = 10 \log_{10}\left(\frac{P_{\text{out}}}{P_{\text{in}}}\right) \tag{4.12}$$

The decibel is also used to state an output power level against an input power standard. We could, for instance, express the output *power* of an amplifier in dB_m, instead of watts. The m signifies that the dB scale is referenced to an input power of 1 mW. An output power of 1 watt could be given as 30 dB_m using this procedure. dB_W is another standard, using 1 watt as the input reference level. A thousand-watt amplifier delivers 30 dB_W at full output. This usage makes some sense when one of the standard input power levels is part of a device's test specifications:

$$\text{dB}_m = 10 \log_{10}\left(\frac{P}{10^{-3}\text{ W}}\right) \qquad \text{dB}_W = 10 \log_{10}\left(\frac{P}{1\text{ W}}\right)$$

Assuming a circuit with equal input and load resistances, the dB measure can also be used to describe the circuit's voltage or current gains:

$$\text{dB} = 10 \log \frac{V_{\text{out}}^2}{V_{\text{in}}^2} = 20 \log \frac{V_{\text{out}}}{V_{\text{in}}} \qquad \textbf{(4.13)}$$

In fact, by convention this is routinely done *even if the input and load resistances are known to be different.* We simply take Equation 4.13 as the definition of a voltage gain measured in decibels, and we do not care that it is not a true power ratio. The main advantage of doing this is that frequency response SLAs may be made on semilog paper, since the dB units provide a logarithmic measure for the vertical gain scale. Magnitude and phase plots can then be made on the same kind of graph paper. At least with current and voltage gains, the logarithm is taken of a dimensionless ratio. Carrying this practice to an extreme, transimpedances or transadmittances could be converted to a decibel measure by taking the logarithm, despite the fact that the ratios are not dimensionless. In these cases we will prefer to revert to log-log paper and give the proper units for the transfer function.

The key to making SLA plots was that the transfer function takes the form Cs^m in any region between break frequencies. On log-log paper, this plots as a straight line with a slope of m. It rises $2m$ in an octave and $10m$ in a decade. On semilog paper with a dB vertical scale, the curve remains a straight line, but it rises at $(20 \log 2)m \approx 6m$ dB per octave or $(20 \log 10)m = 20m$ dB per decade. With this minor change in specifying slopes, the procedure used to draw SLA plots remains the same.

EXAMPLE 4.5

Sketch the decibel magnitude plot and the phase plot for the transfer function

$$A_v = \frac{\vec{V}_o}{\vec{V}_g} = \frac{2(s-2)}{(s+2)^2}$$

Solution

The function has a single break frequency of 2 rad/s and a zero in the RHP:

$$A_v = \begin{cases} -1 = 1\angle + 180^\circ & \omega < 2 \\ 2s^{-1} = \dfrac{2}{\omega}\angle -90^\circ & \omega > 2 \end{cases}$$

The magnitude plot is the same whether or not there are factors present in the RHP. In this case the low-frequency asymptote is horizontal, making it easy to position the plot vertically (Figure 4.9).

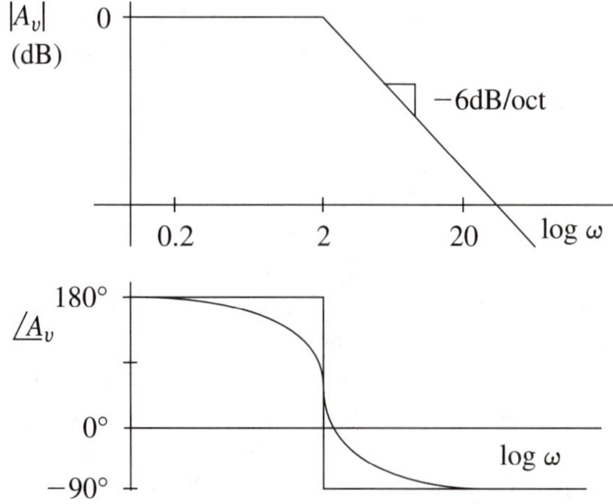

Figure 4.9

To check that we have the correct phase, calculate the transfer function at the break:

$$A_v(s = j2) = \frac{2(-2 + j2)}{(2 + j2)^2} = \frac{-1 + j1}{(1 + j)^2} = \frac{\sqrt{2}\angle 135°}{2\angle 2(45°)} = \frac{1}{\sqrt{2}}\angle 45°$$

In measuring the bandwidth of resonances, we use a criterion based on the impedance being $\sqrt{2}$ or $1/\sqrt{2}$ of its minimum or maximum values, respectively. These numbers convert to $\pm(20 \log \sqrt{2}) = \pm 3$ dB. On this basis, we measure the bandwidth at a series resonance using the frequencies where the gain is 3 dB up from its minimum. Similarly, the bandwidth for a parallel resonance is determined from the frequencies where the gain is down 3 dB from the maximum.

4.4 GENERAL SYSTEMS: *R-L-C* CIRCUITS

In general, system polynomials will factor into a combination of real roots and complex roots but without a dominating resonant condition. In these cases, the SLA approach still provides the most useful information. After completing this section you will be able to:

- Apply the Bode SLA technique to systems with mild resonances.
- Define *damping ratio* and *natural resonant frequency.*

In circuits containing all three passive elements, and where the resistors are not just inherent inductor winding resistances, the circuit polynomials *may* contain some complex roots. When the polynomials have real coefficients, any complex root must appear along with its conjugate. Consider the pair of roots

$$(s + ce^{j\beta})(s + ce^{-j\beta})$$

If we use exactly the same rule as for real roots, we will just keep the larger number. So if $|s| < c$, we keep the constant term; otherwise we keep the s. Since the roots break together, we will be getting a c^2 if $|s| < c$, or we will be getting s^2 when $|s| > c$.

Multiplying out the conjugate roots gives the quadratic that created them. Notice that using the standard rule of keeping the larger number

$$(s + ce^{j\beta})(s + ce^{-j\beta}) = s^2 + (e^{j\beta} + e^{-j\beta})cs + c^2 = s^2 + (2c \cos \beta)s + c^2$$

is equivalent to ignoring the s term in the original quadratic. We therefore adopt the following rule for SLA plots: *If a quadratic would factor into conjugate roots, leave the quadratic unfactored and ignore its center* (s^1) *term. Keep the larger of the two remaining numbers in* ($s^2 + c^2$).

A standard notation has been established for the quadratic based on the undamped natural resonant frequency, ω_n (which is also the break frequency), and the damping ratio, ζ (zeta):

$$s^2 + 2c \cos \beta s + c^2 = s^2 + 2\omega_n \zeta s + \omega_n^2$$

If the damping ratio is 1 or higher, the quadratic factors into real roots. For damping ratios less than 1, conjugate roots result. A damping ratio of zero puts the roots on the $j\omega$ axis. The damping ratio appears only in the center term; therefore it is ignored in obtaining the SLA. However, the actual value of the quadratic at the break is *exactly* the center term:

$$\left| \frac{\text{actual value @ } \omega_n}{\text{SLA value @ } \omega_n} \right| = 2\zeta \qquad \textbf{(4.14)}$$

We conclude that when the standard SLA approximation is applied to complex roots, the actual curve will closely cling to the SLA, for $0.5 < \zeta < 1$, so that the SLA approximation for the quadratic is probably at least as good as, if not better than, the results obtained for real roots (Figure 4.10).

For $\zeta < 0.5$, complex roots show resonant peaking near the SLA corner instead of rounding it off. At the very least, the presence of such roots should serve as an alert that a few calculations are needed at or around the break to ensure an accurate plot. With those calculations, the SLA remains useful as a frequency response analysis tool, despite the presence of slight resonant effects.

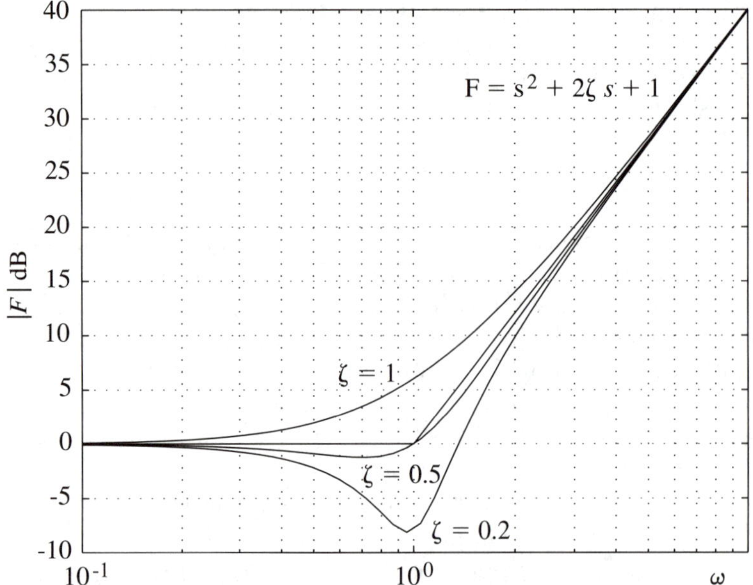

Figure 4.10 Variation of actual and asymptotic response curves in the vicinity of the break frequency of a quadratic factor.

EXAMPLE 4.6

Plot the SLA of the impedance magnitude for the circuit in Figure 4.11a.

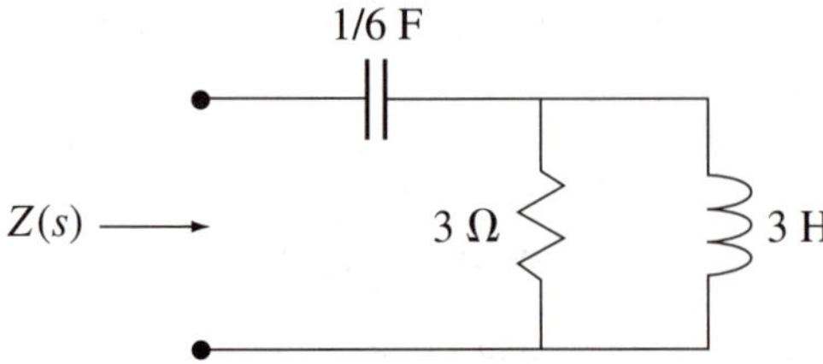

Figure 4.11a

Solution

The impedance is found as

$$Z(s) = \frac{6}{s} + \frac{3(3s)}{3 + 3s} = \frac{18 + 18s + 9s^2}{3s(s + 1)} = 3\frac{s^2 + 2s + 2}{s(s + 1)}$$

The numerator quadratic produces zeros when

$$s = \frac{-2 \pm \sqrt{4 - 4(2)}}{2} = -1 \pm j$$

so we leave the quadratic unfactored and note that

$$2\zeta\omega_n = 2, \qquad \omega_n = \sqrt{2}, \qquad \therefore \zeta = 0.707$$

This value of damping ratio suggests that the SLA will be a good predictor of the actual response. No extra calculations are necessary for this value of ζ

There are two break frequencies, 1 rad/s and $\sqrt{2}$ rad/s (Figure 4.11b):

$$Z(s) \approx 3\frac{(s^2 + 2)}{s(s + 1)} = \begin{cases} 6s^{-1} & \omega < 1 \\ 6s^{-2} & 1 < \omega < \sqrt{2} \\ 3s^0 & \omega > \sqrt{2} \end{cases}$$

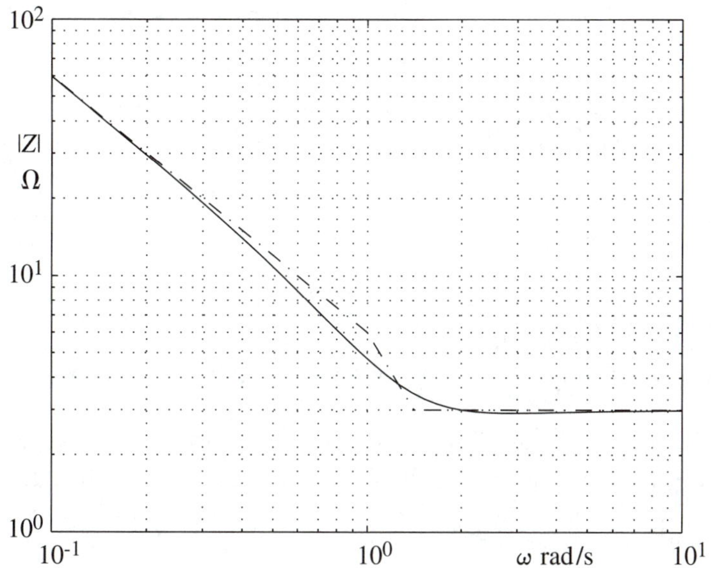

Figure 4.11b Comparison of actual and Bode SLA magnitude response for the impedance of Example 4.6.

Both pole-zero plots and Bode SLAs can be useful in estimating a system's frequency response in appropriate circumstances. Each technique can also provide some insight that could be valuable in design problems. Still, when it comes down to a final design, the ability to obtain a computer-generated frequency response plot is indispensable.

4.5 MATLAB LESSON 4

This lesson will concentrate attention on those MATLAB commands most often used for obtaining the frequency response of continuous-time signals and for graphing this information. After completing this section you will be able to:

- Use the MATLAB frequency response tools.
- Compare SLA and actual response curves.
- Recognize and correct unrealistic decibel scale ranges.
- Capture important data from a graph or from a vector.

MATLAB EXAMPLES

We saw several ways to evaluate a function along some path in a complex plane in Chapters 1 and 2. If the path is the $j\omega$ axis, and the simplified polynomial notation is used, special commands are available to handle this task:

freqs **bode**

We will demonstrate by plotting the frequency response for the impedance of Example 4.2 in the vicinity of its pole.

```
>num=(2/3)*[1 .2  4.008 .3]; % defines the numerator polynomial
>den=[1 .05 1];              % defines the denominator polynomial
>w=linspace(.95,1.05);       % defines a frequency range of about twice the
                               expected bandwidth around parallel
                               resonance
>Z=freqs(num,den,w);         % calculates Z over the specified frequency
                               range if s = jw
>plot(w,abs(Z))              % compare |Z|max and B with estimates from
                               the example
```

freqs(num,den,w) is the equivalent of polyval(num,j*w)./(polyval(den,j*w), where a range of w has been defined. It allows us to calculate a driving point or transfer function over any desired frequency range, and lets us select among appropriate plotting options.

```
>freqs(num,den,w)  % this is easier, but allows less control over the
                     graphs
>bode(num,den,w)   % this apparently does not allow graph changes
```

Now let's plot the magnitude and phase angle for the transfer function of Example 4.4 and compare it to Figure E4.4. The range of frequencies used should usually extend *about* a decade below the lowest break frequency and *about* a decade above the highest break frequency. This ensures that the response will nearly reach the high- and low-frequency asymptotes and that the trend of the graph is not misleading.

```
>num=12*[1 3/2];         % defines the numerator polynomial 12(s + 1.5)
>den=[1 3 0];            % defines the denominator polynomial s^2 + 3s
>w = logspace(-1,1);     % picks 50 (default) equally spaced frequency
                           points along a log scale from 10^-1 to 10^1
>Av=freqs(num,den,w);    % calculates the values of Av over the frequency
                           range specified
```

loglog semilogx semilogy

```
>loglog(w,abs(Av))              % produces a plot of |Av| vs. ω on log-log
                                  axes
>semilogx(w,180*angle(Av)/pi)  % produces a plot of /Av in degrees vs.
                                  log ω
>AvDB=20*log10(abs(Av))        % calculates Av in decibels
>semilogx(w,AvDB)              % plots |Av| in dB vs. log ω
> bode(num,den,w)              % plots |Av| in dB and /Av in degrees vs.
                                  log ω
```

Computers tend to be nonjudgemental about decibel scales. *Any scale exceeding a range of 120 dB should be regarded with skepticism.* Use the **axis** command to limit the axis range to realistic values.

Suppose you have a graph and want to extract numerical information from it. There are a variety of techniques available. The first, and simplest, is to put the graph data in paired points and print it to the screen.

```
>w=linspace(.9, 1.1);
>num=[1 0];
>den = [1 .08 1.1];      % defines a resonant response s/(s^2 + .08s + 1.1)
>z=freqs(num,den,w);     % calculate the impedance (or whatever)
>plot(w,abs(z))          % you should have a graph with a peak of about 12.5
                           @ ω = 1.05
>a=[w' abs(z)']          % print a column array of paired values of ω and
                           |z|.
```

By scrolling through the data table we find the calculated maximum is 12.4982, and it occurs at $\omega = 1.0495$. The impedance would be down to 0.7071 of its maximum when $|z| = 8.838$, which occurs at an ω of about 1.0091 and 1.0899. The bandwidth consequently is 0.0808.

max min

If you don't want to search manually, you can find the maximum or minimum element in an array:

```
>[zmax,row]=max(a(:,2))   % search all rows of column 2 in the array a for
                            a maximum
>wmax=w(row)              % knowing the row allows you to find the
                            corresponding frequency
```

With a little ingenuity, you can also find the bandwidth:

```
>b=abs(1./(zmax*.7071 - abs(z)));   % generate a curve that has a max at
                                      the condition desired
```

```
>[don't_care,k]=max(b)          % find index of half power
                                  frequency.don't_care could be used to
                                  estimate accuracy
>B=abs(wmax-w(k))*2             % calculate bandwidth assuming equal
                                  deviation about wmax
```

zoom ginput

You also can just read data off of a graph. If you want, you can zoom in by positioning the mouse cursor at a point on a graph and using the **zoom** command:

```
>zoom on
    click (left) mouse button to establish region to zoom around and start
    zooming.
    click until done
>zoom off
```

The zoom feature has to be turned on and off so MATLAB will know what command the mouse click goes with. Another command using the mouse click is **ginput.** It allows you to pick up graphical data. If necessary, get back the impedance information.

```
>w=linspace(.9, 1.1);            % these four steps have already been done
                                   unless starting over
>num=[1 0];                      %
>den = [1 .08 1.1];              %
>z=freqs(num,den,w);             % you may already have everything to here
>zmax=max(abs(z));               % determine the maximum
>plot(w,abs(z),w,.7071*zmax)     % plot |z|, and .7071 of max. on same graph
>[freq,ohms]=ginput(3)           % cursor changes to crosshairs and graph
                                   comes forward
```

Click on the intersection of the curves and at the maximum. You have specified three clicks. After you have made three clicks, return to the Command Window to see the results. *Warning: ginput* collects the coordinates at which the clicks were made. It is up to you to click on the curve and, when necessary, to keep track of the order in which the clicks were made. Now the variables *freq* and *ohms* contain the information needed. If, for instance, you clicked on the desired points in succession going left to right, you could calculate bandwidth as B=freq(3)-freq(1).

EXAMPLE 4.7

Use MATLAB to plot the current gain transfer function given by

$$\frac{\vec{I}_o}{\vec{I}_i} = 20\,\frac{s - 4}{s^2 + 4s + 16}$$

Determine the maximum gain in decibels and the frequency where the phase is zero.

Solution

```
>num=20*[1 -4];den=[1 4 16];        % define the transfer function
>w=logspace(-1,2,100);              % run ω from 0.1 to 100 (both breaks
                                      are @ 4)
>Ai=freqs(num,den,w);               % calculate current gain
>semilogx(w,20*log10(abs(Ai)));     % plot gain in dB
>grid
>xlabel('frequency in radians/sec.')
>ylabel('current gain in dB')
```

The first plot is shown in Figure 4.12a.

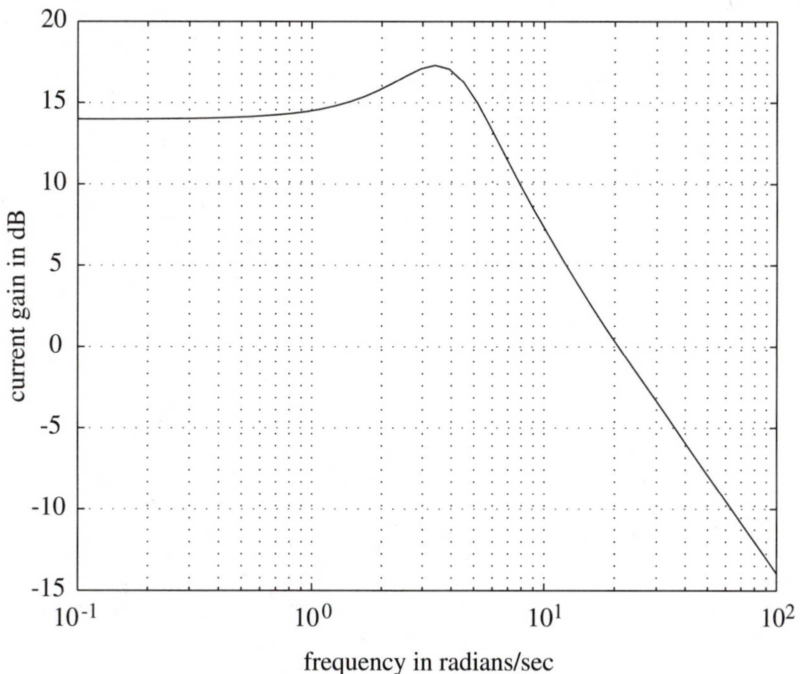

Figure 4.12a

```
>semilogx(w,180*angle(Ai)/pi)
>grid
>xlabel('frequency in radians/sec')
>ylabel('phase in degrees')
```

The second plot is given in Figure 4.12b.

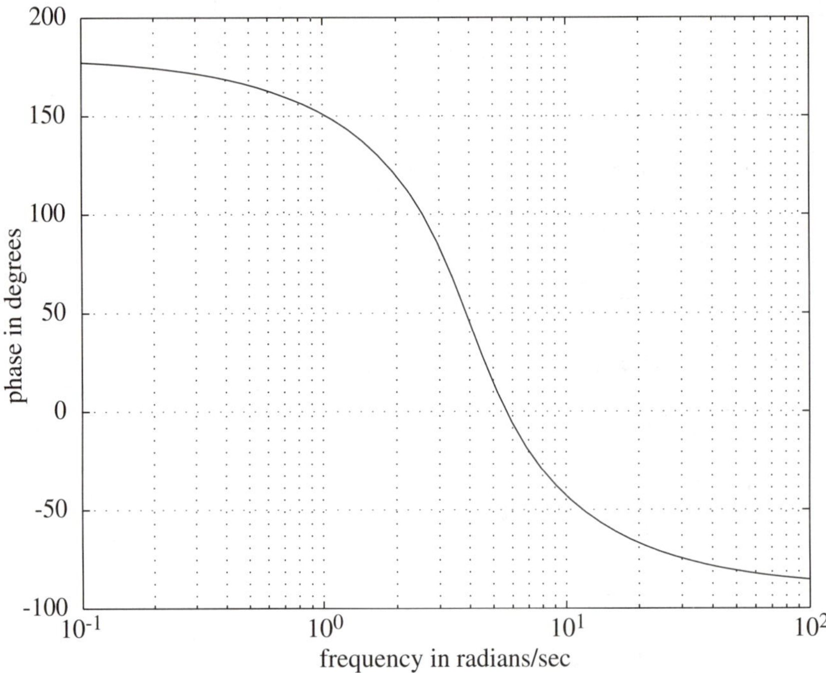

Figure 4.12b

```
>maxgain=max(20*log10(abs(Ai)))   % find the maximum gain in dB
maxgain = 17.3055
>phase=abs(180*angle(Ai)/pi);     % find phase magnitude
>[ang,row]=min(phase)             % Find the min. absolute phase
ang = 1.2345                      % to be 1.2345 degrees at the 59th freq.
row = 59
>w(59)
ans = 5.7224                      % phase data closest to zero is at 5.72
                                    rad/s.
```

CHAPTER SUMMARY

Evaluating an $F(s)$ analog the $+j\omega$ axis reveals its frequency response, which is a standard way of characterizing the capabilities of many systems. Having a software tool for providing a frequency response plot is essential to anyone working in the electrical technologies.

For highly resonant systems, the pole-zero diagram is often useful in predicting the main features of the system response, namely, its maximum or minimum amplitude and its bandwidth.

For systems without dominating resonance effects, using logarithmic graph scales allows an overall picture of a response to be developed with few if any calculations. The frequency response can be represented by a series of straight-line segments on logarithmic graph scales. Often the dB unit is used to give a linear gain scale a logarithmic measure. These Bode SLAs can frequently be generalized to provide essential design information.

PROBLEMS

Section 4.1

1. For the circuits of Figures P4.1a and b, find $Z(s)$ in pole-zero form and sketch its magnitude vs. ω.

Figure P4.1a **Figure P4.1b**

2. For the circuit of Figure P4.2:
   a. Find $Z(s)$ in pole-zero form and sketch its magnitude vs. ω.
   b. Find $\vec{I}_o/\vec{I}_g$ and sketch its magnitude vs. ω.

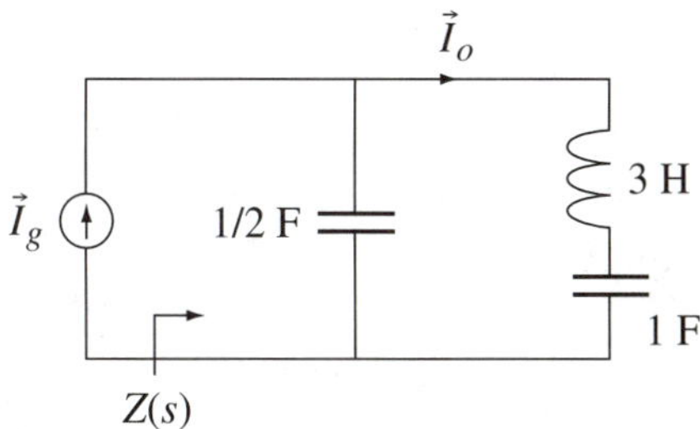

Figure P4.2

3. For the circuit of Figure P4.3:
   a. Find $\vec{V}_o/\vec{V}_g$ and sketch its magnitude vs. ω.
   b. Find $Z(s)$ in pole-zero form and sketch its magnitude vs. ω.

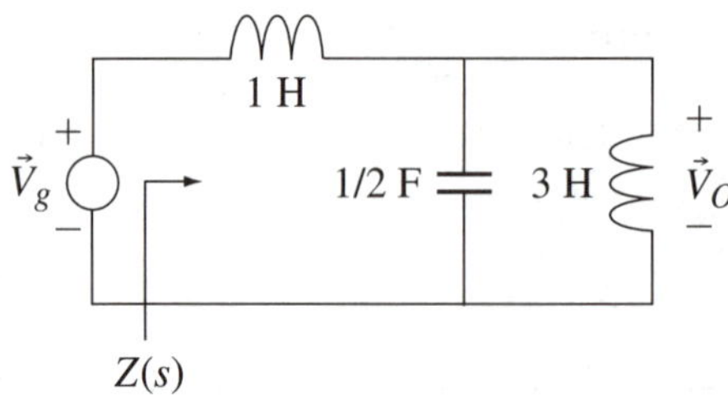

Figure P4.3

4. For the circuit of Figure P4.1b, assume the inductor has an inherent resistance of 0.05 Ω. Find $Z(s)$ in pole-zero form, and estimate the magnitude and bandwidth of the impedance at each resonance.
5. For the circuit of Figure P4.2, assume the inductor has an inherent resistance of 0.1 Ω. Find $Z(s)$ in pole-zero form, and estimate the magnitude and bandwidth of the impedance at each resonance.
6. For the circuit of Figure P4.3, assume each inductor has an inherent resistance of 0.1 Ω.
   a. Find $Z(s)$ in pole-zero form, and estimate the magnitude and bandwidth of the impedance at each resonance.
   b. Find $\vec{V}_o/\vec{V}_g$ in pole-zero form, and estimate the magnitude and bandwidth of the gain at each resonance.

Section 4.2

7. For the circuits of Figures P4.7a and b, sketch the SLA of the impedance magnitude and phase. Identify all important levels and frequencies.

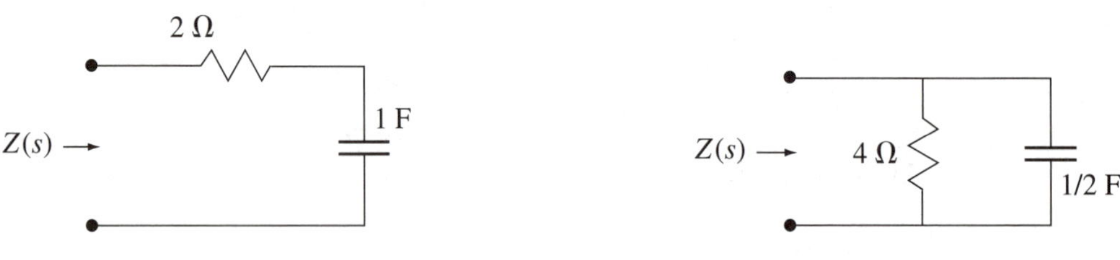

Figure P4.7a

Figure P4.7b

8. For the circuits of Figures P4.8a and b, sketch the SLA for the magnitude of the transfer function suggested, and obtain formulas for the break frequency values and gain levels.

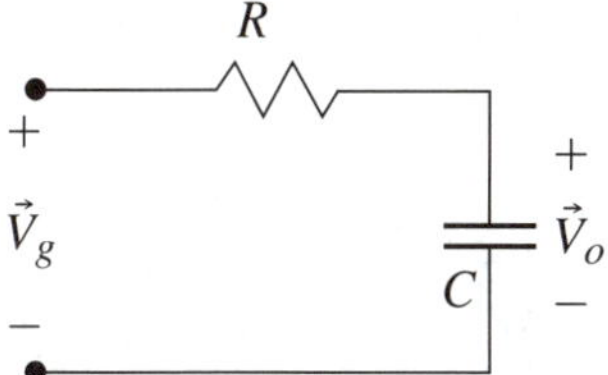

Figure P4.8a

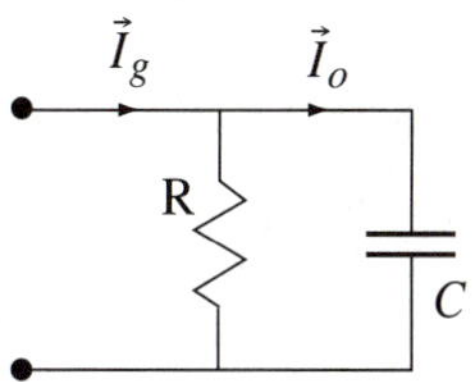

Figure P4.8b

9. For the circuits of Figures P4.9a and b, sketch the SLA of the impedance magnitude and phase. Identify all important levels and frequencies. Identify the frequency at which the phase has the largest magnitude.

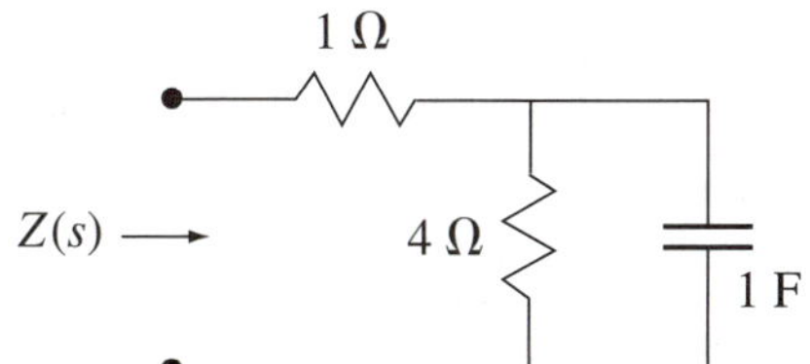

Figure P4.9a

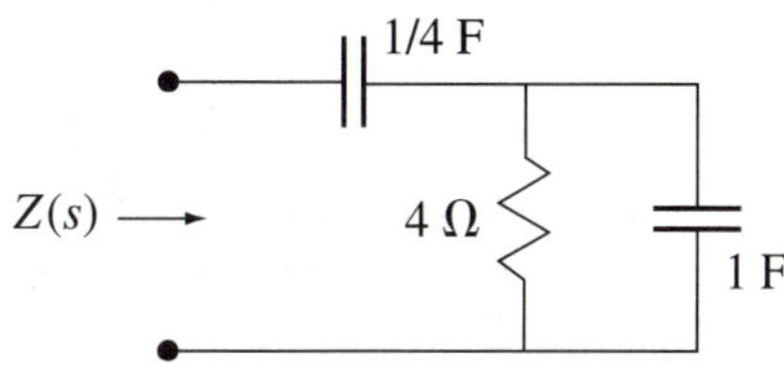

Figure P4.9b

10. For the circuits of Figures P4.10a and b, sketch the SLA of the voltage gain magnitude, and obtain formulas for the break frequency values and gain levels.

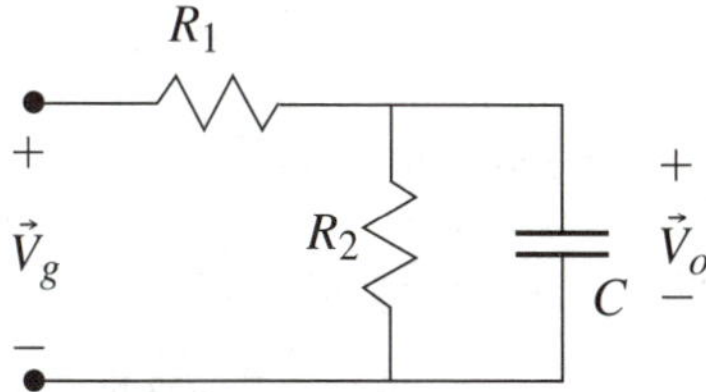

Figure P4.10a

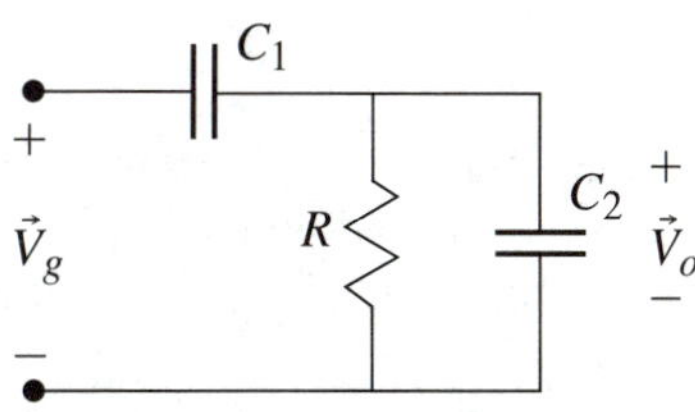

Figure P4.10b

Section 4.3

11. Express a power of 7 watts in dB_m and dB_W.
12. An amplifier rated at 100 W has a noise output of 1 μW. Express the output to noise power ratio in decibels.

Section 4.4

13. For the circuit of Figure P4.13, sketch the SLA dB voltage gain given that $R = 1\ \Omega$. Do you expect the actual curve to stay close to the SLA at the break frequency? Why?

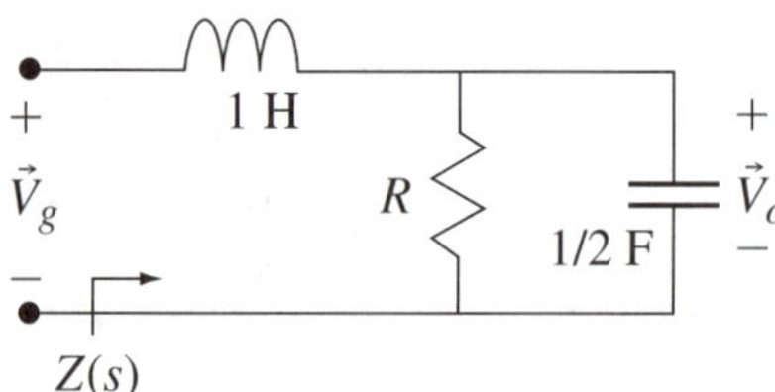

Figure P4.13

14. For the circuit of Figure P4.14, sketch the SLA for the magnitude of the impedance. Do you expect the actual curve to stay close to the SLA? Why?

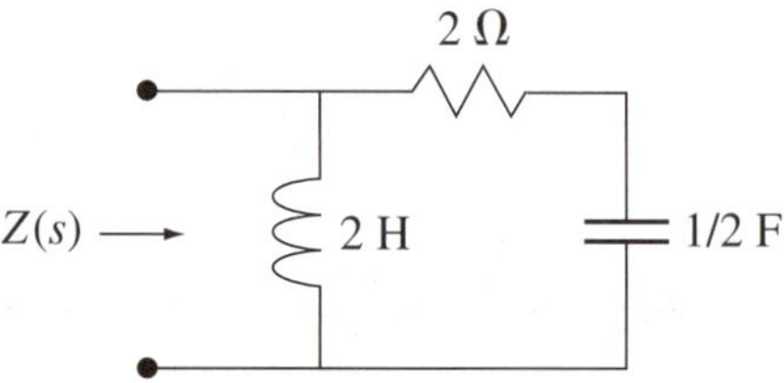

Figure P4.14

15. Sketch the SLA, magnitude and phase, for the gain $G(s)$. Calibrate the gain scale in decibels and the phase scale in degrees. Determine the actual maximum gain achieved and the frequency where the phase is zero in each case.

a. $G(s) = \dfrac{12s}{(s + 2)(s + 8)}$ b. $G(s) = \dfrac{12s}{(s^2 + 2s + 16)}$ c. $G(s) = \dfrac{12s}{(s + 4)^2}$

16. Each of the functions *F*, *G*, and *H* have the same SLA plots.
    a. Addressing the magnitude response curves, which will have an actual response closest to the SLA? Which will have an actual response furthest from the SLA?
    b. Addressing the phase response curves, which will have an actual response closest to the SLA? Which will have an actual response furthest from the SLA?

$$F(s) = \frac{4}{s^2 + 4s + 4} \qquad G(s) = \frac{4}{s^2 + 2s + 4} \qquad H(s) = \frac{4}{s^2 + 4}$$

Section 4.5

17. For the circuit of Figure P4.17, obtain an actual frequency response plot for the magnitude of $Z(j\omega)$ on log-log paper. Use a frequency range of 0.01–10 rad/s. Superimpose the SLA on top of the plot.

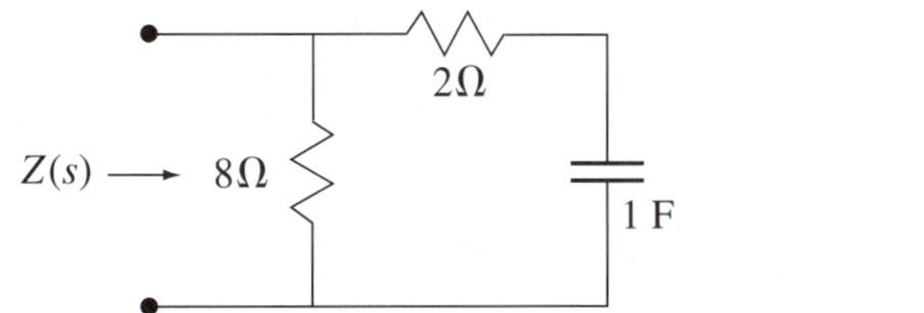

Figure P4.17

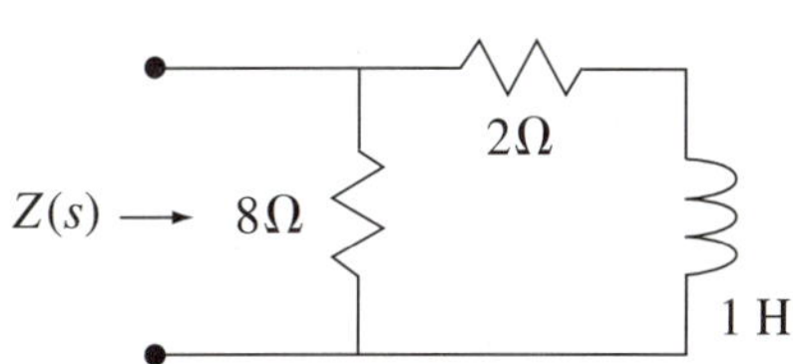

Figure P4.18

18. For the circuit of Figure P4.18, obtain an actual frequency response plot for the magnitude of $Z(j\omega)$ on log-log paper. Use a frequency range of 1–100 rad/s. Superimpose the SLA on top of the plot.
19. Obtain the actual impedance magnitude and phase responses for the circuit of Figure P4.13 for $R = 2\ \Omega$. Draw the SLA for each on top of the actual response curves. What is the value of the damping ratio for the complex root?
20. A graph of a transfer function has been generated over the frequency range of 0.1–10 rad/s using the MATLAB command >freqs(num,den,w). The gain magnitude needs to be limited to a range from 1 to 10. Show the MATLAB command sequence required to accomplish this.

Advanced Problem

21. A scope using a properly compensated $\times 10$ probe is shown in Figure P4.19. Sketch the input impedance (magnitude and phase) of the scope as a function of frequency. Determine the break frequency in hertz.

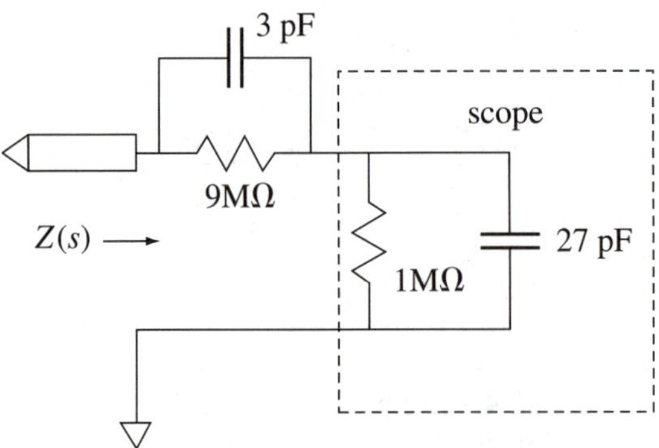

Figure P4.19

5

STANDARD FILTERS

OUTLINE

OBJECTIVES

1. Define characteristics of ideal low-pass and bandpass filters.
2. Discuss the characteristics of standard filter prototypes.
3. Denormalize prototype filter polynomials.
4. Transform a prototype to any desired filter type.
5. Develop simple prototype circuits.
6. Use MATLAB to find, denormalize, and transform filter polynomials.

INTRODUCTION

A common problem in signal processing is to isolate a desired signal from an interfering signal. When the signals occupy different frequency ranges, a filter is used to accomplish this task. Before we can implement a filter, we must first identify the properties that an ideal filter should have. Then we can search for a transfer function that will best approximate as many of those ideal characteristics as possible. This process has led to the Butterworth, Chebyshev, and other standard filter polynomials detailed in design handbooks.

In the past, the next stage in the design process would be to match the filter transfer function to a circuit that could implement it. In power applications that is still a necessary step. Today, however, the filtering of low-power signals can be accomplished with software rather than hardware, and with superior results. Some filter circuits are still needed to get the signal into and out of the digital processor, but these can be relatively simple and inexpensive.

The filter design handbook has also been made obsolete by software. The standard filter polynomials can be defined easily in the s domain, and denormalized or transformed to the particular filter type desired. Although these transformations involve only algebra, they are very difficult to accomplish in high-order filters without the assistance of appropriate computer software.

5.1 IDEAL FILTERS

An ideal filter cannot be built, but it establishes the properties to be strived for in real filters. After completing this section you will be able to:

- Define characteristics of ideal low-pass and bandpass filters.
- Define *cutoff frequency, passband,* and *stopband.*
- Explain the importance of constant gain in the passband.
- Explain the importance of linear phase in the passband.

In the frequency domain, the properties of an ideal filter seem obvious. Every sinusoid in the desired signal should emerge from the filter unchanged, while any other sinusoids should not emerge from the filter at all. The ideal filter frequency response is divided into passbands and stopbands, with the boundaries between them identified as cutoff frequencies.

Actually, there is no objection to signals in the passband emerging amplified, so long as all emerging sinusoids receive exactly the same gain. If this is the *only* change experienced by the signal, the output is an amplified duplicate of the input. If sinewaves in the passband are not amplified equally, the output suffers from *amplitude distortion.*

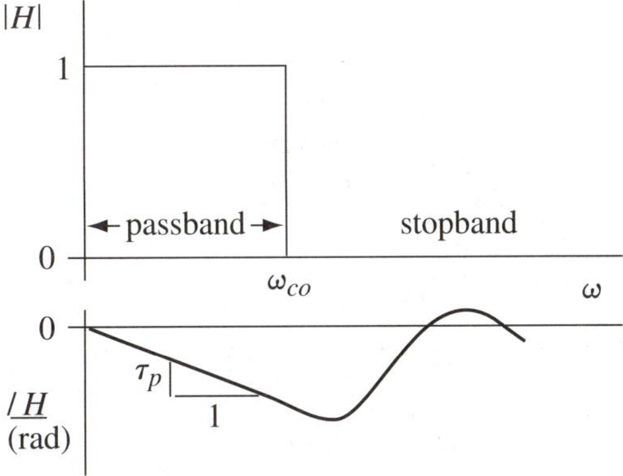

Figure 5.1 Ideal low-pass filter characteristics: constant amplitude gain and constant phase delay over the passband, no signal passage in the stopband. The edge of the passband is defined by the cutoff frequency ω_{co}.

It is impossible to build a frequency-selective circuit that does not shift the phase of sinewaves passing through it. Phase shift is also not a problem, provided the passband sinusoids are all delayed the same amount *in time.* This requires a phase shift that decreases linearly with frequency. The output is still an exact replica of the input, but it experiences a *phase delay,* τ_p, given by Equation 5.1. Any other phase shift relationship will introduce *phase distortion* in the filter output.

$$\cos \omega(t - \tau_p) = \cos(\omega t - \omega \tau_p) = \cos(\omega t + \theta)$$

$$\tau_p = -\frac{\theta}{\omega} \quad \text{(phase delay)} \tag{5.1}$$

An ideal low-pass filter introduces neither amplitude nor phase distortion and has the characteristic shown in Figure 5.1.

The characteristics of an ideal bandpass filter (Figure 5.2) are essentially the same as for the low-pass case, except it is only necessary for the time delays experienced in the passband to be constant relative to the other sinusoids in that region. Bandpass-type signals usually arise from using low-frequency information to modulate a high-frequency carrier. The demodulation process will eventually discard the carrier, so it is unimportant what absolute phase shift it experiences. The modulation envelope at the output experiences a *group delay,* τ_g, determined by the relative phase shift over the passband as given by Equation 5.2.

$$\tau_g = -\frac{d\theta}{d\omega} \quad \text{(group delay)} \tag{5.2}$$

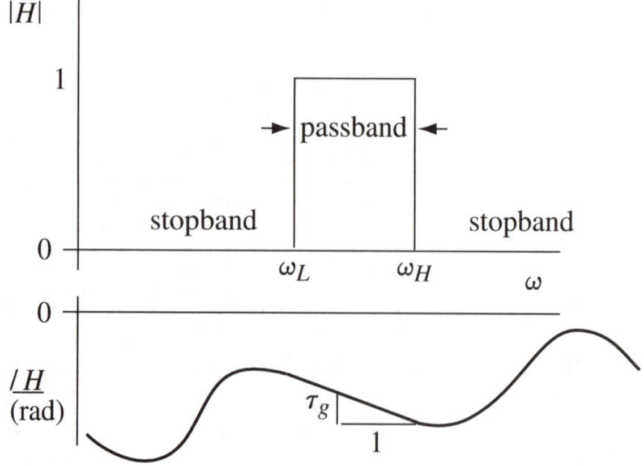

Figure 5.2 Ideal bandpass filter characteristics: constant amplitude gain and constant group delay over the passband, no signal passage in the stopband. The edges of the passband are defined by the cutoff frequencies ω_L and ω_H.

In defining these ideal filter characteristics, it is well to remember that the natural response of the system has been ignored. It is possible that the ideal filter for steady-state sinusoids may not be the best indicator of performance in applications where abrupt changes in the input signal occur regularly and excite the circuit's natural response. Standard filters arise from compromising one or another aspect of the ideal filter to achieve a practical result. The "best" filter is the one that achieves the most acceptable compromises in performance, as well as in cost, for the application intended.

5.2 FILTER PROTOTYPES

Mathematicians have identified polynomials that provide an optimal match to some feature of the ideal filter. To provide a unified design procedure, their results are summarized in a prototype filter, which is a low-pass filter with a cutoff frequency of 1 rad/s. After completing this section, you will be able to:

- Discuss the characteristics of standard filter prototypes.
- State the definition of *cutoff* used in the standard filters.
- Explain the effect of polynomial degree on filter implementation.
- Calculate the stopband attenuation of a Butterworth filter.
- Identify which aspects of the ideal filter are best approximated by which polynomials.

Real filters do not have an abrupt transition from the passband to the stopband. The definition used for *cutoff* varies with the prototype filter. Two that will be emphasized here are all-pole filters having the transfer function of Equation 5.3. Although a detailed SLA cannot be drawn until the polynomial coefficients have been determined, the high- and low-frequency asymptotes for this transfer function show the gross features of the response in Figure 5.3. Clearly, the higher the filter order n, the closer the high-frequency asymptote approximates the infinite drop-off rate of the ideal low-pass filter. One may also assume that higher-order polynomials would, because of their greater number of terms, permit a tighter fit to the ideal characteristics. Actual filters are named for the investigators who specified the polynomial coefficients that best approximate some feature of the ideal filter.

$$H(s) = \frac{a_0}{s^n + b_{n-1}s^{n-1} + \cdots + b_2 s^2 + b_1 s + b_0} \tag{5.3}$$

The *Butterworth* filter is one of the most popular of the all-pole filters. It does the best job of approximating the flat gain desired in the passband. An nth-order Butterworth prototype has a "maximally flat" passband and a transfer function that reduces to the simple form of Equation 5.4. This equation shows that cutoff for the Butterworth filter is defined by the −3-dB (0.7071) point, which is the standard definition of cutoff used in most circuits. The only parameter to specify for a Butterworth prototype is the filter order, n.

$$|H(\omega)|^2 = \frac{1}{1 + (\omega/\omega_{co})^{2n}} \tag{5.4}$$

Since the Butterworth transfer function takes this simple form, it may be solved to find the required n to meet a particular minimum attenuation in the stopband. It is the only one that takes a simple form, which may account for part of its popularity.

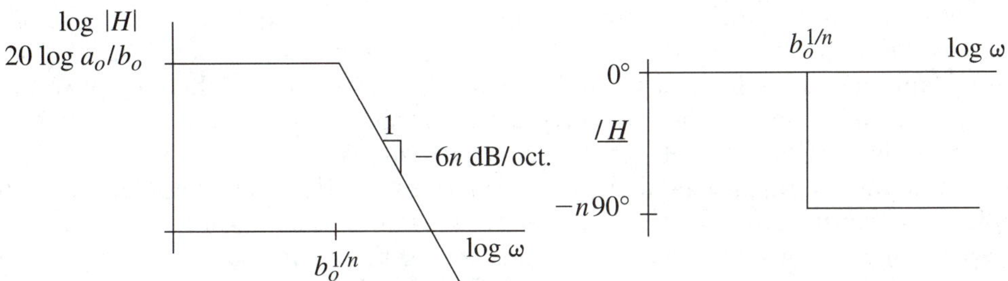

Figure 5.3 High- and low-frequency asymptotes for the all-pole filter characteristic of Equation 5.3. The frequency at which these asymptotes intersect does not necessarily have any physical significance.

EXAMPLE 5.1

What is the required order for a Butterworth low-pass filter if the gain must be down 20 dB at $\omega = 2\omega_{co}$?

Solution

A gain of −20 dB is equivalent to an $|H|$ of

$$-20 = 20 \log |H| \qquad \text{or} \qquad |H| = 0.1$$

Using Equation 5.4,

$$\begin{aligned} 1 + \left(\frac{\omega}{\omega_{co}}\right)^{2n} &= \frac{1}{(0.10)^2} = 100 \\ (2)^{2n} &= 99 \\ 2n \log 2 &= \log 99 \\ n &= 3.31 \end{aligned}$$

A fourth-order filter is required.

The frequency response for a third-order Butterworth is shown in Figure 5.4. The poles of a Butterworth prototype can be shown to be equally spaced in the LHP along a unit circle centered on the origin of the s plane.

Another popular all-pole prototype is the *Chebyshev* type 1 filter. By adjusting the polynomial coefficients, Chebyshev was able to trade off flatness in the passband for a higher *initial rate* of attenuation in the stopband. The Chebyshev polynomials create an *equiripple passband* in which the gain is allowed to vary a specified amount. The more variation allowed in the passband, the higher the initial attenuation in the stopband. The concept of an "equiripple passband" suggests that the appropriate definition for cutoff would be the frequency at which the gain finally exceeds the variation specified. In the prototype, this always occurs at 1 rad/s.

An nth-order Chebyshev (type 1) prototype response has a total of n maximums and minimums in the passband. All the maximums are the same height, and all the minimums are the same depth, which is the reason for the *equi*ripple designation. The poles of a Chebyshev (type 1) prototype are in the LHP and are equally spaced along an ellipse whose eccentricity is related to the allowed passband ripple. Specifications for a Chebyshev prototype include the filter order, n, and the allowed passband ripple, in decibels. A third-order Chebyshev prototype with 1-dB passband ripple is shown in Figure 5.5.

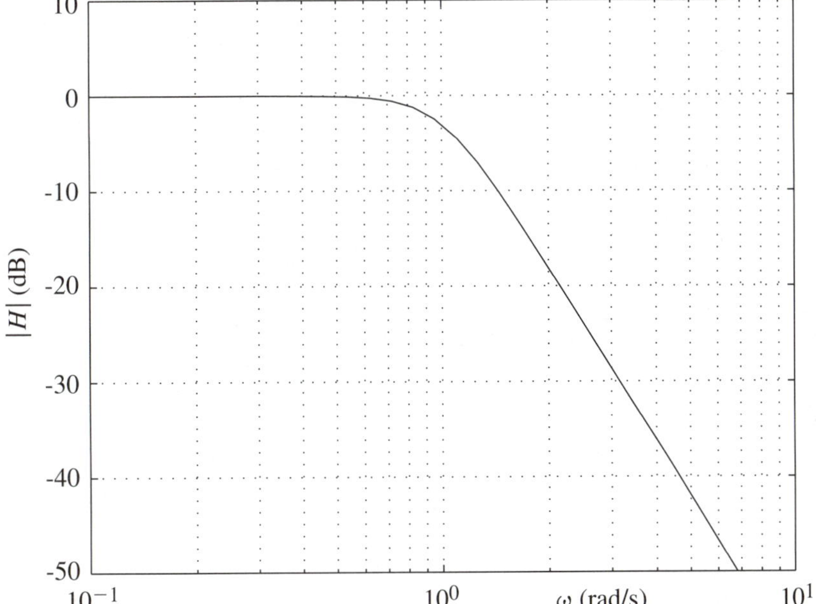

Figure 5.4 Response of a third-order Butterworth prototype.

The Bessel prototype is another all-pole filter, but its polynomial coefficients are chosen to best approximate the linear phase characteristic of the ideal filter. It, like the Butterworth, is fully specified by the filter order, n, and the design process is identical for both. Although the phase response of a filter may be more important than the amplitude response in some situations, we will find that a linear phase characteristic is easily achieved in digital filtering. As a result, the Bessel polynomials will be of little future interest to us, and we will not discuss them further.

Placing zeros in the stopband region of a low-pass transfer function leads to another interesting class of filters. One scheme that results in such a transfer function is to start with a Chebyshev all-pole prototype, transform it into a high-pass filter, and subtract it from unity. The result is a low-pass prototype with the ripple moved from the passband into the stopband. This prototype retains the abrupt transition from the passband to the stopband that the Chebyshev polynomials provide, and it is called a Chebyshev type 2 filter. Its polynomial coefficients are directly related to the type 1 filter. The specifications for this filter type include the order, n, and the minimum attenuation allowed in the stop band. The cutoff frequency of the type 1 Chebyshev prototype translates into the starting frequency of the type 2 stopband. Although it is a viable filter option, it is not particularly popular, primarily because the design does not define the end of the pass region. When we refer to a

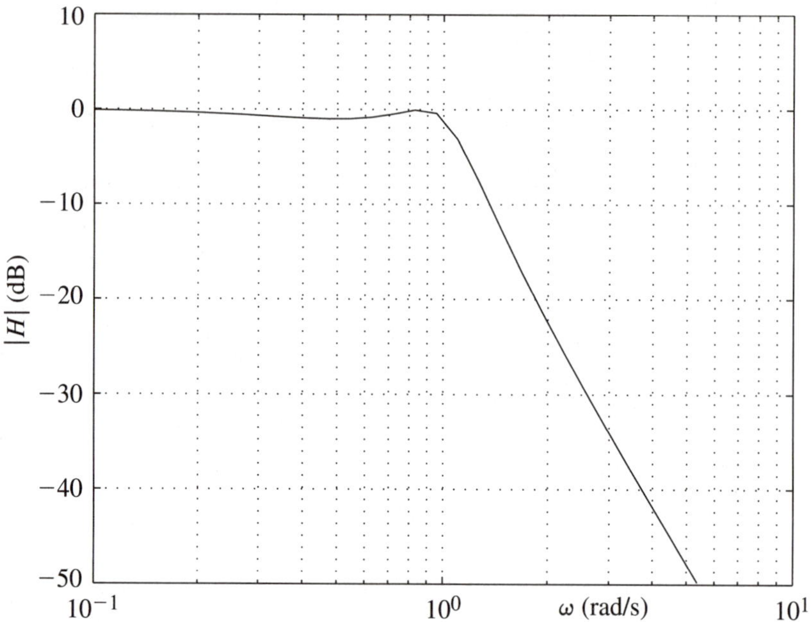

Figure 5.5 A third-order Chebyshev (type 1) prototype with 1-dB passband ripple. Cutoff is defined as the frequency where the gain drops below −1 dB. Although equal-order all-pole filters have the same ultimate *rate* of attenuation, the high-frequency asymptote of the Chebyshev filter has a lower break frequency than the equivalent Butterworth filter. As a result, the Chebyshev response approaches its high-frequency asymptote from above, while the Butterworth approaches its asymptote from below. This gives the Chebyshev filter a higher initial stopband attenuation.

Chebyshev filter, it will mean the type 1 filter unless otherwise stated. A third-order type 2 Chebyshev filter with a minimum attenuation of 30 dB at $\omega = 1$ rad/s is shown in Figure 5.6.

Zeros may be placed right on the $j\omega$ axis by using, for example, series resonant traps to shunt the transmission path. Equation 5.5 shows the transfer function of such a low-pass filter.

$$H(s) = \frac{a_m s^m + a_{m-1} s^{m-1} + \cdots + a_0}{s^n + b_{n-1} s^{n-1} + \cdots + b_2 s^2 + b_1 s + b_0} \tag{5.5}$$

The high-frequency asymptote is $a_m s^{m-n}$, where m is an even integer if resonant traps provide the zeros. The high-frequency asymptote will provide increasing attenuation with increasing frequency as long as $n > m$. For the limiting case of

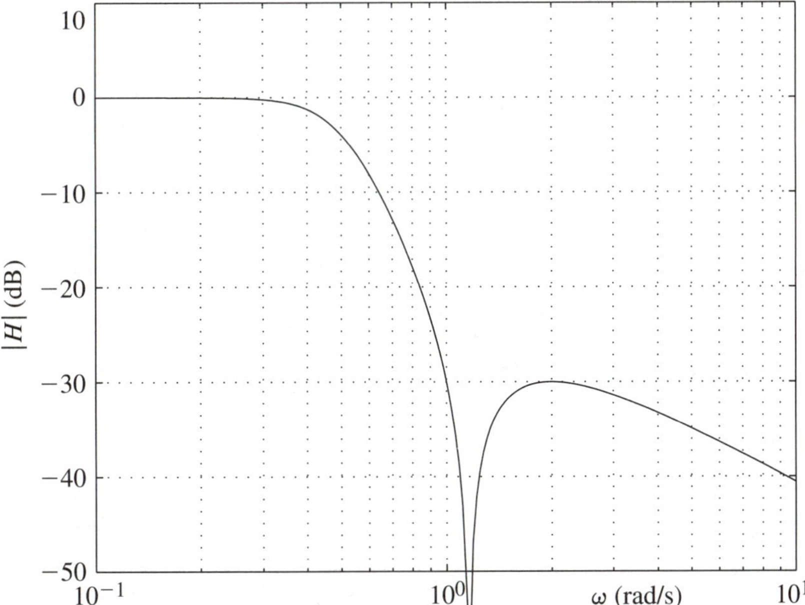

Figure 5.6 A third-order Chebyshev (type 2) prototype with a 30-dB minimum stopband attenuation. The prototype cutoff specification is replaced by the frequency where the minimum stopband attenuation is first achieved.

$n = m$, the high-frequency asymptote is horizontal, but the successive zeros keep pulling the gain back to zero to create the stopband. By adjusting a_m and properly spacing the zeros, it should be possible to achieve an equiripple stopband. Allowing ripple in both the passband and the stopband provides a filter with the most abrupt transition at cutoff.

The polynomials that provide equiripple pass and stopbands are elliptical functions, of which the Chebyshev polynomials are a special case. The filters are often referred to as *Cauer filters* in honor of another prominent investigator in this area. The specifications for a Cauer filter includes n, the allowed passband ripple, in decibels, and the minimum stopband attenuation, in decibels. A third-order Cauer filter with 1-dB passband ripple and 30-dB minimum stopband attenuation is shown in Figure 5.7.

Selecting among Butterworth, Chebyshev, Bessel, and Cauer filters is also influenced by the sensitivity each polynomial has to errors in circuit component values. Some designers favor the Butterworth because its flaws tend to be less obvious than an "unequiripple" Chebyshev response. In digital filters, however, the polynomials can be made very precise, and this criterion does not enter the filter selection process.

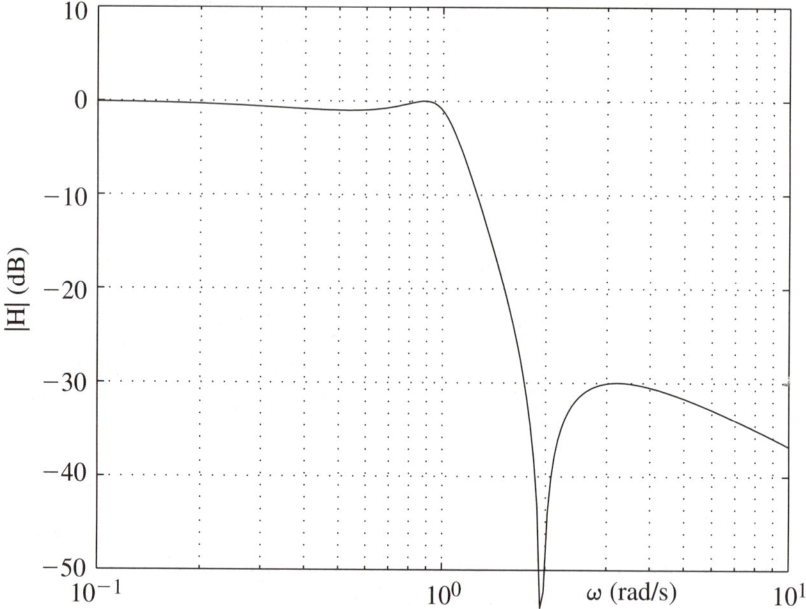

Figure 5.7 A third-order Cauer prototype with a 1-dB passband ripple and a minimum of 30 dB stopband attenuation. Cutoff is defined as the frequency where the gain leaves the equiripple passband, as in the case of a Chebyshev filter.

5.3 DENORMALIZATION

Denormalization is the process of converting a prototype filter to a low-pass filter with the desired cutoff frequency. After completing this section you will be able to:

- Denormalize prototype filter polynomials.
- Obtain the transfer function of a low-pass filter with a desired ω_{co}.
- Modify a circuit to have the cutoff frequency desired.

A filter handbook will give the prototype polynomials for a reasonable selection of filters. The Butterworth and Bessel prototype polynomials, for example, depend only on the polynomial degree, *n,* so they would require, at most, a page for each. Chebyshev prototypes, however, require a table of polynomials for every possible passband ripple. Even worse, Cauer prototypes would require a full set of tables for each Chebyshev table. A limited selection of prototype polynomials is given in Tables 5.1–5.3. The frequency variable is represented by p instead of s in these tables to reduce the chance for error during denormalization. If the prototype polynomial is to be denormalized for a cutoff frequency of ω_{co}, the p variable must be

Table 5.1 Butterworth Polynomials

$(p + 1)$

$p^2 + 1.4142p + 1$

$(p + 1)(p^2 + p + 1) = p^3 + 2p^2 + 2p + 1$

$(p^2 + 0.7654p + 1)(p^2 + 1.8478p + 1) =$
$p^4 + 2.6131p^3 + 3.4142p^2 + 2.6131p + 1$

$(p + 1)(p^2 + 0.6180p + 1)(p^2 + 1.6180p + 1) =$
$p^5 + 3.2361p^4 + 5.2361p^3 + 5.2361p^2 + 3.2361p + 1$

For cutoff frequency of 1 rad/s.
$a_o = 1$ for a maximum gain of 0 dB (see Equation 5.3).

Table 5.2 Chebyshev Polynomials: 0.5-dB Ripple; $\omega_{co} = 1.0$ rad/s

$p + 2.8628$

$p^2 + 1.4256p + 1.5162$

$(p + 0.6265)(p^2 + 0.6265p + 1.1424) = p^3 + 1.2529p^2 + 1.5349p + 0.7157$

$(p^2 + 0.3507p + 1.0635)(p^2 + 0.8467p + 0.3564) =$
$p^4 + 1.1974p^3 + 1.7169p^2 + 1.0255p + 0.3791$

$(p + 0.3623)(p^2 + 0.5862p + 0.4768)(p^2 + 0.2239p + 1.0358) =$
$p^5 + 1.1725p^4 + 1.9374p^3 + 1.3096p^2 + 0.7525p + 0.1789$

To set the maximum gain to 0 dB, make $a_o/b_o = 0.9441$ if n is even and $a_o/b_o = 1$ if n is odd (see Equation 5.3).

Table 5.3 Chebyshev Polynomials: 2.0-dB Ripple; $\omega_{co} = 1.0$ rad/s

$p + 1.3076$

$p^2 + 0.8038p + 0.8231$

$(p + 0.3689)(p^2 + 0.3689p + 0.8861) = p^3 + 0.7378p^2 + 1.0222p + 0.3269$

$(p^2 + 0.2098p + 0.9287)(p^2 + 0.5064p + 0.2216) =$
$p^4 + 0.7162p^3 + 1.2565p^2 + 0.5168p + 0.2058$

$(p + 0.2183)(p^2 + 0.3532p + 0.3932)(p^2 + 0.1349p + 0.9523) =$
$p^5 + 0.7065p^4 + 1.4995p^3 + 0.6935p^2 + 0.4593p + 0.0817$

To set the maximum gain to 0 dB, make $a_o/b_o = 0.7943$ if n is even and $a_o/b_o = 1$ if n is odd (see Equation 5.3).

replaced everywhere by the expression (s/ω_{co}). The denormalization process can also be described in terms of functional notation as

$$H_{LP}(s) = H_p\left(\frac{s}{\omega_{co}}\right) \tag{5.6}$$

where H_p denotes the prototype transfer function polynomials. The denormalized filter polynomial will have the same value at $s = j\omega_{co}$ that the prototype has at $p = j1$ rad/s.

EXAMPLE 5.2

Determine the transfer function for a third-order 0.5-dB ripple Chebyshev low-pass filter with a cutoff frequency of 1 kHz.

Solution

The prototype filter transfer function is, from Table 5.2,

$$H_p(p) = \frac{0.7157}{p^3 + 1.2529p^2 + 1.5349p + 0.7157}$$

To move cutoff to 1 kHz, each p in the prototype is replaced by the expression s/ω_{co} to obtain the denormalized low-pass filter polynomial. The desired filter is

$$H_{LP}(s) = \frac{0.7157}{(s/2\pi \cdot 10^3)^3 + 1.2529(s/2\pi \cdot 10^3)^2 + 1.5349(s/2\pi \cdot 10^3) + 0.7157}$$

Multiplying through by $(2\pi \cdot 10^3)^3$ gives a final form of

$$H_{LP}(s) = \frac{178 \cdot 10^9}{s^3 + 7.872 \cdot 10^3 \cdot s^2 + 60.60 \cdot 10^6 \cdot s + 178 \cdot 10^9}$$

There is, of course, a great deal of difference between knowing what the filter transfer function should be and having a circuit with that transfer function. Often, filter handbooks will give circuits that implement a particular prototype, and the designer will have to denormalize the circuit. Circuits also have an impedance level, which is independent of the transfer function polynomial coefficients. The load will usually be normalized to 1 Ω. In past chapters we have also normalized our circuits to impedance levels and frequencies that keep the numbers simple so that we could concentrate on the ideas presented without becoming bogged down with unwieldy powers of 10. Denormalizing our own or handbook circuits for a practical design is straightforward. The first step is to select new capacitors and inductors that will have the same impedance at the desired cutoff frequency as they presently have at the normalized cutoff frequency. Using primes to indicate the new values:

$$\omega'_{co}\, C' = \omega_{co}\, C \qquad \text{and} \qquad \omega'_{co}\, L' = \omega_{co}\, L$$

$$C' = \omega_{co}\frac{C}{\omega'_{co}} \qquad L' = \omega_{co}\frac{L}{\omega'_{co}} \tag{5.7}$$

where ω_{co} is usually 1. Next, the impedance level is increased by some arbitrary factor k. The value of k is selected to make one component, usually the load resistor, a specified value:

$$R' = kR, \qquad L'' = kL', \qquad \text{and} \qquad C'' = \frac{C'}{k} \tag{5.8}$$

Naturally, (5.7) and (5.8) could be combined into a single operation.

EXAMPLE 5.3

The circuit of Figure 5.8 is a third-order 2-dB Chebyshev prototype. Modify it to provide a 1-kHz cutoff frequency while driving a 600-Ω load.

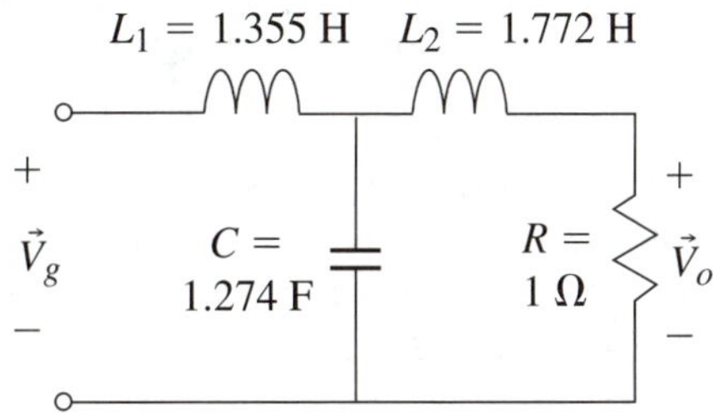

Figure 5.8

Solution

The new cutoff frequency is $\omega' = 2\pi \cdot 10^3$, so applying Equation 5.7 gives

$$L_1' = \frac{1.355}{2\pi \cdot 10^3} = 0.215 \text{ mH}$$
$$L_2' = \frac{1.772}{2\pi \cdot 10^3} = 0.282 \text{ mH}$$
$$C' = \frac{1.274}{2\pi \cdot 10^3} = 202.8\ \mu\text{F}$$

Now the impedance level must be raised by a factor of 600 to give the load resistance specified. Applying Equation 5.8 gives

$$R' = 600\ \Omega$$
$$L_1'' = 600(0.215 \text{ mH}) = 129 \text{ mH}$$
$$L_2'' = 600(0.282 \text{ mH}) = 169 \text{ mH}$$
$$C'' = \frac{202.8\ \mu\text{F}}{600} = 338 \text{ nF}$$

5.4 FILTER TRANSFORMATIONS

Prototype transfer functions may also be converted to high-pass, bandpass, or stopband filters. The transformation may be made to a prototype circuit or just to the prototype polynomials. After completing this section you will be able to:

- Transform a prototype transfer function to any desired filter type.
- Relate bandwidth and center frequency to passband cutoff frequencies.

Figure 5.9 demonstrates the filter varieties available. To find the polynomials for the new filter, start with the prototype and replace each p with the indicated argument. In the case of bandpass or stopband filters, the bandwidth and center frequency need to be determined. These are related to the cutoff frequencies, ω_H and ω_L, by

$$B = \omega_H - \omega_L \qquad \text{and} \qquad \omega_c = \sqrt{\omega_H \omega_L} \tag{5.9}$$

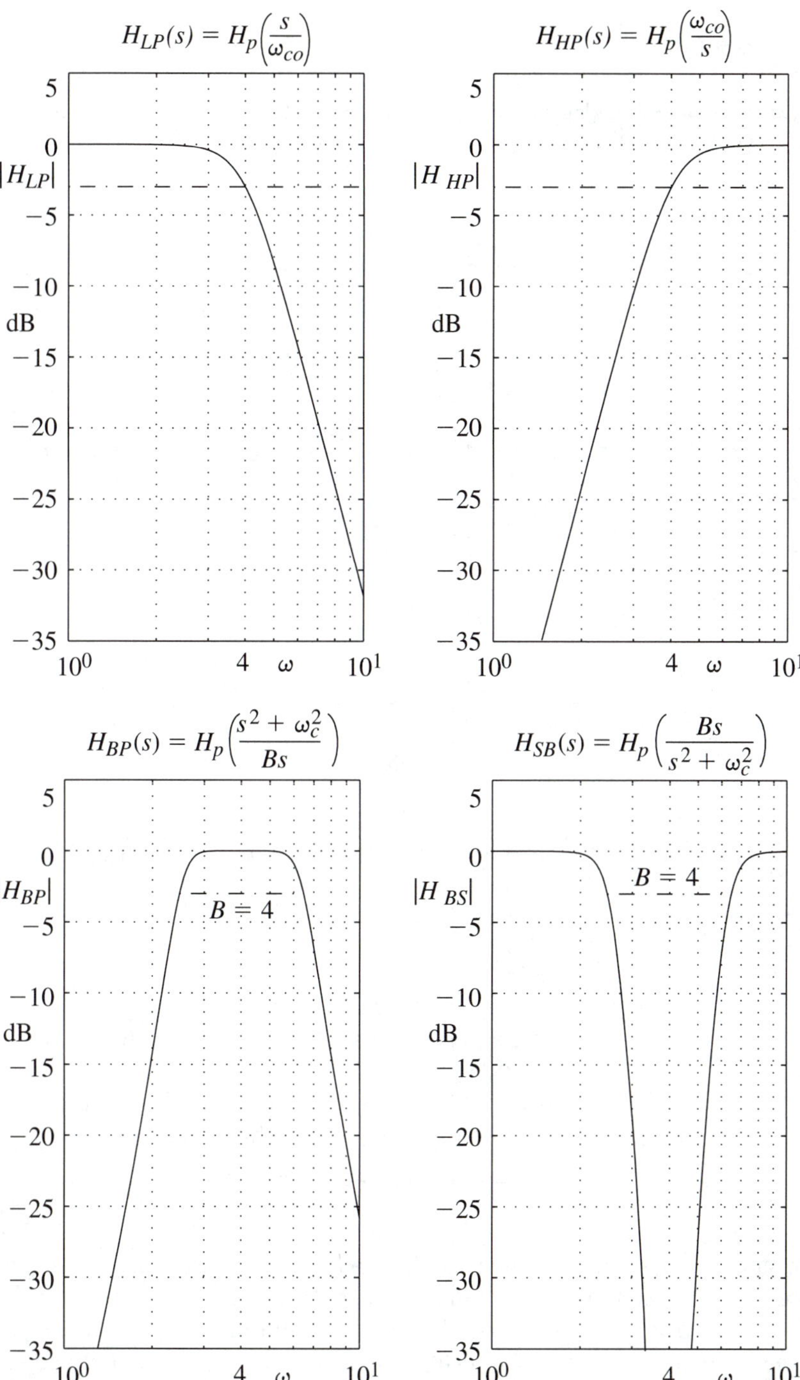

Figure 5.9 Characteristics of low-pass, high-pass, bandpass, and bandstop filters derived from a fourth-order Butterworth prototype. Cutoff for the low- and high-pass units is 4 rad/s. Bandpass and stopband units were designed for a center frequency of 4 rad/s and a bandwidth of 4 rad/s.

Find $H(s)$ for a second-order Chebyshev bandpass filter with 2 dB of passband ripple and cutoff frequencies of 5 and 8 rad/s.

Solution

We start with the appropriate prototype. From Table 5.3:

$$H(p) = \frac{0.6538}{p^2 + 0.8038p + 0.8231}$$

We calculate $B = 8 - 5 = 3$ and $\omega_c = \sqrt{40} = 3.62$, giving a transformation of

$$p \to \frac{s^2 + 40}{3s}$$

Then

$$H_{BP}(s) = \frac{0.6538}{\left(\frac{s^2 + 40}{3s}\right)^2 + 0.8038\left(\frac{s^2 + 40}{3s}\right) + 0.8231}$$

$$= \frac{0.6538(9s^2)}{(s^2 + 40)^2 + 0.8038(3s)(s^2 + 40) + 0.8231(9s^2)}$$

After a little bookkeeping,

$$H_{BP}(s) = \frac{5.884s^2}{s^4 + 2.411s^3 + 87.41s^2 + 96.46s + 1600}$$

Although the process is routine, it is clear that with realistic frequencies and higher-order prototypes, the process becomes very tedious and prone to error!

Prototype circuits may also be transformed directly to, for instance, a bandpass circuit. Figure 5.10 demonstrates this process with a second-order prototype. It is of interest primarily because it shows the origin of the bandpass transformation. To achieve the correct bandwidth, the prototype must first be denormalized to a cutoff frequency equal to the bandwidth eventually desired. The inductor that passes low

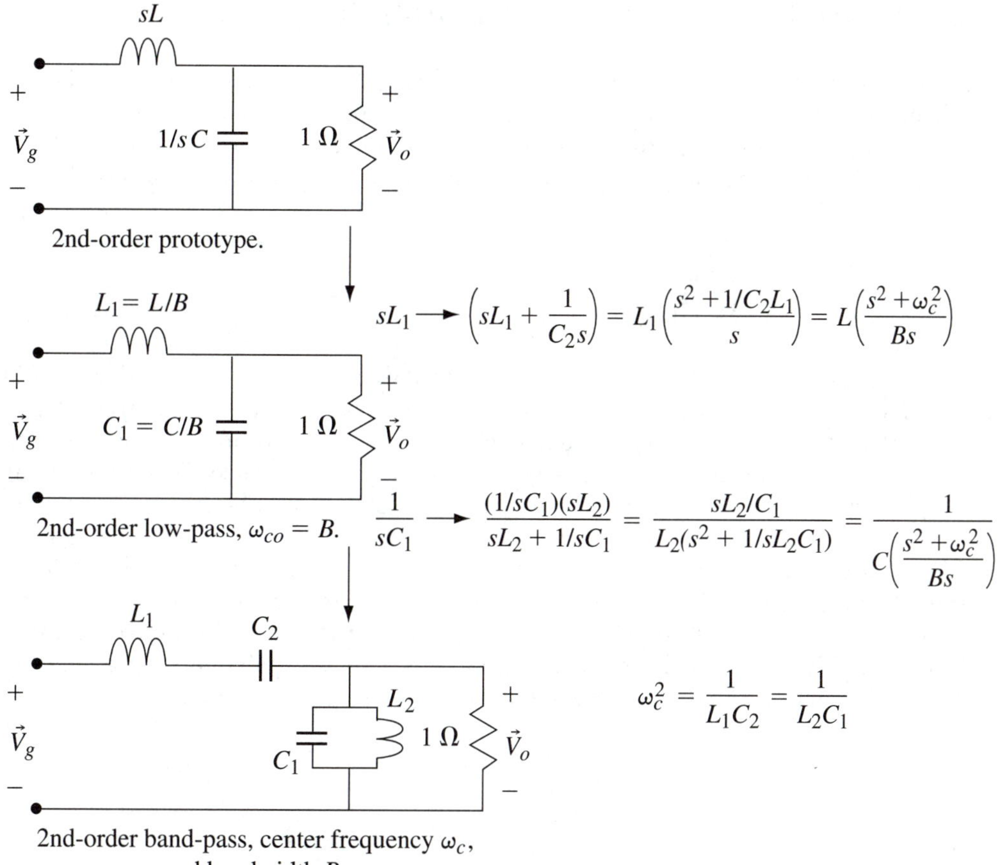

Figure 5.10 Prototype-to-bandpass transformation. First the prototype is denormalized to the proper bandwidth. Then the series path is made a series resonant circuit ($sL_1 + 1/sC_2$) and the shunt path is made a parallel resonant circuit (sL_2 in parallel with $1/sC_1$). Mathematically, the final result involves only replacing the s variables in the prototype with the bandpass transformation.

frequencies must then be replaced with a series resonant circuit that only passes signals around ω_c. Similarly, the capacitor that shunts the signal to ground at high frequencies must be replaced by a parallel resonant circuit that shunts the signal at all frequencies except those around ω_c.

The same process is used to obtain a stopband filter except that the series path must become a parallel resonant circuit while the shunt path becomes a series resonant circuit. The prototype should again be denormalized to the proper bandwidth first.

5.5 PROTOTYPE CIRCUIT DEVELOPMENT

We have been able to modify prototype polynomials or circuits to meet a design requirement. While prototype polynomials can be computer generated, prototype circuits have to be described in a handbook. We will develop our own mini-handbook to suggest the process and to provide a few prototype circuits. After reading this section you will be able to:

- Develop simple prototype circuits.
- State the advantages of active filters.

The general process of finding a circuit that implements a particular $G(s)$ is known as *circuit synthesis,* and it is usually a graduate-level course for circuit theorists. Fortunately, we do not need to be general and can just propose a low-pass circuit topology. As long as we can achieve the desired polynomial coefficients without having to provide negative resistors, inductors, or capacitors, the design can be considered successful.

The all-pole prototype filter is shown in Figure 5.11. The order of the filter equals the number of independent inductors and capacitors present. The design procedure is to obtain the transfer function for the filter order desired and to determine the relationship between the circuit components and the polynomial coefficients. This task becomes more difficult as higher-order filters are attempted.

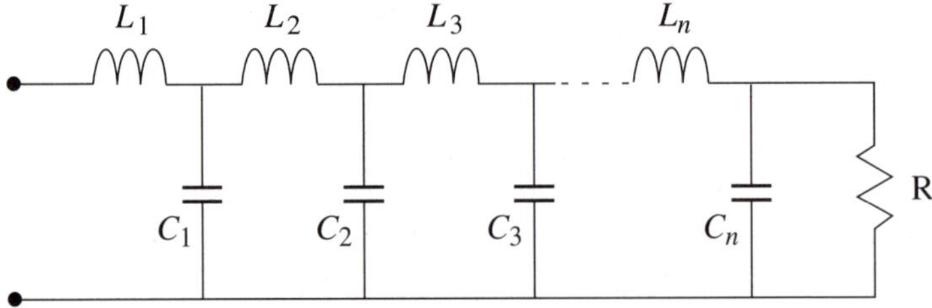

Figure 5.11 Ladder implementation of an all-pole prototype of order $2n$.

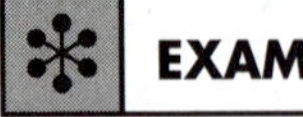

EXAMPLE 5.5

Determine the circuit components needed for a second-order Butterworth prototype filter.

Solution

The circuit of Figure 5.12 will give a second-order all-pole filter. The transfer function can be found with the voltage divider:

$$\frac{\vec{V}_o}{\vec{V}_g} = \frac{\dfrac{R/sC}{R + 1/sC}}{sL + \dfrac{R/sC}{R + 1/sC}} = \frac{\dfrac{R}{sCR + 1}}{sL + \dfrac{R}{sCR + 1}} = \frac{1/LC}{s^2 + s/CR + 1/LC}$$

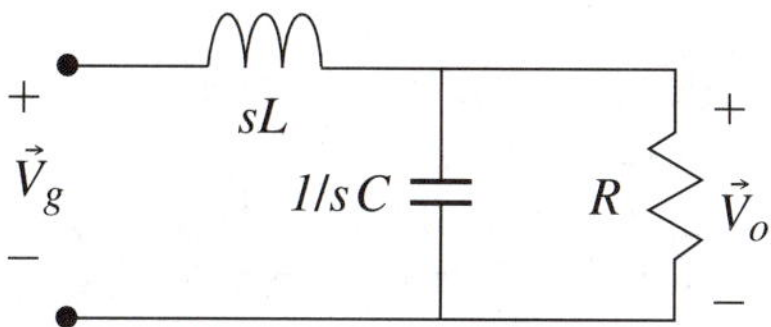

Figure 5.12

For a Butterworth prototype, $1/CR = \sqrt{2}$ and $1/LC = 1$. One of the many ways this could be accomplished is $R = 1$, $C = 0.7071$, and $L = 1.4142$. (See Problem 11, page 154, for a third-order prototype.)

The results of Example 5.5 might be incorporated in a handbook either as a prototype circuit or as a sequence of design steps as shown in Figure 5.13a and b. Neither approach reflects the full range of options available for the design, but they keep things simple for unsophisticated audiences. We prefer to keep all of the design options open.

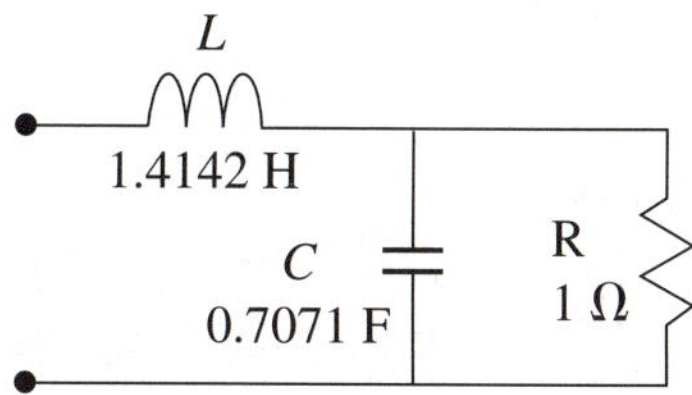

Figure 5.13a Second-order Butterworth prototype circuit.

1. Select R.

2. $C = \dfrac{1}{\sqrt{2}R}$

3. $L = \dfrac{1}{C}$

Figure 5.13b Second-order Butterworth prototype design equations.

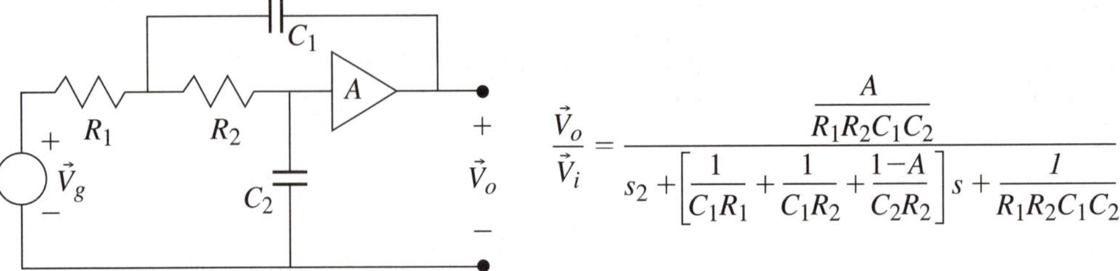

Figure 5.14 An active filter quadratic section.

The classic all-pole filters have complex poles. We have seen that R-C and R-L circuits have only real poles and zeros, so both inductors and capacitors must be present in passive circuits to create Butterworth or Chebyshev filters. This limitation can be overcome by adding the gain or feedback capabilities of an op amp to the circuit. Since inductors are generally more expensive and less ideal than capacitors, especially at low frequencies, schemes that allow the synthesis of complex roots with R-C circuits are very attractive. An op amp circuit that provides this capability is shown in Figure 5.14.

There are five components to adjust and only two polynomial coefficients to set. This means that several arbitrary decisions can be made in the design. The op amp also prevents loading on the circuit, so filter coefficients are independent of loading effects, and higher-order filters may be created by simply cascading second-order sections. Circuits using op amps to achieve these advantages are called *active filters*.

EXAMPLE 5.6

Design an active fourth-order Butterworth prototype filter.

Solution

Table 5.1 shows that the required polynomials are

$$(p^2 + 0.7654p + 1)(p^2 + 1.8478p + 1)$$

One popular option is to let $A = 1$ and $R_1 = R_2 = 1$. This reduces the transfer function of a quadratic section to

$$\frac{\vec{V}_o}{\vec{V}_g} = \frac{1/C_1C_2}{s^2 + (2/C_1)s + 1/C_1C_2}$$

which requires $C_1 = 2/0.7654 = 2.613$ F and $C_2 = 0.3827$ F for the first section. The values for the second section are $C_1 = 2/1.8478 = 1.0824$ F and $C_2 = 0.9239$ F.

Active circuits are limited by the frequency capability of the op amps used and are suitable only for low-power applications. They also require power supplies for biasing the op amp, which may make them uneconomical unless other electronic circuits are already present. Notice from Figure 5.14 that a large value of A could result in poles in the RHP! (Test #1 of the Routh-Hurwitz criterion.)

Cauer filters require a different topology, since they have both poles and zeros. A third-order Cauer filter requires a minimum of five components, because it has to be able to set five polynomial coefficients independently. Its transfer function is given as Equation 5.9.

$$H(p) = \frac{a_2 s^2 + a_0}{s^3 + b_2 s^2 + b_1 s + b_0} \tag{5.9}$$

A couple of circuits that might serve as the prototype are shown in Figure 5.15. The transfer function for circuit (a) is provided as Equation 5.10. It matches the desired polynomials provided that a_o and b_o can be the same. This is a common requirement imposed by circuits, and simply sets the d-c gain at unity. This could be changed by breaking the load resistor into two pieces and only taking the output across part of it, but there is no point to doing that. An entire transfer function can be multiplied by any constant without changing the nature of its frequency response. In the case of a third-order Cauer filter, $a_o = b_o$ anyway.

$$\frac{\vec{V}_o}{\vec{V}_g} = \frac{\frac{1}{L_3} s^2 + \frac{1}{C}\left(\frac{1}{L_1 L_2} + \frac{1}{L_1 L_3} + \frac{1}{L_2 L_3}\right)}{s^3 + \left(\frac{1}{L_2} + \frac{1}{L_3}\right) s^2 + \frac{1}{C}\left(\frac{1}{L_1} + \frac{1}{L_2}\right) s + \frac{1}{C}\left(\frac{1}{L_1 L_2} + \frac{1}{L_1 L_3} + \frac{1}{L_2 L_3}\right)} \tag{5.10}$$

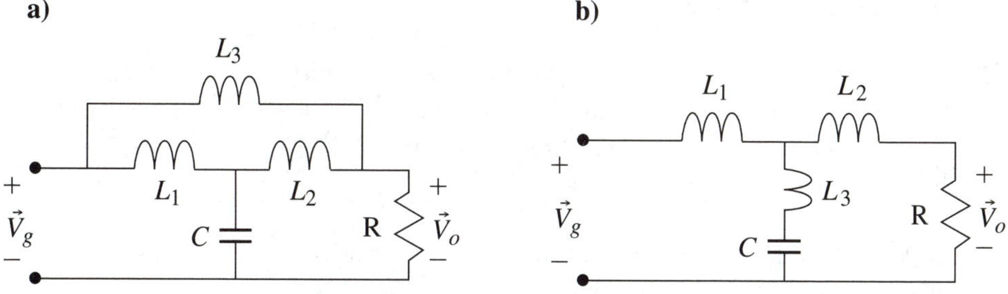

Figure 5.15 Possible circuit topologies for a Cauer third-order prototype.

Relating the circuit components to the polynomial coefficients provides the design equations.

$$L_3 = \frac{1}{a_2}$$

and

$$\frac{1}{L_2} + a_2 = b_2 \qquad \text{or} \qquad L_2 = \frac{1}{b_2 - a_2}$$

Dividing b_0 by b_1 eliminates C, leaving L_1 as the only unknown. After some algebra, the result is

$$L_1 = \frac{b_1 b_2 - b_0}{(b_2 - a_2)(b_0 - a_2 b_1)}$$

With all the inductors known, C can be found from

$$C = \frac{1}{b_1}\left(\frac{1}{L_1} + \frac{1}{L_2}\right)$$

Notice that there is a potential for some of the inductors to come out with negative values. Should that happen, another circuit topology would have to be tried or the filter specifications relaxed.

Cauer filters are not generally used in simple filtering applications and have been discussed only for the sake of completeness. They are often implemented as digital filters, however.

5.6 MATLAB LESSON 5

MATLAB will serve as our filter polynomial handbook and save us from transformation drudgery. It also allows us to reduce any self-imposed drudgery by creating scripts to execute a sequence of commands repeatedly. After completing this section you will be able to:

- Use MATLAB to find, denormalize, and transform filter polynomials.
- Write and execute MATLAB scripts.

MATLAB EXAMPLES

Matlab commands exist for finding prototype filter polynomials and for transforming them into the filter types desired. The *ap* ending to the following commands indicates *analog prototype.*

buttap **cheb1ap** **cheb2ap** **ellipap**

```
>[z,p,k]=buttap(6);   % returns the zeros, poles, and multiplier for a 6th-
                        order Butterworth
>num=k*poly(z);       % creates the Butterworth numerator, z is an empty
                        set
>den=poly(p);         % creates the Butterworth denominator
>w=logspace(-1,1);    % go a decade down and up from cutoff
>freqs(num,den,w);    % display the frequency response of the Butterworth
                        prototype
```

This sequence may be used to plot the response of any of the prototypes. The **ellipap** command is for the Cauer prototype. Command formats are

```
[z,p,k]=cheb1ap(order, ripple in dB)
[z,p,k]=cheb2ap(order, minimum stopband attenuation in dB)
[z,p,k]=ellipap(order, ripple in dB, minimum stopband attenuation in dB)
```

Try one of your choice.

The task of denormalizing or transforming to a different filter type, which gets very messy for realistic filter specs, can also be turned over to MATLAB.

lp2lp **lp2hp** **lp2bp** **lp2bs**

```
>[tnum,tden]=lp2bp(num,den,2*pi*1000, 2*pi*500)   % returns the polynomials
                                                    for a bandpass filter
                                                    centered at 1 kHz and
                                                    500 Hz wide.
>format short e                                   % will need to change
                                                    format to see all the
                                                    polynomial terms
>w=logspace(3,5);                                 % make a frequency
                                                    response run
>h=freqs(tnum,tden,w);                            % use this format to
                                                    retain control of what
                                                    is graphed
>semilogx(w,20*log10(abs(h)))
>axis([2000 20000 -80 10])                        % curve shows straight-
                                                    line segments, need
                                                    more points
>w=logspace(3.5,4.5, 100);                        % use shorter range and
                                                    more points
>h=freqs(tnum,tden,w);
>semilogx(w,20*log10(abs(h)))
>axis([2000 20000 -80 10])
```

In many situations a particular command sequence is used repeatedly. It can quickly become tedious retyping the same commands or even recovering them with the up-arrow key. An alternative is to create a text file consisting of the command sequence, just as you would normally enter them in the Command Window. The file is named, and given an **.m** extension. Once it has been saved, it may be called by typing its name in the Command Window.

We may create an **.m** file with any text editor, or by selecting **New** under the MATLAB **File** menu. Once the desired command sequence has been entered, we must **Save** the file either to the hard drive or to a floppy, and give it a name. When that name is used in the Command Window, MATLAB goes searching for an **.m** file of that name, and if it finds it, executes the instructions it contains. If the file is saved in the MATLAB directory, it will be found. If it is saved in another directory or on a floppy, MATLAB must be directed to include that directory or floppy in its search path. This is accomplished by following the **Set Path** procedure under the **File** menu. On a Microsoft Windows system, entering **cd a:** while in the MATLAB Command Window will direct MATLAB to search the floppy before going through its usual search path.

Now that MATLAB can find your script, you can call it and edit it if necessary. Remember to resave it after any editing.

Variables in a script become part of the workspace and must not conflict with variable names already in the Command Window. Scripts basically are used for a special purpose and then discarded. MATLAB also allows users to create their own functions, which are similar to scripts but have many advantages. We will discuss function **.m** files later, but if a script has fairly widespread application, it should probably be rewritten as a function. If a script is not discarded after use, make sure it contains enough commentary to remind you what the file does and what information has to be provided before it can be called.

EXAMPLE 5.7

Your company sells 1-dB Chebyshev low-pass filters. Prepare a family of curves showing the stopband attenuation from cutoff out to four times cutoff, for even-ordered filters out to the tenth order.

Solution

We will start by creating an **.m** file **File** **New** **M-File**

```
% This script plots the roll-off of Chebyshev 1 dB filters
% call after defining filter order and frequency range w
```

```
[z,p,k]=cheb1ap(n,1); num=k*poly(z); den=poly(p);
h=freqs(num,den,w);
semilogx(w,20*log10(abs(h)))
```

File **Save** (to floppy) **As** **ez1.m**

Back in the Command Window (Microsoft system)

```
>cd a:                          % put the floppy in the search path
>w=logspace(0,.7); n=2; ez1     % define n and w and execute the scrip.
>hold
>n=4; ez1
>n=6; ez1
>n=8; ez1
>n=10; ez1
>axis([1 4 -100 10])
>grid                           % complete the labeling, and use gtext to
                                  label individual curves.
```

See Figure 5.16 for the result.

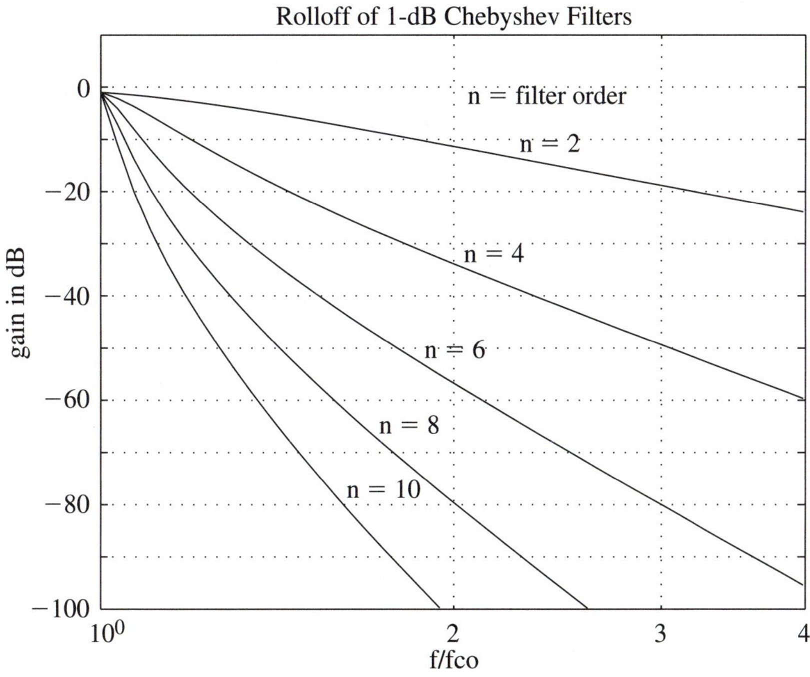

Figure 5.16

CHAPTER SUMMARY

For a signal to pass through a filter without distortion, the filter must have a constant gain and linearly decreasing phase over its passband. For interfering signals to be completely eliminated by the ideal filter, an abrupt transition to the stopband and zero gain in the stopband is needed. Butterworth, Bessel, Chebyshev, and Cauer filters all seek to approximate one aspect of these ideal characteristics by compromising the others.

Prototype filters are low-pass filters with a cutoff frequency of 1 rad/s. Actual filter polynomials are obtained from the prototype by denormalization and transformation. For implementation, circuit transfer functions are compared to the filter polynomials to arrive at design equations.

In recent years, handbooks of filter polynomials have been replaced with computer software that can generate the prototypes directly. In addition, low-power signals are being filtered by a digital program rather than a circuit. This technique allows much higher-order filters than are practical with circuits and thus much closer approximations to the ideal.

PROBLEMS

Section 5.1

1. A signal having frequency components below 2 rad/s is passed through a filter with the characteristics shown in Figure P5.1. Compare the input and output signals.

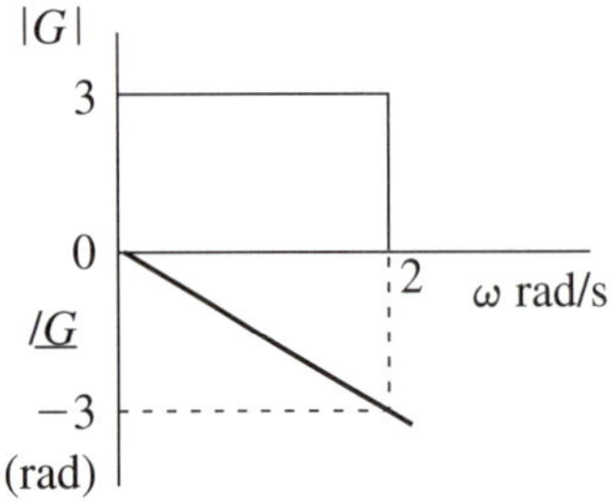

Figure P5.1

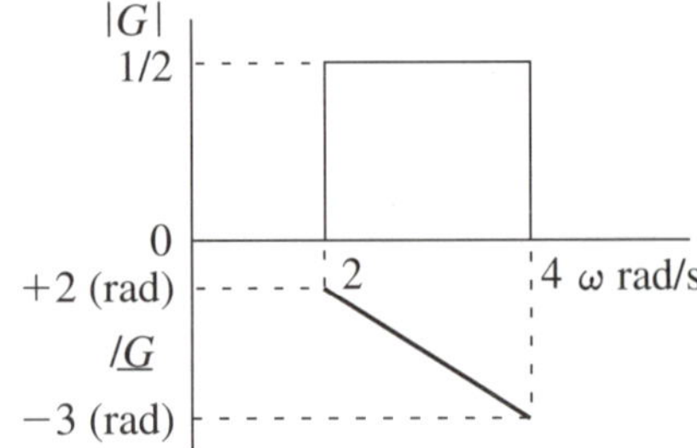

Figure P5.2

2. A signal having frequency components between 2 and 4 rad/s passes through the filter of Figure P5.2. Compare the output and input signals.

3. To demonstrate the effects of phase and group delay, create the function $f = \cos 9t + \cos 10t - \cos 10.5t$ for $0 \geq t \geq 12$, using about 500 points. This signal is input to a bandpass filter with a constant phase delay of 2 seconds. The output is $g = \cos(9t - 18) + \cos(10t - 20) - \cos(10.5t - 21)$.
   a. Verify that the expression for g corresponds to f with a phase delay of 2 seconds. Plot f and g to show they are *identical* except for the 2-second delay.
   b. Create a new signal, x, by adding any arbitrary number (say, 5) to each phase term in the expression for g. This signal no longer has a constant phase delay, but it does still have a constant group delay of 2 seconds. Create two more new signals, y and z, by adding different arbitrary constants. Plot g, x, y, and z on the same graph. Observe that all the functions are different but that they all fit in the same envelope and that the envelope is delayed by 2 seconds.
4. Using the identity $2 \cos A \cos B = \cos(A + B) + \cos(A - B)$, show analytically that if the amplitude-modulated signal $f(t) = (1 + m \cos \omega_m t) \cos \omega_c t$ passes through a unity-gain bandpass filter whose phase plot is as shown in Figure P5.4, then the output is $g(t) = [1 + m \cos(\omega_m(t - \tau_g)] \cos \omega_c(t - \tau_p)$.

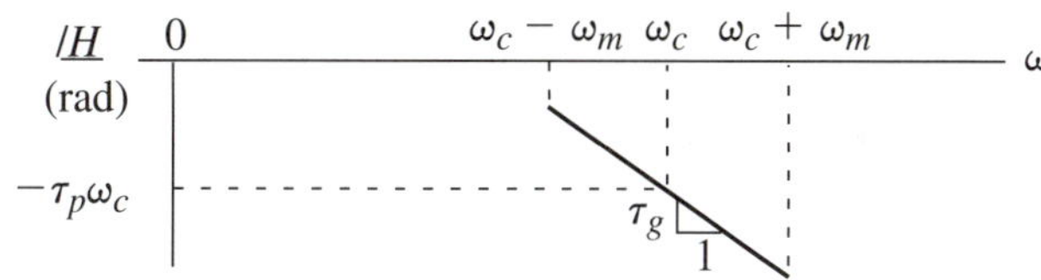

Figure P5.4

Section 5.2

5. Insert the third-order Butterworth prototype coefficients into Equation 5.3 and show that it reduces to the form of Equation 5.4.
6. Determine the order of a Butterworth low-pass filter required if the gain is to be down 80 dB at $f = 2f_{co}$.
7. What order Butterworth low-pass filter can be used if the gain needs to be down 20 dB at $f = 1.5f_{co}$?
8. From consideration of Figure 5.3, show that the extra attenuation achieved by the Chebyshev filter high-frequency asymptote over an equal-order Butterworth high-frequency asymptote is $20 \log b_o$, assuming $a_o = b_o$.

Section 5.3

9. Starting with the information in Table 5.1, provide the $H(s)$ for a third-order Butterworth low-pass filter with a cutoff frequency of 4 kHz.

10. Starting with the information in Table 5.2, provide the $H(s)$ for a third-order Chebyshev 0.5-dB low-pass filter with a cutoff frequency of 1 kHz.
11. Starting with the third-order 2-dB Chebyshev prototype filter of Figure 5.8, provide a schematic of a third-order 2-dB Chebyshev low-pass filter with a cutoff frequency of 20 kHz and a 4.7-kΩ-load resistor.
12. The circuit of Figure P5.12 is a notch filter whose notch is at 2 rad/s. Show how you would modify the circuit so that it notches at 120 Hz using a 10-H inductor.

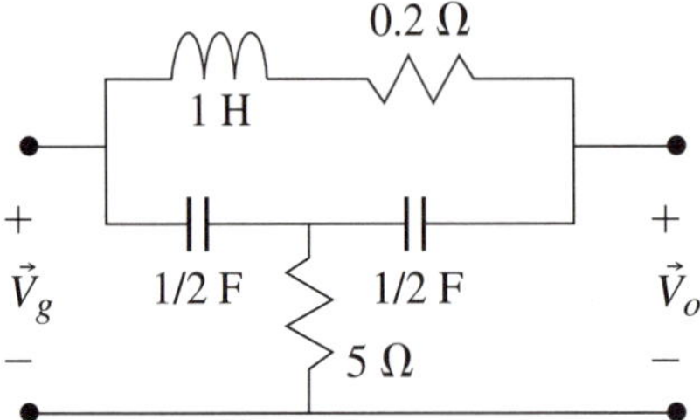

Figure P5.12

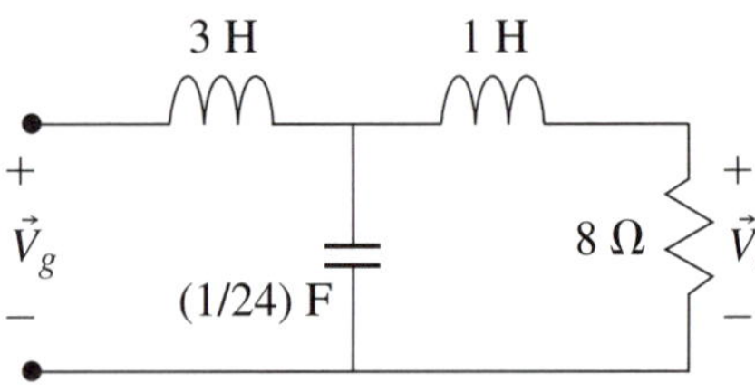

Figure P5.13

13. The circuit of Figure P5.13 is a Butterworth low-pass filter with a cutoff frequency of 4 rad/s. Modify the circuit so that it cuts off at 4.0 kHz and supplies a 1.2-kΩ load.

Section 5.4

14. Starting with the $H(s)$ for a third-order 0.5-dB Chebyshev prototype, obtain the transfer function for a high-pass filter with a cutoff frequency of 1 kHz.
15. Starting with the $H(s)$ for a second-order Butterworth prototype, obtain the transfer function for a bandpass filter with cutoff frequencies of 3 and 5 rad/s.
16. Provide a schematic for a second-order 2-dB Chebyshev bandpass filter having cutoff frequencies of 1 kHz and 4 kHz and a 1-kΩ-load resistor (see Figure P5.16).

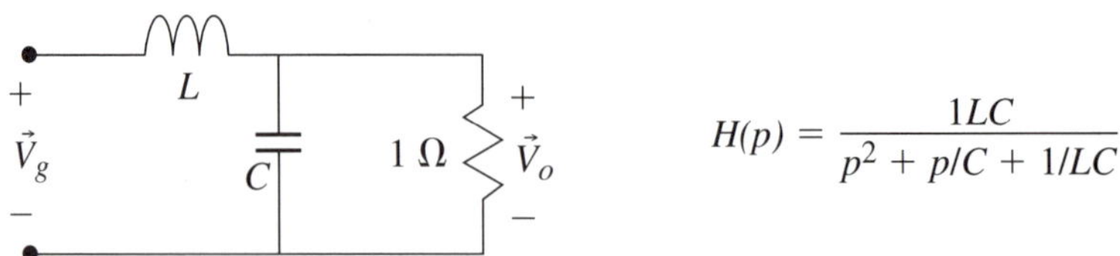

Figure P5.16 Second-order prototype.

17. Starting with a second-order 2-dB Chebyshev prototype, obtain the transfer function for a stopband filter with cutoff frequencies of 1 and 4 rad/s.
18. Provide the schematic of a third-order high-pass Butterworth filter with cutoff at 400 Hz and a 1-kΩ load.
19. Provide the schematic of a second-order Butterworth bandpass filter with cutoff frequencies of 0.5 and 2 rad/s.

Section 5.5

20. Provide component values for the circuit of Figure P5.20 to serve as a third-order prototype 2-dB Chebyshev low-pass filter.

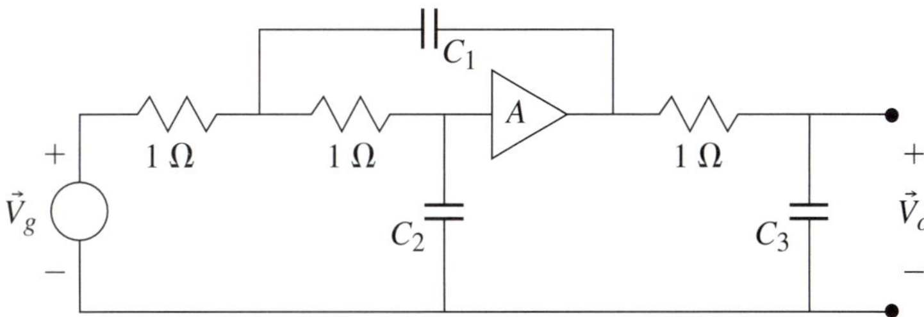

Figure P5.20

21. Provide a prototype circuit for a third-order Butterworth filter using only passive circuit elements.
22. Provide a prototype circuit for a third-order Cauer filter with 1-dB passband ripple and 30-dB minimum stopband attenuation.

Section 5.6

23. Obtain the polynomials of $H(s)$ for a Butterworth sixth-order low-pass filter with cutoff at 10 kHz. Plot $|H(s)|$ vs. f.
24. Plot the transfer function, magnitude and phase, for a Butterworth bandpass filter derived from an eighth-order prototype and with cutoff frequencies of 8 kHz and 12 kHz. Estimate the group delay for the filter.
25. Prepare a family of curves showing the stopband rolloff of even-ordered Chebyshev 2-dB filters similar to that of Figure 5.16. Use a linear frequency range of $1 \leq f/f_{co} \leq 2$, and include filters out to the fourteenth order.

6

SPECTRAL ANALYSIS

OUTLINE

OBJECTIVES

1. Mathematically state the complex form of Fourier's series.
2. Determine a signal's spectral power distribution.
3. Compare the spectra of a pulse and a repetitive pulse.
4. Improve partial sums of the Fourier series.
5. Create MATLAB functions.

INTRODUCTION

By specializing in e^{st} forcing functions, system differential equations become algebraic equations and are easily solved. Work originating with Joseph Fourier (1768–1830) and subsequently expanded upon by many others has shown that any signal waveform may be created by superimposing appropriate amounts of e^{st} signals (Laplace transform) and/or $e^{j\omega t}$ signals (Fourier transform). There are two major implications of these facts:

1. If we can decompose an arbitrary forcing function into its $e^{j\omega t}$ components, we can apply each one individually to a system, use the system frequency response to determine its magnitude and phase at the output, and superimpose the output signals to find the resulting output signal waveform. This removes almost all restrictions on the types of input signals for which we can find the output.
2. The decomposition of a signal into $e^{j\omega t}$ components describes the signal's spectrum. The spectrum tells us the magnitude and phase of the $e^{j\omega t}$ signal present at each frequency. That, in turn, tells us how the signal's power is distributed versus frequency, and will influence the design of circuits intended to detect or pass such signals.

6.1 FOURIER SERIES

Fourier's original work claimed that an infinite series of sinusoids could be used to represent any periodic function, even one with square corners. The idea that smoothly varying functions like sinewaves could be used to create a square corner was greeted with much skepticism by his colleagues, but Fourier was right. After reading this section you will be able to:

- Mathematically state the complex form of the Fourier series.
- Identify the period and fundamental frequency of a periodic waveform.
- Find the Euler phasors in a periodic waveform.

A periodic function is shown in Figure 6.1. One cycle of the waveform occurs in the *period T*. As a result of Fourier's work, his successors showed that this type of periodic waveform could be expressed in an infinite sum of the form

$$f(t) = \sum_{m=-\infty}^{m=+\infty} c_m e^{jm\omega_o t} \qquad \textbf{(6.1)}$$

where $\omega_o = 2\pi/T$, m is an integer, and

$$c_m = \frac{1}{T}\int_T f(t)e^{-jm\omega_o t}\,dt \qquad \textbf{(6.2)}$$

This is known as the complex form of the Fourier series.

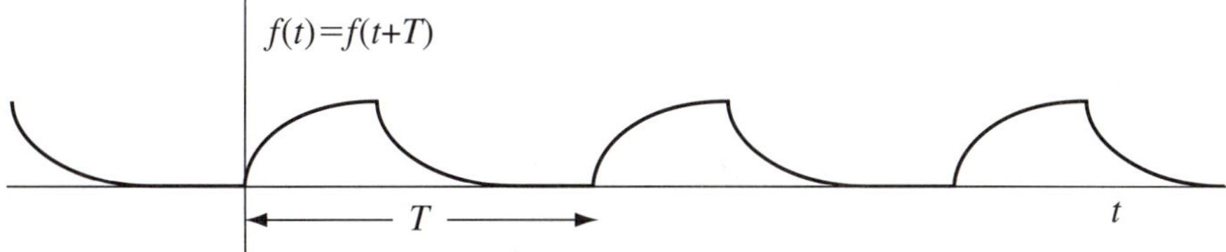

Figure 6.1 A periodic signal with period T has the property that $f(t) = f(t + T)$ for all t. It has a fundamental frequency or repetition rate of $\omega_o = 2\pi/T$.

The frequencies present are $\pm\omega_o$, called the *fundamental* frequency, and all integer multiples of the fundamental, called *harmonics*. The integer m gives the harmonic number; and the c_m are, in general, complex numbers that describe how much of each exponential time function is present and what phase it has at $t = 0$. In fact, the last time we encountered these numbers, we called the c_{+m} values *Euler phasors.* (Since $f(t)$ is generic and the study of the Fourier series spans many disciplines where the term *phasor* is meaningless, the phasor arrow notation is not used here.)

Equation 6.2 shows that

$$c_{-m} = c_m^* \tag{6.3}$$

which is exactly the condition required for the positive and negative frequency exponentials to combine to give a real, sinusoidal $f(t)$. We can determine the magnitude and phase of that sinusoid either by taking equal $\pm m$ terms together or by just using the $+m$ value and the Euler phasor transformation.

There are two reasons for evaluating the Fourier series. One is to obtain an expression for $f(t)$ that applies everywhere, rather than only over a single period. The other is to determine the phasors, which indirectly tell how much power is available at each harmonic of the waveform.

EXAMPLE 6.1

Find c_m for the periodic waveform of Figure 6.2.

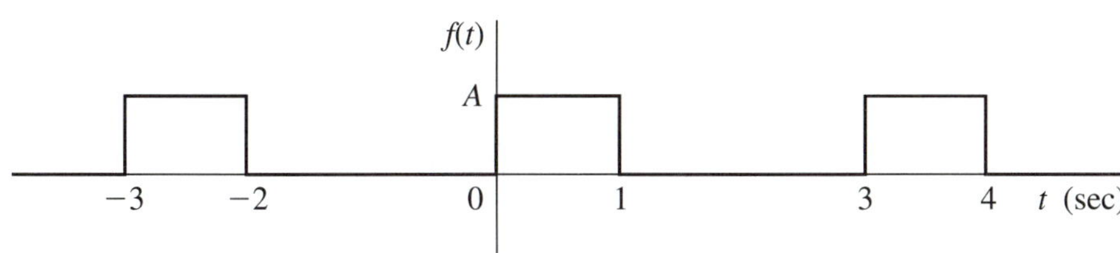

Figure 6.2

Solution

There are basically four steps involved in evaluating Equation 6.2.

1. *Identify the period.* In this case $T = 3$. (Take *any* 3-second interval, duplicate it, concatenate the duplicates, and the waveform of Figure 6.2 will result.)
2. *Find an expression for f(t) that is good over any one period.* In this example we choose

$$f(t) = \begin{cases} A & 0 < t < 1 \\ 0 & 1 < t < 3 \end{cases}$$

   (The function of this example is extremely easy to describe in a piecewise fashion. The Fourier series will provide us with a single equation that applies for any value of t.)
3. *Evaluate the integral.* (Use a good set of integral tables, and any "tricks" you know to simplify the process.)

$$c_m = \frac{1}{3}\int_0^3 f(t)e^{-jm\frac{2\pi}{3}t}\,dt = \frac{1}{3}\left[\int_0^1 Ae^{-jm\frac{2\pi}{3}t}\,dt + \int_1^3 0\,dt\right]$$

$$c_m = \frac{A}{3}\left.\frac{e^{-jm\frac{2\pi}{3}t}}{-jm\dfrac{2\pi}{3}}\right|_0^1 = \frac{A}{3}\left[\frac{e^{-jm2\pi/3} - 1}{-jm2\pi/3}\right]$$

$$c_m = \frac{A}{m\pi}e^{-jm\pi/3}\sin\frac{m\pi}{3}$$

4. *Resolve any indeterminate cases.* (It is not unusual for the resulting expression for c_m to be indeterminate at one or two m values. Check to see if there is any m value for which the denominator goes to zero. A finite-sized wave cannot possibly have an infinite component, so it must be indeterminate at that m value.)

$$c_0 = \frac{0}{0}$$

This can be resolved using L'Hopital's rule or from

$$\lim_{m\to 0}\sin\left(\frac{m\pi}{3}\right) = \frac{m\pi}{3}$$

or by returning to the integral specifically for c_0, which is

$$c_0 = \frac{1}{3}\int_0^3 f(t)\,dt = \frac{1}{3}\left[\int_0^1 A\,dt + \int_1^3 0\,dt\right] = \frac{A}{3}$$

Note that c_0 is the d-c or average value of the function, which is also often obvious by inspection.

The resulting Fourier series is

$$f(t) = \sum_{-\infty}^{\infty}\left(\frac{A}{m\pi}\sin\frac{m\pi}{3}\right)e^{jm(\omega_o t - \pi/3)}$$

or, if you prefer,

$$f(t) = \frac{A}{3} + \sum_{1}^{\infty}\left(\frac{2A}{m\pi}\sin\frac{m\pi}{3}\right)\cos\left(m\omega_o t - \frac{m\pi}{3}\right)$$

Generalizing waveforms, when possible, gives the best return on the time invested. The results can be collected in tables and applied to specific numerical problems without rederiving them from scratch. The period sets the fundamental frequency and consequently tells what frequencies are potentially present. The position of the waveform relative to the time origin affects only the phase of the Euler phasors. It can be chosen to simplify the integration and then adjusted to match the waveform position desired. For many applications, the phase is unimportant.

EXAMPLE 6.2

Determine the Euler phasors for the waveform of Figure 6.3.

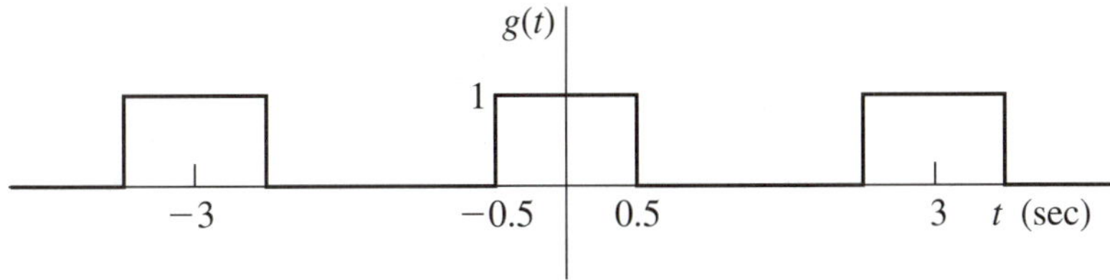

Figure 6.3

Solution

If $g(t)$ is delayed by 1/2 second, it would be identical to the $f(t)$ of Example 6.1, with $A = 1$. From the properties of ideal filters we know that a pure delay is represented by a phase shift of

$$e^{-j\tau_p\omega} = e^{-j\tau_p m\omega_o} = e^{-j2\pi m\tau_p/T}$$

With $\tau_p = 1/2$ and $T = 3$, multiplying each Euler phasor of $g(t)$ by the factor $e^{-j\pi m/3}$ will delay $g(t)$ by 1/2 second; or, equivalently, multiplying the Euler phasors of $f(t)$ by $e^{+j\pi m/3}$ will move it ahead in time by 1/2 second. Since the Fourier series for $f(t)$ is

$$f(t) = \sum_{-\infty}^{\infty}\left(\frac{A \sin m\pi/3}{m\pi}\right)e^{jm(\omega_o t-\pi/3)}$$

the Fourier series for $g(t)$ must be

$$g(t) = \sum_{-\infty}^{\infty}\left(\frac{\sin m\pi/3}{m\pi}\right)e^{jm\omega_o t}$$

The spectrum of an $f(t)$ is a plot of its c_m vs. m or $m\omega_o$. For theoretical work we prefer to allow m to range over the positive and negative integers, which gives the amplitudes and phases of exponential signals. If we are only given the c_m for positive m values, we know the rest are just the conjugate of their positive frequency mate.

Nontechnical personnel are more comfortable with real signals than with complex exponentials, so commercial spectrum analyzers modify the c_m values to give the amplitudes and phases of the corresponding sinusoids. The different representations are shown in Figures 6.4a, b, and c. The phase of each amplitude spectrum

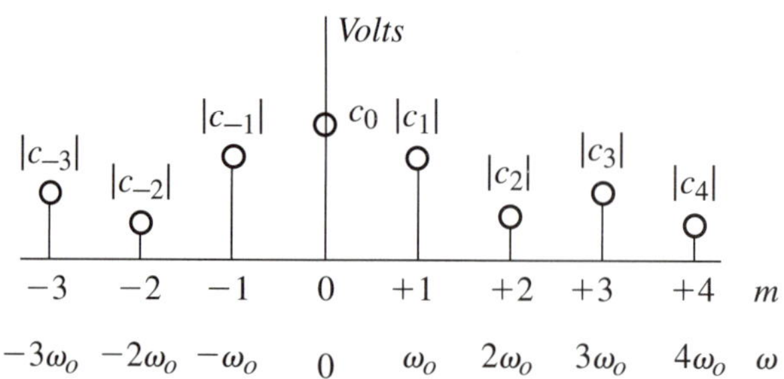

Figure 6.4a The amplitude spectrum of the exponential components of a periodic waveform.

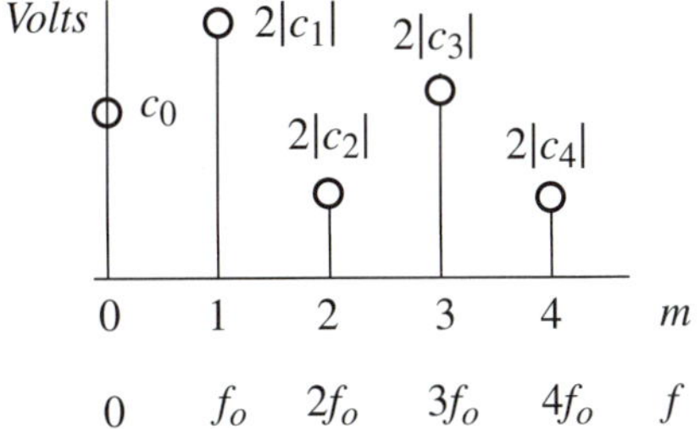

Figure 6.4b The amplitude spectrum of sinusoids as it would be displayed on a commercial spectrum analyzer.

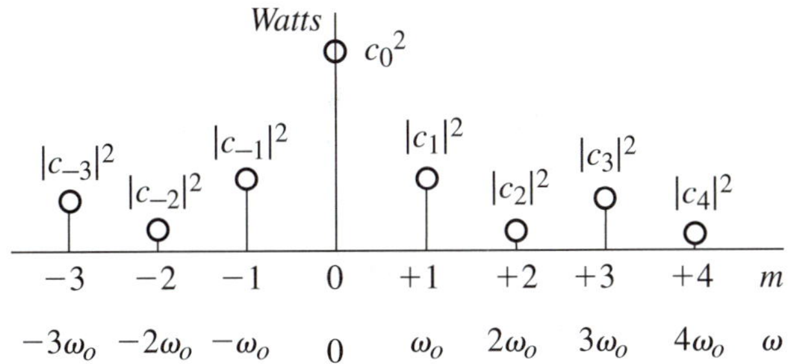

Figure 6.4c The power spectrum of a periodic waveform as it would be displayed for exponential components.

component may also be provided in a second graph. The power spectrum is entirely real and has no phase.

6.2 PARSEVAL'S THEOREM

The average power of a periodic waveform may be found in either the time domain or the frequency domain. After completing this section you will be able to:

- Determine a signal's spectral power distribution.
- Find the percentage of total power in each frequency component.

In the time domain, we define the average power of the function $f(t)$ as

$$P_{ave} = \frac{1}{T}\int_T f(t)^2\,dt$$

If $f(t)$ were a voltage or current applied to a 1-Ω resistor, this definition of average power would be familiar. However, we will use this definition for the power of the signal whether or not its units are known or any resistor is implied. This way we will be able to discuss the "power" of a signal that consists of just a sequence of numbers in a computer.

Expressing one $f(t)$ in the form of its Fourier series gives

$$P_{ave} = \frac{1}{T}\int_T f(t)^2\,dt = \frac{1}{T}\int_T f(t)\left[\sum_{-\infty}^{\infty} c_m e^{jm\omega_o t}\right]dt$$

and interchanging the order of integration and summation results in

$$P_{ave} = \sum_{-\infty}^{\infty} c_m \underbrace{\left[\frac{1}{T}\int_T f(t)e^{jm\omega_o t}\,dt\right]}_{c_{-m}} = \sum_{-\infty}^{\infty} c_m c_{-m} = \sum_{-\infty}^{\infty} c_m c_m^* = \sum_{-\infty}^{\infty} |c_m|^2$$

A less elegant but more familiar approach is to add the powers of the sinusoids in Figure 6.4b. Double each c_m to get the peak value of the sinusoids, divide by the square root of 2 to convert to rms values, and square the rms values to get power. The power provided by sinusoids of different frequency does superimpose, so the components can just be added up. The d-c term is already rms and only needs to be squared. The simplest equation comes from using the spectrum of the exponential signals instead of the spectrum for sinusoids.

$$P_{ave} = c_0^2 + \sum_{m=1}^{\infty}\left(\frac{2|c_m|}{\sqrt{2}}\right)^2 = \sum_{m=-\infty}^{\infty} |c_m|^2$$

That the signal power can be determined in either the time or the frequency domain is called Parseval's theorem:

$$P_{ave} = \frac{1}{T}\int_T f(t)^2\,dt = \sum_{m=-\infty}^{\infty} |c_m|^2 \qquad \text{Parseval's Theorem} \tag{6.4}$$

EXAMPLE 6.3

Determine the percentage of the total power represented by the fourth harmonic in the waveform of Figure 6.5 given that

$$c_m = \frac{1}{T}\int_0^{T/4} \sin\left(\frac{8\pi t}{T}\right)e^{-jm2\pi t/T}\,dt = \frac{2}{\pi}\frac{1 - e^{-jm\pi/2}}{16 - m^2}$$

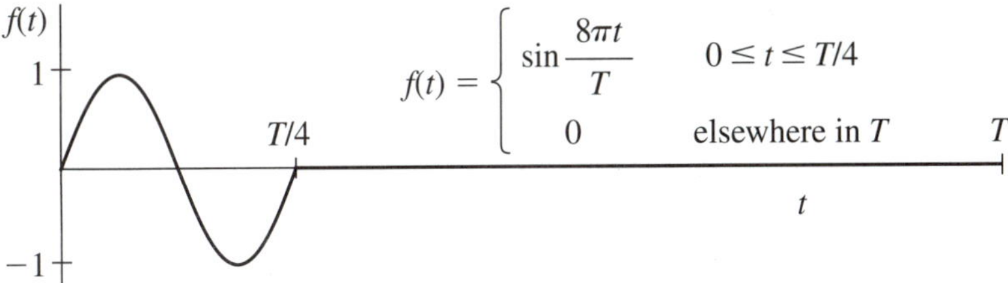

Figure 6.5

Solution

The expression for c_m becomes indeterminate at $m = 4$. Using L'Hopital's rule, we get

$$c_4 = \frac{2}{\pi}\frac{(j\pi/2)e^{-jm\pi/2}}{-2m} = -j\frac{1}{\pi}\frac{\pi}{8} = \frac{-j}{8}$$

$$P_4 = \left(2\frac{1/8}{\sqrt{2}}\right)^2 = \frac{1}{32}$$

The total power in the waveform is found from the time domain as

$$P_{ave} = \frac{1}{T}\int_T f(t)^2\,dt = \frac{1}{T}\int_0^{T/4} \sin^2\!\left(\frac{8\pi}{T}t\right)dt = \frac{1}{2T}\int_0^{T/4}\left[1 - \cos\!\left(\frac{16\pi t}{T}\right)\right]dt = \frac{1}{8}$$

$$\therefore\ 100\frac{P_4}{P_{ave}} = 100\frac{1/32}{1/8} = 25\%$$

The total power could also be found in the frequency domain by adding up the contributions from all of the significant harmonics.

6.3 THE FOURIER TRANSFORM

Signals that consist of a single pulse have a finite energy but zero average power. The Fourier transform is a spectral distribution function from which the signal's energy can be determined in the frequency domain. After reading this section you will be able to:

- State the Fourier transform and the inverse Fourier transform.
- Compare the spectra of a pulse and a repetitive pulse.
- Determine the Fourier series from the Fourier transform.

The Fourier transform gives the magnitudes and phases of a *continuous sum* (integral) of $e^{j\omega t}$ signals that combine to form an *aperiodic* (not periodic) pulse:

$$f(t) = \frac{1}{2\pi}\int_{\omega=-\infty}^{\infty} F(\omega)e^{j\omega t}\,d\omega \qquad \text{Inverse Fourier Transform} \tag{6.5}$$

where

$$F(\omega) = \int_{t=-\infty}^{\infty} f(t)e^{-j\omega t}\,dt \qquad \text{Fourier Transform} \tag{6.6}$$

If $f(t)$ is a voltage, Equation 6.6 shows $F(\omega)$ has the units of volt second, or volts per unit bandwidth, which describes a signal's spectral density. The energy provided by $f(t)$ can be found from

$$W = \int_{t=-\infty}^{\infty} f^2(t)\,dt = \frac{1}{2\pi}\int_{\omega=-\infty}^{\infty} |F(\omega)|^2\,d\omega \tag{6.7}$$

which is derived in the same fashion as for the periodic case, and represents Parseval's theorem for an energy signal.

EXAMPLE 6.4

Find the Fourier transform of the unit impulse $\delta(t)$.

Solution

The unit impulse is zero everywhere, except when its argument is zero. It is infinite at $t = 0$ but has an area of unity:

$$\delta(t) = \begin{cases} 0 & t \neq 0 \\ \infty & t = 0 \end{cases}$$

$$\int_{-\infty}^{\infty} \delta(t)\,dt = \int_{0-}^{0+} \delta(t)\,dt = 1$$

Inserting the impulse in Equation 6.6 gives

$$F(\omega) = \int_{t=-\infty}^{\infty} f(t)e^{-j\omega t}\,dt = \int_{0-}^{0+} \delta(t)e^{-j\omega t}\,dt = \int_{0-}^{0+} \delta(t)e^{-j\omega 0}\,dt = 1$$

The unit impulse makes integrations easy because it exists only when its argument is zero. Over that infinitesimal range, the other integrand terms may be considered constant.

Notice that the spectral density of a unit impulse is a constant, independent of frequency. Consequently, if a unit impulse acts as the input signal to a system, the output frequency response must be exclusively the natural response of the system. An impulse input provides equal amounts of energy at all frequencies.

The Fourier transform is actually the most general way of describing the frequency domain characteristics of signals. Even the Fourier transform of a periodic waveform can be defined:

$$F(\omega) = 2\pi \sum_{m=-\infty}^{\infty} c_m \delta(\omega + m\omega_o) \qquad \text{Fourier Transform of a Periodic Function}$$

It shows the expected infinite energy at multiples of the fundamental frequency; if inserted into the inverse Fourier transform, it reproduces the Fourier series.

Most of the theorems developed for signals come from the Fourier transform. Our interest is primarily in the Fourier series, but we want to explore further what Fourier's series and transform have in common.

Consider the waveforms and relationships summarized in Figure 6.6. The integrals for $c_m T$ and $F(\omega)$ are identical except for the way their parameter is

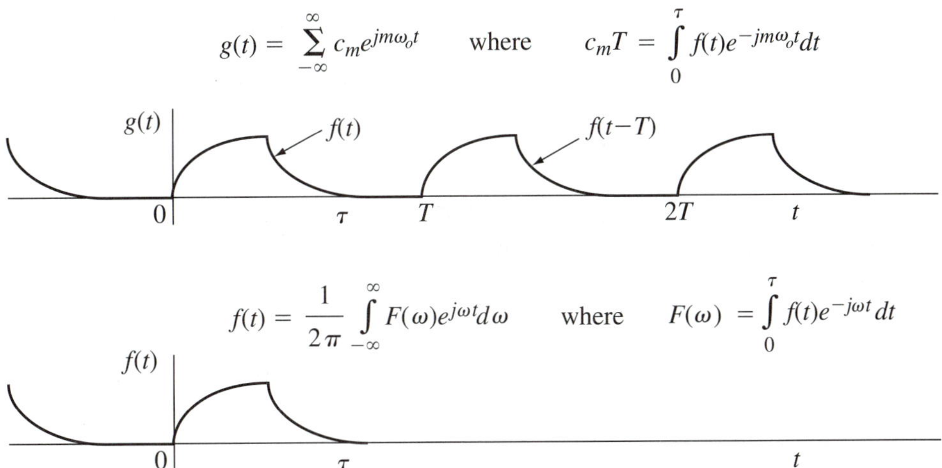

Figure 6.6 A periodic signal $g(t)$ is generated when the single pulse $f(t)$ is repeated every T seconds. The integrals for $F(\omega)$ and $c_m T$ are identical.

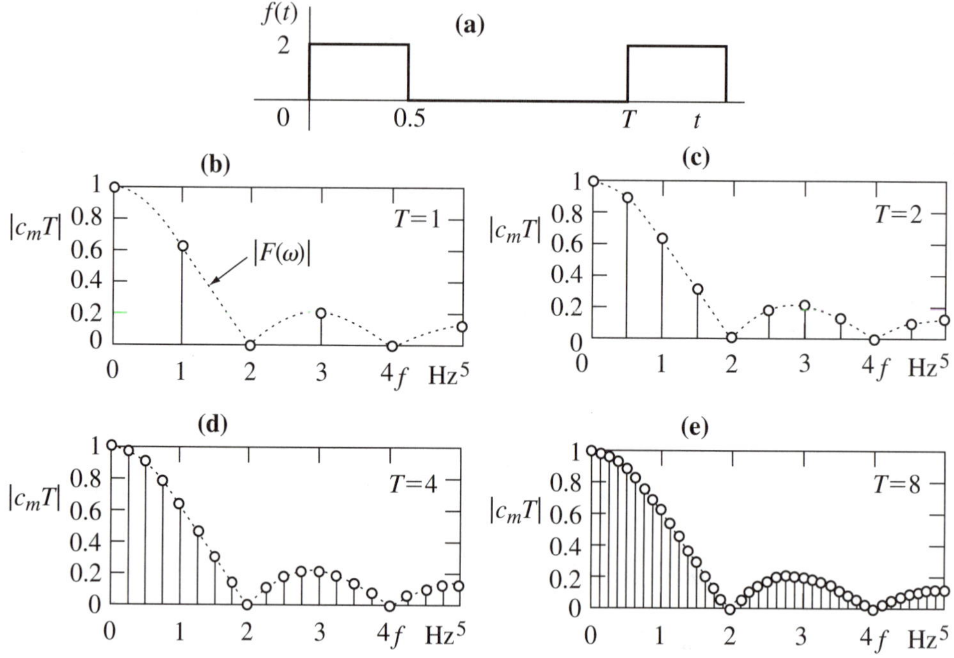

Figure 6.7 The spectra resulting from a fixed 0.5-second rectangular pulse is observed as the period T increases. The $c_m T$ values of the Fourier series are seen to be samples of the Fourier transform $F(\omega)$ taken at frequencies m/T.

stated: $m\omega_o$ for one and ω for the other. Neither of these parameters affects the integration with respect to time, so the way we label them can be altered before or after the integration is carried out. The conclusion is that

$$c_m T = F(\omega)|_{\omega = m\omega_o} = F(m\omega_o) \tag{6.8}$$

The significance of this result is that $F(\omega)$, which is based only on the pulse shape, establishes all possible amplitudes of the Fourier series components. The period determines only where the $F(\omega)$ is sampled to obtain a specific series component. Moreover, if we want to see $F(\omega)$ for a particular pulse, we can get it from the Fourier series by making the pulse periodic, with a very large period. Both of these concepts are demonstrated for the rectangular pulse of Figure 6.7a.

EXAMPLE 6.5

Determine the Fourier series coefficients of a waveform consisting of the pulse of Figure 6.8 repeated every 2 seconds.

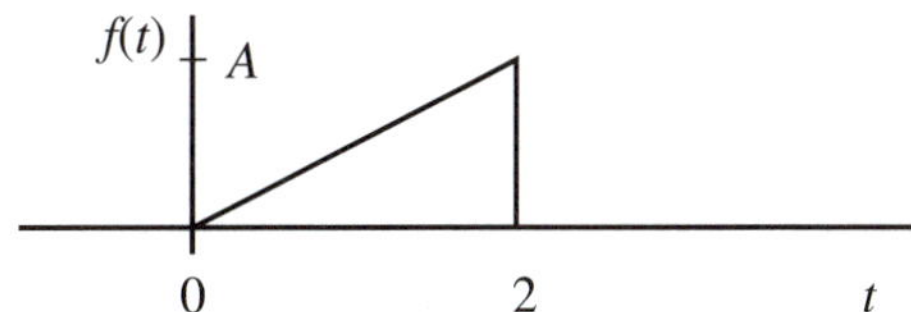

Figure 6.8

Solution

$$f(t) = \begin{cases} \dfrac{At}{2} & 0 < t < 2 \\ 0 & \text{elsewhere} \end{cases}$$

$$F(\omega) = \int_{-\infty}^{\infty} f(t)e^{-j\omega t}\,dt = \frac{A}{2}\int_0^2 te^{-j\omega t}\,dt = \frac{A}{2}\left[\frac{e^{-j\omega t}}{-\omega^2}(-j\omega t - 1)\right]_0^2$$

$$F(\omega) = \frac{A}{2}\left[\frac{e^{-j\omega t}}{\omega^2}(j\omega t + 1)\right]_0^2 = \frac{A}{2\omega^2}[e^{-j2\omega}(j2\omega + 1) - 1]$$

Now from Equation 6.8, the effect of the period is entered: $(T = 2,\ \omega_o = \pi)$

$$2c_m = F(\pi m) = \frac{A}{2(\pi m)^2}[e^{-j2(\pi m)}(j2\pi m + 1) - 1]$$

$$c_m = \frac{1}{2}F(\pi m) = \frac{A}{4\pi^2 m^2}[j2\pi m] = \frac{jA}{2\pi m} \qquad m \neq 0$$

The c_0 value appears to be infinite in the final result because the simplifications that were made have hidden the fact that it originally was indeterminate. The easiest way to obtain c_0 is by inspection of the waveform's average value, which is $A/2$.

6.4 WINDOWS

One application of the Fourier series is to provide an equation to represent a particular shape. To make it practical to use, the series must be limited to a finite number of terms. The Fourier series coefficients may need to be tweaked with a window function to give best results in these partial sums. After reading this section you will be able to:

- Improve partial sums of the Fourier series.
- Discuss how the Fourier series converges.

- Identify several window functions.
- Recommend when a window function should be used.

While many of Fourier's contemporaries scoffed at the notion that smooth functions like sinusoids could be added up to provide a sharp corner, Dirichlet showed that indeed the Fourier series converges to $f(t)$ everywhere $f(t)$ is continuous, and to the mean of $f(t)$ at the discontinuities. A *discontinuity* is any point where the function abruptly jumps from one value to another.

For functions without discontinuities, the Fourier series coefficients need no modification. The only decision is how many terms to use in the partial sum. The waveform of Figure 6.5 is continuous. By writing a program to evaluate and plot its Fourier series, the results of Figure 6.9 are obtained for a 21-term sum. The highest error in the partial sum occurs where the curve suddenly changes direction. High-frequency terms are needed for corners, and they are what partial sums limit.

Discontinuous functions behave less well when approximated by a Fourier partial sum. In order to produce the jumps required at discontinuities, the Fourier

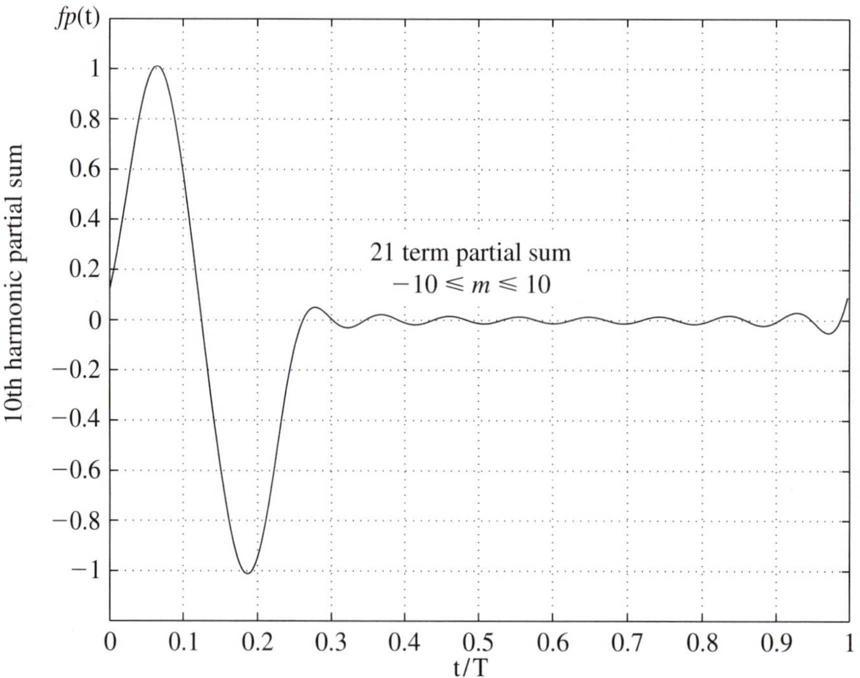

Figure 6.9 Partial sum representation of the waveform of Figure 6.5, including terms out to the tenth harmonic.

series retains excessive amplitudes in its high-frequency terms. When a partial sum is used, the discarded high-frequency terms are no longer present to cancel out the excess amplitudes of terms retained in the partial sum. As a result, the partial sum waveform shows rippling, as if it has too much high-frequency content. As more terms are included in the partial sum, the rippling does not disappear or even decrease substantially, but instead it is pushed closer and closer to the point of discontinuity. In the limit, as the partial sum becomes the infinite sum, the rippling is all pushed to the discontinuity, leaving the waveform with "ears" known as *Gibb's phenomenon* (Figure 6.10).

The process of obtaining a partial sum approximation for a function is equivalent to putting the original function through an ideal low-pass filter. This rectangular *window* keeps all of the Fourier coefficients out to the Mth harmonic, but multiplies all the harmonics higher than the Mth harmonic by zero. We have seen that this works fine for continuous functions, but results in rippling in discontinuous functions. Since this rippling is the result of having excess amplitude in the high-frequency terms that are retained, it is reasonable to seek a way to systematically taper off the high frequencies. In other words, for discontinuous functions, the Fourier series coefficients will be taken as a starting point; but instead of passing them through a rectangular window, we will pass them through a tapered window that gradually reduces each successively higher harmonic out to the last one retained.

The result of doing this for the sawtooth waveform of Example 6.5 is demonstrated in Figure 6.11. The windowed partial sum appears to be identical to the desired straight line over 80% of the period, but does have a wider transition region than the unwindowed series.

The windowing process is pictured in Figures 6.12a and b for rectangular and triangular windows. There are a wide variety of window functions. They all multiply the d-c term by unity. The d-c term only sets the level of the waveform but does not affect its shape. At this point the only criterion we have for a window function is that it systematically taper off the Fourier coefficients and that it be easily calculated for any arbitrary number of harmonics to be retained in the partial sum. A few popular window functions are given in Table 6.1.

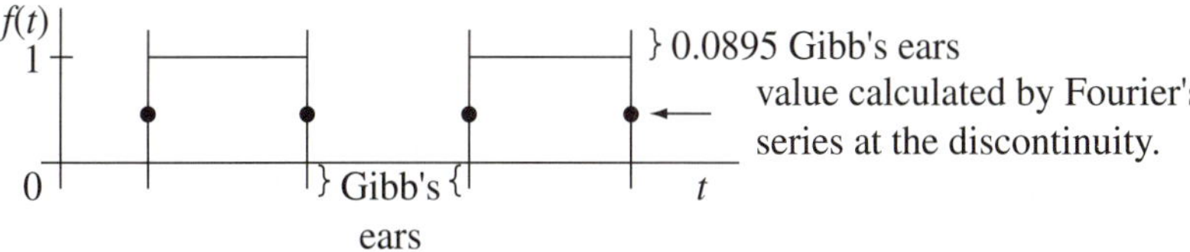

Figure 6.10 Gibb's ears at points of discontinuity in a square wave. The waveform still averages to the mean at the jumps, satisfying Dirichlet's conditions.

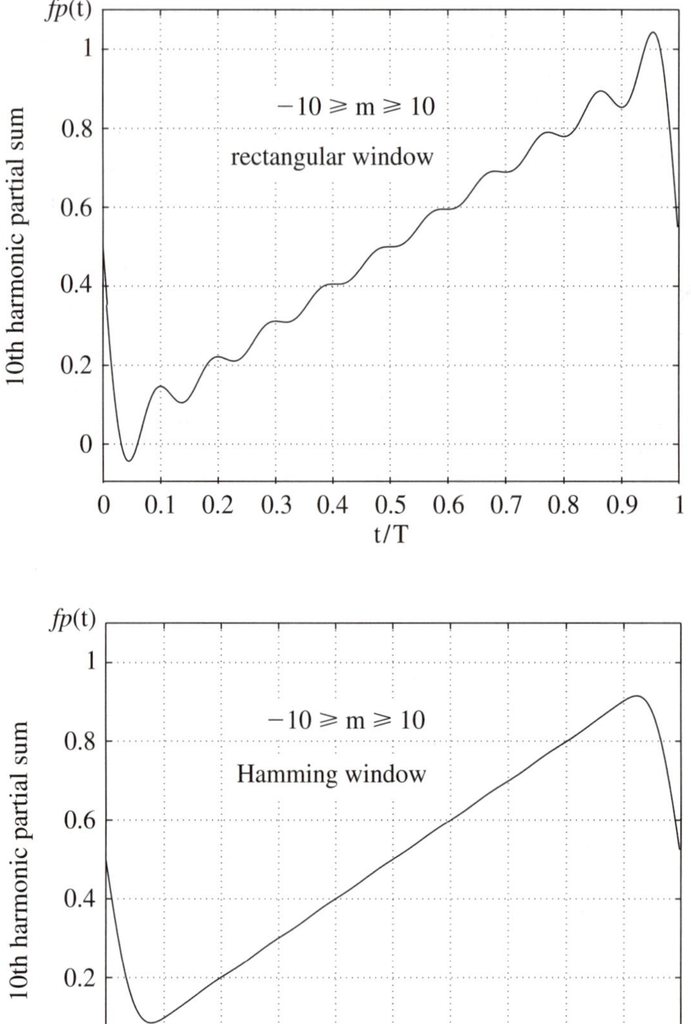

Figure 6.11 The partial Fourier series of a sawtooth waveform is demonstrated with and without windowing.

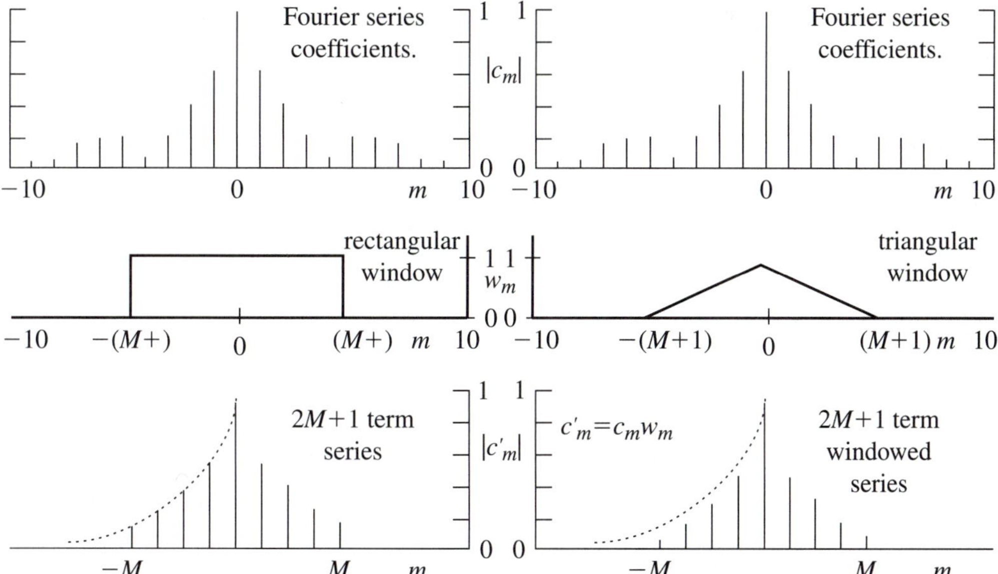

Figure 6.12a Terminating the Fourier series after $2M + 1$ terms is equivalent to multiplying the spectrum by a rectangular window.

Figure 6.12b Reducing the series coefficients to zero by a linearly increasing percentage is equivalent to multiplying the spectrum by a triangular window.

Table 6.1 Common Window Functions

Name	***Window Function***
Triangular	$w(m) = 1 - \|m\|/(M + 1)$
Hamming	$w(m) = 0.54 + 0.46\cos(m\pi/M)$
Hanning (Von Hann)	$w(m) = 0.5[1 + \cos(m\pi/(M + 1))]$

$M =$ highest harmonic number; $-M \leq m \leq M$ ($2M + 1$ terms).

EXAMPLE 6.6

The Fourier series of a discontinuous function is

$$f(t) = \cdots 0.2e^{-j4t} + 0.6e^{-j3t} + e^{-j2t} + 2e^{-jt} - 2 + 2e^{+jt} + e^{+j2t} + 0.6e^{+j3t} + 0.2e^{+j4t} + \cdots$$

The series is to be approximated by a five-term partial sum and modified with a triangular window to reduce rippling. Determine the required partial sum.

Solution

The unmodified (rectangular window) five-term partial sum is

$$f_p(t) = \underbrace{1}_{c_{-2}}e^{-j2t} + \underbrace{2}_{c_{-1}}e^{-jt} \underbrace{-2}_{c_o} + \underbrace{2}_{c_1}e^{+jt} + \underbrace{1}_{c_2}e^{+j2t}$$

The triangular window of Table 6.1 gives, for $M = 2$ (five terms)

$$w(m) = 1 - \frac{|m|}{3} = \underbrace{\frac{1}{3}}_{w_{-2}} \quad \underbrace{\frac{2}{3}}_{w_{-1}} \quad \underbrace{1}_{w_o} \quad \underbrace{\frac{2}{3}}_{w_1} \quad \underbrace{\frac{1}{3}}_{w_2}$$

The triangular windowed five-term partial sum consequently is

$$f_p(t) = \underbrace{c_{-2}w_{-2}}_{c'_{-2}}e^{-j2t} + \underbrace{c_{-1}w_{-1}}_{c'_{-1}}e^{-jt} + \underbrace{c_0w_0}_{c'_0} + \underbrace{c_1w_1}_{c'_1}e^{+jt} + \underbrace{c_2w_2}_{c'_2}e^{+j2t}$$

or

$$f_p(t) = \frac{1}{3}e^{-j2t} + \frac{4}{3}e^{-jt} - 2 + \frac{4}{3}e^{+jt} + \frac{1}{3}e^{+j2t}$$

There are some annoying inconsistencies in the window functions provided by different references. If we are going to compare a rectangular window with some other window, we ought to at least be comparing equal-length series. The more terms we include in a partial sum, the better we expect the function to be approximated by the series. If a series has a nonzero c_M term, the window function should not force that Mth harmonic term to zero, or we would be comparing an Mth harmonic series to a windowed $(M - 1)$th harmonic series. The equations in Table 6.1 have been written so that this does not happen.

6.5 MATLAB LESSON 6

It would be useful to have a program that would plot a Fourier partial sum to demonstrate issues that have arisen in this chapter. Creating a MATLAB function is one way to accomplish this. After completing this section you will be able to:

- Create MATLAB functions.
- Graph a Fourier partial sum with or without a window function.
- Perform a matrix multiplication.
- Find the output of a circuit with a periodic input signal.

MATLAB EXAMPLES

hamming hanning triang

MATLAB provides a variety of window functions for partial sums of the complex Fourier series. Two that we will use most often are the Hamming and Von Hann (Hanning) windows. Suppose we want to window the Fourier coefficients of Example 6.1 with a Hamming window, including out to the fifth harmonic. The MATLAB calculations would look like this:

```
>m=-5:5;
>cm=exp(-j*pi*m/3).*sin(m*pi/3)./(m*pi);
```

This last statement stipulates the following calculation:

$$c_m = \frac{e^{-j\pi m/3}}{m\pi} \sin \frac{m\pi}{3}$$

```
>cm(6)=1/3;             % correct the NaN
>win=hamming(11);       % m=-5:5 gives 11 coefficients
>cwin=cm.*win';         % note the transpose
```

MATLAB's window functions are delivered as column vectors instead of the usual row vectors, for reasons best known to them. *win'* turns the column vector back into a row vector. Since *win* is entirely real, either *win.'* or *win'* will give the same results.

A **matrix multiply** is defined for a row vector and a column vector of the same length:

$$a = (a_1 \quad a_2 \quad a_3) \qquad b = \begin{pmatrix} b_1 \\ b_2 \\ b_3 \end{pmatrix} \qquad a^*b = a_1b_1 + a_2b_2 + a_3b_3$$

The order is important for $a^*b \neq b^*a$. Both are legitimate operations, but the one we will want requires that the second vector be the column vector and the result is a single constant. There are two applications we have for matrix multiplies.

APPLICATION 1

Suppose we have the Fourier series coefficients as one vector and the exponentials as a second vector. We will have to specify a value for t before defining the e vector. If $t = 0.2$, and cm $= [c_{-3}\, c_{-2}\, c_{-1}\, c_0\, c_1\, c_2\, c_3]$ while

$$e = [e^{-j3*t} \quad e^{-j2*t} \quad e^{-j*t} \quad 1 \quad e^{j*t} \quad e^{2j*t} \quad e^{3j*t}]$$

then $f = c*e.'$ calculates $f(0.2)$ for the partial sum:

```
>t=0.2            % any arbitrary value of t
>e=exp(j*m*t);    % creates the e vector for t=0.2 (and ωo =1)
>f=cm*e.'         % calculates the following:
```

$$f(0.2) = \sum_{m=-3}^{3} c_m e^{jm(0.2)}$$

Notice the *e.'* term, which is an *array transpose.* The row vector is turned into a column vector, but no other changes are made. If this sequence of steps were performed for a whole range of t, we could plot f vs t.

APPLICATION 2

To calculate the power contained in the d-c thru the third harmonic, we need Equation 6.4

```
>Pave=cm*cm';
```

Notice the *cm'* term, which is a *matrix transpose.* The *cm* row vector is turned into a column vector *and* the complex conjugate is taken. This calculates

$$P_{ave} = \sum_{m=-3}^{3} c_m c_m^* = \sum_{m=-3}^{3} |c_m|^2$$

stem

Stem plots are ideal for functions of a discrete variable, such as the spectrum of a periodic signal. The format is identical to the plot function:

```
>stem(m,abs(cm));
```

We have created *.m* file scripts so that a sequence of commands could be executed repeatedly. MATLAB *function* calls are special versions of *.m* files that allow us to create a toolbox for our own area of specialization. A function file is created with any text editor and has the following characteristics:

1. The file and function names must be identical, but the file name has the *.m* extension.
2. The first nonblank line of the function file *must be*

   ```
   function [out1, out2...] = function _ name(in1,in2....)
   ```

   where out1, out2... are output variables to be calculated by the function using the input variables in1, in2.... If there is only one output variable, the square bracket may be omitted.
3. Successive comment lines following the function line constitute the information brought to the screen if *help function_name* is entered. The first line of this sequence is the only line searched by the *Find* command.
4. The first noncomment line begins the programming, which describes how the output variables are calculated from the input variables. Blank lines are ignored, and may help isolate programming sections.

There are hundreds of examples of function files within the MATLAB directory of any computer on which it is installed, and they are just as accessible as any other text file. Viewing a few will give you an idea of how the professionals do it. The function file reacts with the workplace only through the variables passed to it or returned back from it. This prevents workplace variables from being duplicated and overwritten by the function, which can be a problem with ordinary script files.

We have also used enough function calls by now to appreciate that the names given to input and output variables used by the function are immaterial. It will assign the numerical values of the symbols we use to call the function to the symbols defined in its function file according to their position in the function argument. We have also seen functions that accept a variable number of input variables or that either deliver an output listing or a graph, depending on how the function call is made. These options can be made available by the programmer's using MATLAB inherent *nargin* and *nargout* variables. They count the number of input or output arguments specified in the function call. The function file can include branching to accommodate the instructions implied by the number of input or output arguments that have been specified.

Function files may call other functions. When the function file is first run, it is compiled along with any other function calls included within the function file. Functions consequently execute very efficiently. As with other *.m* files, functions must be located within the MATLAB search path. If your *.m* files are on a floppy, remember to add the floppy to the MATLAB search path.

The following function file demonstrates the format. It should be duplicated and copied to a floppy. It will graph a Fourier partial sum with or without a window function. The user is required to define the series coefficients before calling the function, and provide the window coefficients.

① File and function names must agree

file: **FS.m**

```
function [yy,t]=FS(c,w,T)
% FS Fourier Partial Sum Plotter
%     FS calculates the Fourier series partial
%     sum given the complex Fourier series
%     coefficients c, the window w, and the
%     period T. Default values are a rectangular
%     window and T = 1. The command
%     format is [part_sum,time]=FS(c,w,T).
%     FS(c,w,T) used alone plots the partial sum.
```

② This is the *function* declarative; it must be on the first file line used.

③ This is the *help FS* output. Its first line is the only one searched by the *Find* command. Remember your own experiences and make this information *helpful!*

④ The program starts here. If two input arguments are provided, the period is set to 1. If only one argument is provided, a rectangular window is used.

```
if nargin==2
      T=1;                  % if omitted, set T = 1
      elseif nargin==1      % if omitted, set w = 1
      w=1; T=1;
end

cm=c.*w';                   % calculate windowed coefficients
n=size(c);                  % determine the number of harmonics
M=(n(2)-1)/2; m=-M:M;
for t=1:400                 % use 400 points for the plot
     tt=(t-1)/400;
     arg=j*2*pi*m*tt;       % calculate the exponents
     y(t)=exp(arg)*cm.';    % note the matrix multiply
     x(t)=tt;
end
if nargout==0               % decide what output format is desired
     plot(x*T,real(y))
     title('Fourier series partial sum')
     ylabel('f(t)')
     xlabel('t (seconds)')
elseif nargout==1
     yy=real(y);
elseif nargout==2
     yy=real(y); t=x*T;
end
```

EXAMPLE 6.7

Retaining terms out to the fifth harmonic, plot the partial sum of the waveform in Figure 6.13 using a rectangular window. Also, plot the exponential spectrum and determine the percentage of the total power in these harmonics.

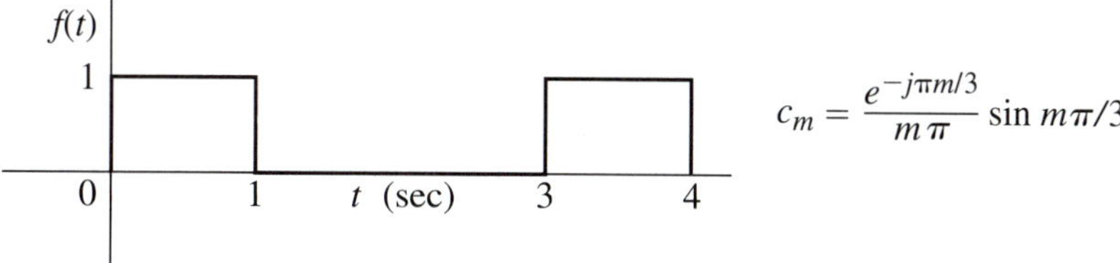

Figure 6.13

Solution

(Assumes the *FS.m* function is on a floppy and has been put in the search path.)

```
>m=-5:5; cm=exp(-j*m*pi/3).*sin(m*pi/3)./(m*pi);
>cm(6)=1/3;     % fix indeterminate values
>FS(cm,1,3)     % call the function
```

See the resultant graph in Figure 6.14.

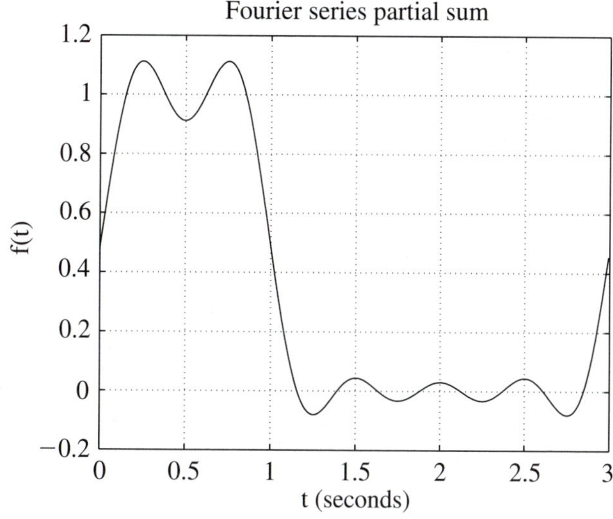

Figure 6.14 Eleven term partial sum Fourier series representation of the waveform of Figure 6.13 using a rectangular window.

To plot the spectrum:

```
>stem(m,abs(cm))
>grid; title('amplitude spectrum');ylabel('|cm|');xlabel('m')
```

See the resultant plot in Figure 6.15.

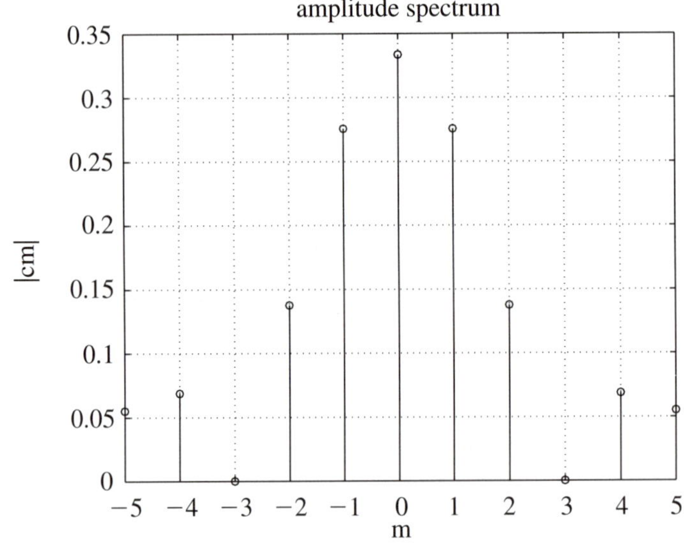

Figure 6.15 The exponential signal amplitude spectrum containing 95% of the power in the waveform of Figure 6.13.

```
>Pave=cm*cm'
Pave = 0.3167     % total power=1 joule spread over 3 sec = 1/3 watt
>PartP =3*Pave
PartP = 0.95      % 95% of the waveform power is in the first five harmonics.
```

EXAMPLE 6.8

The rectangular waveform of Example 6.1 is input to the circuit of Figure 6.16. Determine the output using a 21-term partial sum.

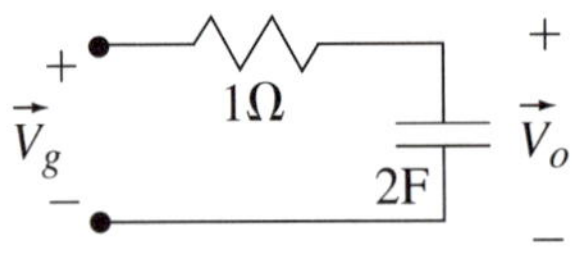

+ $\vec{V}_g$ − 1Ω 2F + $\vec{V}_o$ −

$$\frac{\vec{V}_o}{\vec{V}_g} = \frac{1/2}{s + 1/2}$$

Figure 6.16

Solution

The circuit transfer function must be expressed in terms of the harmonic frequencies:

$$\frac{\vec{V}_o}{\vec{V}_g} = \frac{1/2}{s + 1/2} = \frac{1/2}{jm\omega_o + 1/2} = \frac{1/2}{1/2 + jm2\pi/3}$$

Since the circuit is a low-pass filter, it will tend to remove the rippling, so no window will be used on the Fourier coefficients of $v_g(t)$.

```
>m=-10:10;in=exp(-j*m*pi/3).*sin(m*pi/3)./(m*pi);
>in(11) = 1/3;
>FS(in,1,3)          % plot v_g(t)
>hold
>h=0.5./(.5 + 2j*pi*m/3);% calculate transfer function
>out=in.*h;
>FS(out,1,3)         % plot v_o(t)
>gtext('input')
>gtext('output')
```

The resultant output is displayed in Figure 6.17.

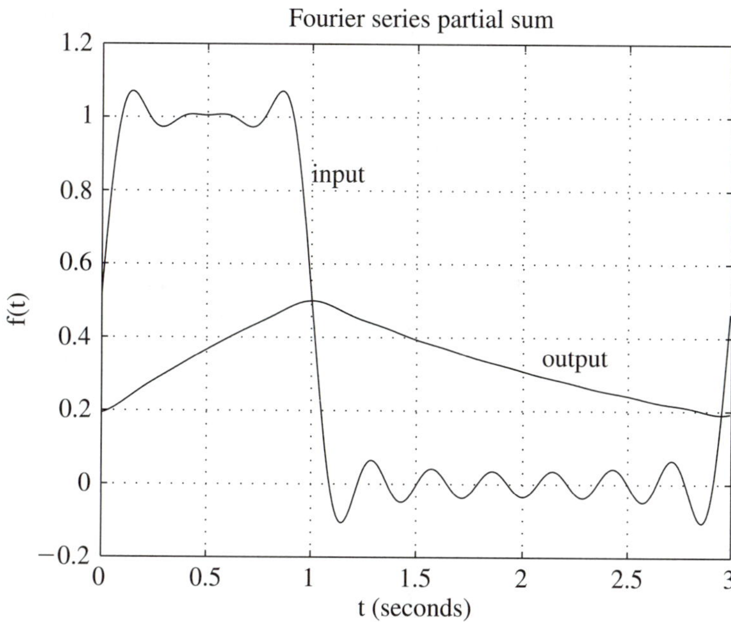

Figure 6.17 Tenth harmonic Fourier partial sum representations of the input and output waveforms in the circuit of Figure 6.16.

CHAPTER SUMMARY

Fourier showed how any repetitive waveform can be created by superimposing sinusoids with the proper amplitude and phase. The sinusoids will have frequencies that are integer multiples of the repetition rate, called the fundamental frequency. The Fourier series converges to the function where it is continuous and to the average value where it is discontinuous.

The spectrum of a repetitive signal is a display of the amplitude, and possibly phase, present at each frequency. The symmetrical display of exponential frequencies is preferable mathematically, but the positive frequency display of sinusoidal components is provided in commercial equipment. A power spectrum is a similar display but gives the average power of the signal at each frequency.

In many applications, the Fourier series is terminated to provide a practical approximation to a desired function. If that function contains a discontinuity, the partial sum will tend to oscillate around the ideal function because the Fourier series of a discontinuous function retains excessive amounts of high-frequency components. By tapering off the amplitudes of the Fourier coefficients with a window function, a smoother result is obtained for the partial sum. Windows are not used with continuous functions.

The Fourier transform is the general tool for determining the frequency spectrum of a pulse. Such signals have a finite energy and therefore zero average power. They also have a continuous spectrum, with infinitesimal components available at all frequencies but with a relative distribution given by $F(\omega)$. If the pulse is repeated so that a repetitive signal is generated, the c_m will show the identical relative frequency distribution as for the single pulse.

Computer-based tools for plotting a Fourier series partial sum make it possible to observe firsthand many of the issues discussed in this chapter. MATLAB functions can be created to perform this task, and they allow users to personalize their toolboxes. The FS MATLAB function created in this chapter is essential to verifying the concepts presented.

PROBLEMS

Section 6.1

1. Show that the waveform of Figure P6.1 has the Euler phasors indicated.
2. The Euler phasors of a waveform periodic in 2π seconds are

$$c_m = \frac{2\cos(m\pi/2)}{1 - m^2}$$

   a. Write the five-term partial sum of the complex Fourier series for this waveform.

$$c_m = A\frac{\tau}{T}\frac{\sin \pi m\tau/T}{\pi m\tau/T}$$

Figure P6.1

b. Write out the first three nonzero terms of the cosine Fourier series for this waveform.

3. The Euler phasors of a waveform periodic in 3 seconds are

$$c_m = \frac{-j}{m\pi}\left[\cos\left(\frac{2\pi m}{3}\right) - 1\right]$$

a. Write out the five-term partial sum ($-2 \le m \le 2$) of the complex Fourier series for this waveform.
b. Write out the first three nonzero terms of a trigonometric Fourier series for this waveform.

4. For the waveforms of Figure P6.4, set up the integral expression for c_m, including all specific numerical information, but do not integrate.

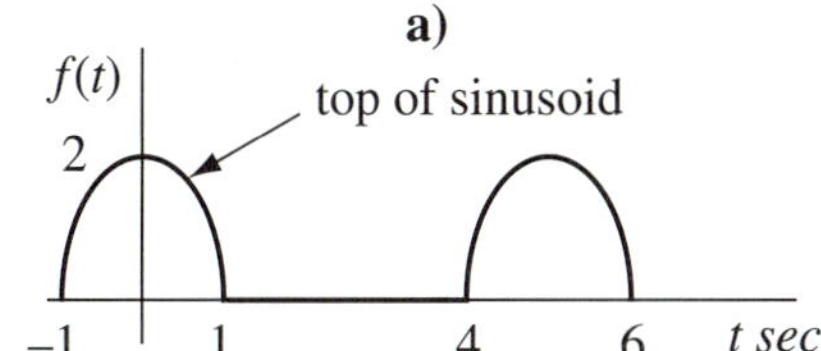

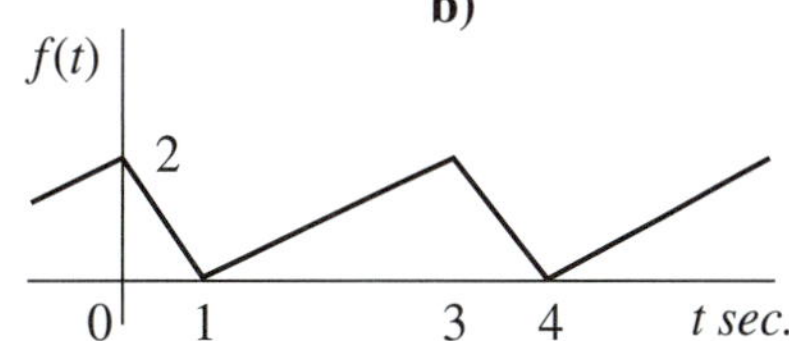

Figure P6.4

5. For the waveforms of Figure P6.5, set up the integral expression for c_m, including all specific numerical information, but do not integrate.

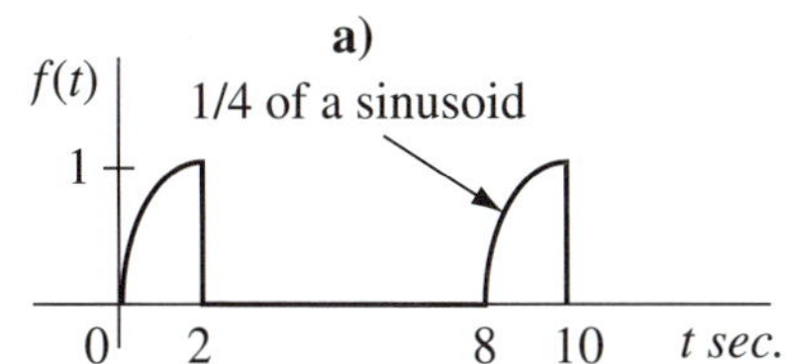

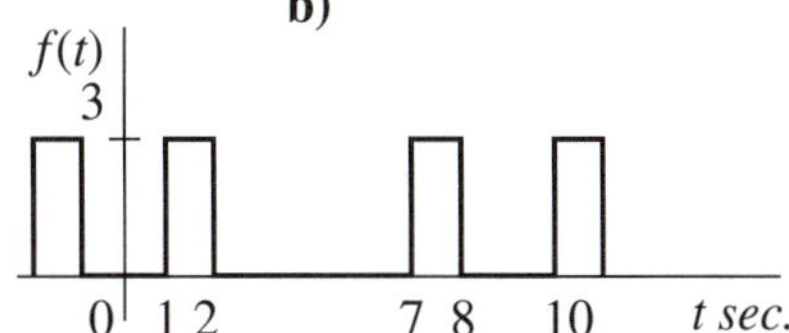

Figure P6.5

6. From the results of Problem 1, determine the Euler phasors for the waveform of Figure P6.6.

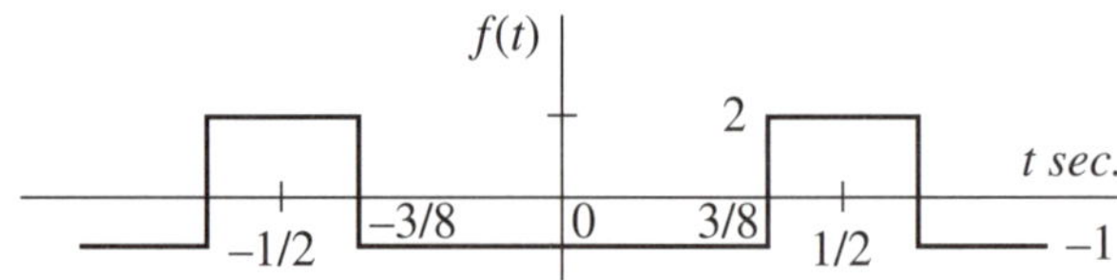

Figure P6.6

7. Show that the waveform of Figure P6.7 has the Euler phasors indicated.

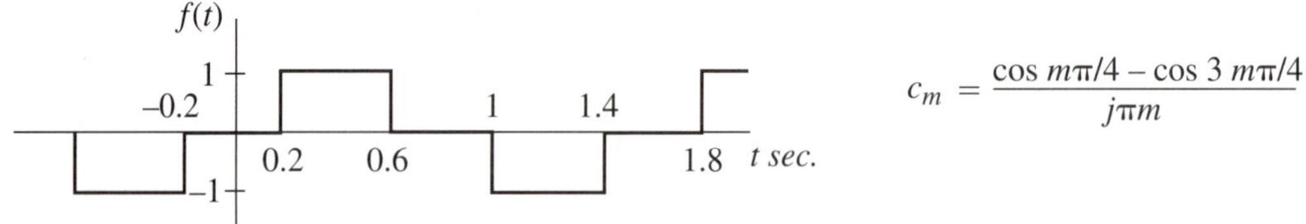

$$c_m = \frac{\cos m\pi/4 - \cos 3\,m\pi/4}{j\pi m}$$

Figure P6.7

Section 6.2

8. a. Sketch the amplitude spectrum of the waveform of Figure P6.1 out to the fifth harmonic if $A = 1$, $\tau = 1$, and $T = 4$.
   b. Determine the percentage of the total power contained in the first five harmonics.
9. a. Sketch the amplitude spectrum for the signal of Example 6.3 out to the fifth harmonic.
   b. Sketch the power spectrum for the signal out to the fifth harmonic.
10. Using the results of Example 6.5, determine the percentage of the total waveform power that exists in the first three harmonics for the waveform of Figure P6.10.

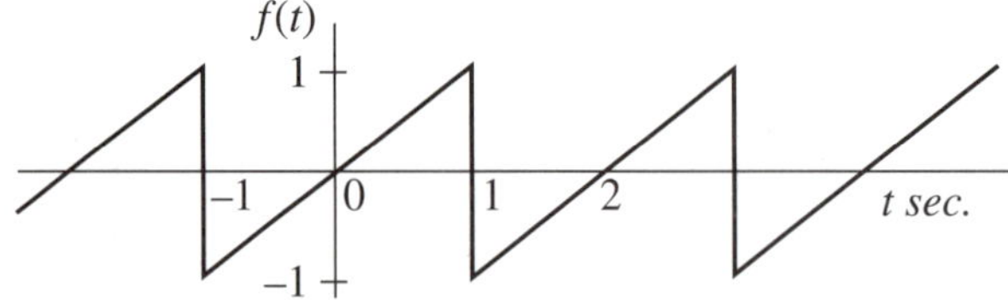

Figure P6.10

Section 6.3

11. The pulse shown in Figure P6.11 has the $F(\omega)$ indicated.
    a. Verify the expression for $F(\omega)$.
    b. Determine the complex Fourier series ($-2 \le m \le 2$) for the waveform that results if the pulse repeats every 3 seconds.

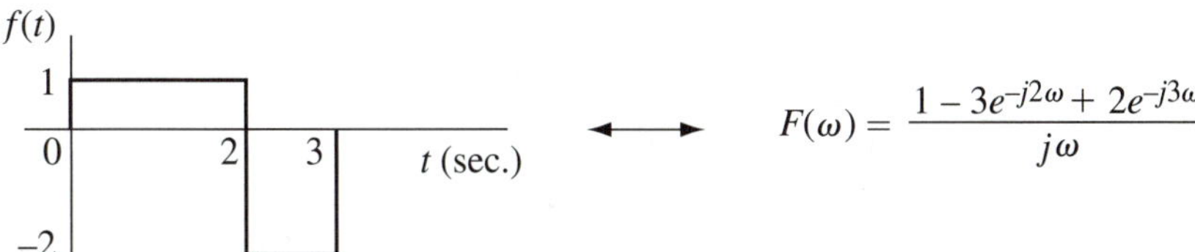

Figure P6.11

12. The pulse described here has the Fourier transform indicated:

$$f(t) = \begin{cases} \sin t & 0 \le t \le 2\pi \\ 0 & \text{elsewhere} \end{cases} \quad \leftrightarrow \quad F(\omega) = 2je^{-j\pi\omega}\frac{\sin \pi\omega}{1 - \omega^2}$$

Determine a few terms ($-2 \le m \le 2$) of the Fourier series of the periodic waveform that results if the pulse repeats in a period of
a. 2π seconds. b. 4π seconds.

13. Find the Fourier transform $X(\omega)$ for the pulse of Figure P6.13.

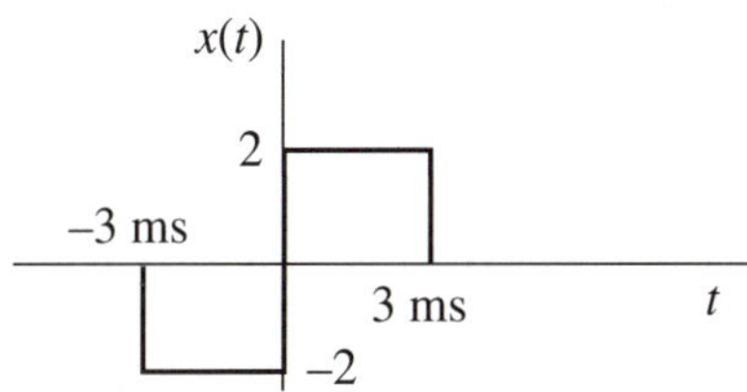

Figure P6.13

14. Find the Fourier transform $X(\omega)$ for the sinusoidal pulse of Figure P6.14.

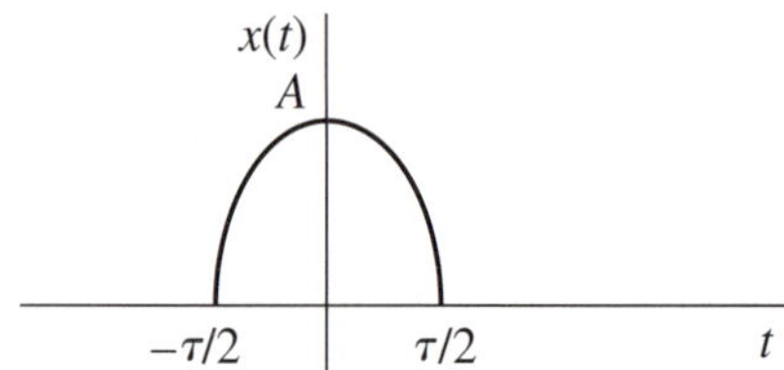

Figure P6.14

Section 6.4

15. The coefficients of a Fourier series are $c_m = 2/m$ $(m \neq 0)$ and $c_0 = 1$. List the coefficients of the 11-term Fourier partial sum using a Hamming window.
16. The coefficients of a Fourier series are $c_m = \sin(m)/2m$. List the values of coefficients $c_0 - c_4$ if a 21-term partial sum is used with a Hanning window.
17. State which of the five waveforms in Figure P6.17 would best be approximated by a nonrectangular windowed Fourier partial sum.

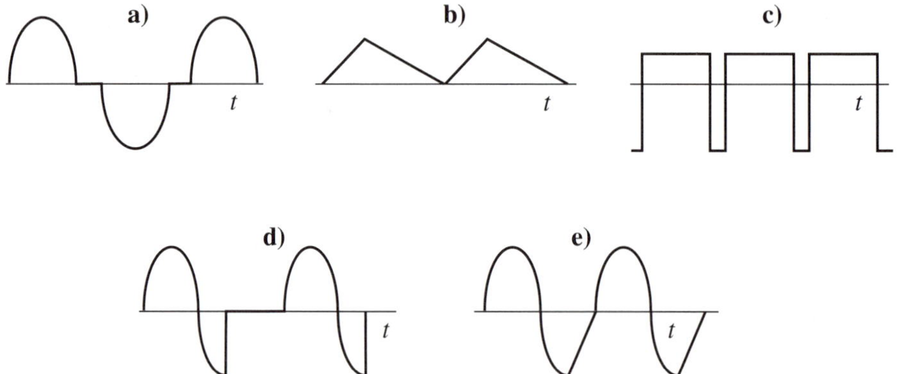

Figure P6.17

Section 6.5

18. Use a 31-term partial sum to produce a plot of one cycle of the $f(t)$ whose c_m are given in Problem 2. Use a rectangular window.
19. Use a 21-term partial sum to produce a plot of one cycle of the $f(t)$ whose c_m are given in Problem 3. Use a rectangular window.
20. a. Plot on the same graph the 21-term partial sum of the waveform in Figure P6.1 with and without a Hanning window for $\tau = 1, T = 3$, and $A = 2$.
    b. Plot the exponential spectrum of the windowed signal.
    c. Determine the average power of the windowed signal.

21. a. Plot on the same graph the 21-term partial sum of the waveform in Figure P6.7 with and without a Hanning window.
    b. Plot the exponential spectrum of the windowed signal.
    c. Determine the average power of the windowed signal.
22. Plot on the same graph the window functions of Table 6.1. Use 41 terms for each window, and identify each curve.
23. For the MATLAB vector $x = [1-j \quad 2 \quad j]$, determine the result of the operations indicated. (One will be an error statement.)
    a. `a=x*x'` b. `b=x.*x` c. `c=x.*x.'` d. `d=x*x.'`

Advanced Problems

24. Carry out the inverse Fourier transform to determine $f(t)$ given

$$F(\omega) = \delta(\omega - \omega_o) + \delta(\omega + \omega_o)$$

25. The signal of Problem 6.1 has $A = 2$, $\tau = 1$, and $T = 6$. It is input to the circuit shown in Figure P6.25. Use a 31-term Fourier partial sum to determine the output waveform.

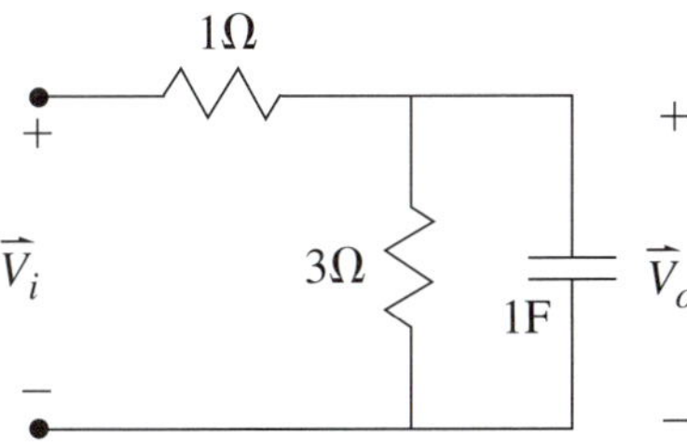

Figure P6.25

26. The signal of Example 6.5 repeats every 4 seconds and has an amplitude of 12 V. It is input to the circuit of Figure P6.26. Determine the output if a 31-term Fourier partial sum with a Hamming window is used to represent the input.

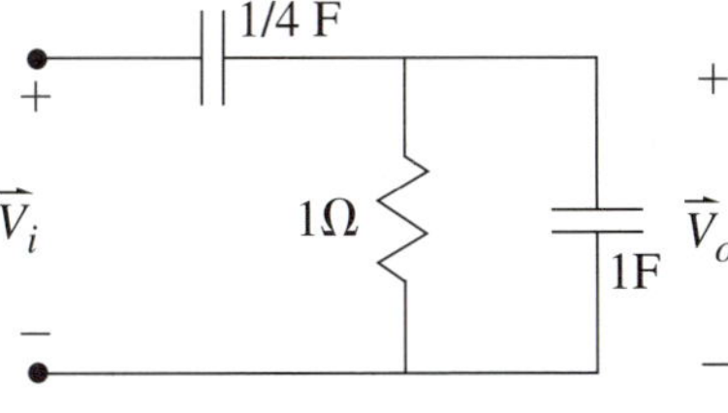

Figure P6.26

7

SAMPLED SIGNALS

OUTLINE

OBJECTIVES

1. Define the discrete Fourier transform
2. State the sampling theorem.
3. Find the spectrum of a discrete-time pulse.
4. Use and interpret the fast Fourier transform (FFT) and inverse fast Fourier transform (IFFT) MATLAB commands.

INTRODUCTION

To have a computer graph an equation $y = f(x)$, we first have it calculate y at a large number of x values. In this way it obtains samples of y that correspond to the samples of x specified. If we do not take enough samples to accurately characterize the $f(x)$, its graph will miss any rapid variations and give a false impression of the original equation.

For the remainder of this text, we will be sampling signals so that they may be processed by computer, and we will be receiving the processed samples back from the computer. We need to know how fast we must sample to avoid losing information. We also need to know what must be done to properly reassemble a continuous-time signal from an output sample sequence.

A sequence of samples constitutes a discrete-time signal. We know the value of only the samples and have no information about the signal at other instants. Still, such a signal has a spectrum, and the spectrum can answer many of the processing questions we may have. An important computer program called the *Discrete Fourier Transform* can be used to calculate the spectrum of a periodic sample sequence, and is it also used to estimate the spectrum of periodic continuous-time signals. Special MATLAB functions will be developed to apply the DFT to applications of interest.

7.1 DISCRETE FOURIER TRANSFORM (DFT)

Evaluating the integral to determine the coefficients of the Fourier series is rarely simple, and it often requires that we resolve indeterminate values for some of the coefficients. By having a computer calculate these coefficients using numerical integration, we can avoid both of these problems. After completing this section you will be able to:

- Define the discrete Fourier transform.
- State one property of a sampled signal's spectrum.
- Compare the DFT and the fast Fourier transform (FFT).

The Fourier series coefficients are given by (Eq. 6.2):

$$c_m = \frac{1}{T}\int_T f(t)e^{-jm\omega_o t}\,dt$$

To have a computer do this integration by numerical methods, we will have to describe the *continuous-time* periodic function, *f(t),* to the computer. Whenever we must represent a function of a continuous variable to a computer, we do so by *sampling* it at integer multiples of some increment, $t = nT_s$, where T_s is the sampling interval. This set of samples constitutes a *discrete-time* signal.

To accomplish our numerical integration problem, we will break the period up into N equal T_s intervals,

$$T = NT_s \tag{7.1}$$

and find the "area under the $f(t)e^{-jm\omega_o t}$ curve," ignoring any conceptual problems that arise because the area is a complex number. We expect that the result will accurately approximate the original integral if we make N very large and, therefore T_s very small. The operation is shown in Figure 7.1. The equation for c_m becomes

$$c_m = \frac{1}{T}\int_T f(t)e^{-jm\omega_o t}\,dt \approx \frac{1}{NT_s}\sum_{n=0}^{N-1}\left[f(nT_s)e^{-jm\frac{2\pi}{NT_s}nT_s}\right]T_s$$

$$c_m \cong \frac{1}{N}\sum_{n=0}^{N-1} f(nT_s)e^{-j\frac{2\pi mn}{N}} \tag{7.2}$$

There are many interesting aspects to Equation 7.2. One important aspect is that the frequency spectrum has become periodic in N harmonics. This is shown from

$$c_{m+N} = \frac{1}{N}\sum_{n=0}^{N-1} f(nT_s)e^{-j\frac{2\pi(m+N)n}{N}} = \frac{1}{N}\sum_{n=0}^{N-1} f(nT_s)e^{-j\frac{2\pi mn}{N}}\underbrace{e^{-j2\pi n}}_{1} = c_m$$

The N harmonics of the periodic waveform span a frequency range of f_s.

$$NT_s = T \qquad or \qquad Nf_o = f_s$$

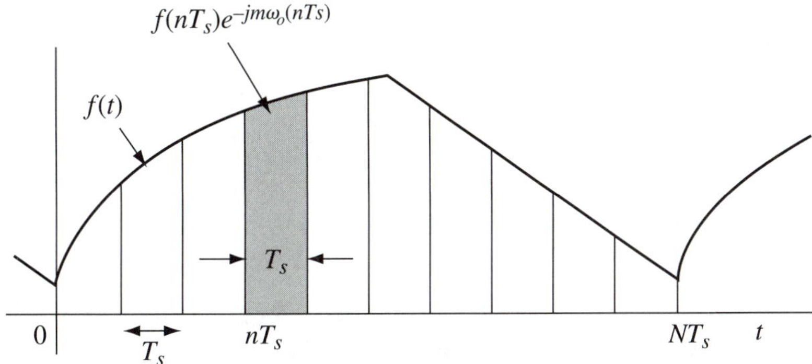

Figure 7.1 Breaking up a period into N equal segments, the integral for c_m may be approximated by a summation of terms of the form $f(nT_s)e^{-jm\omega_o(nTs)}Ts$.

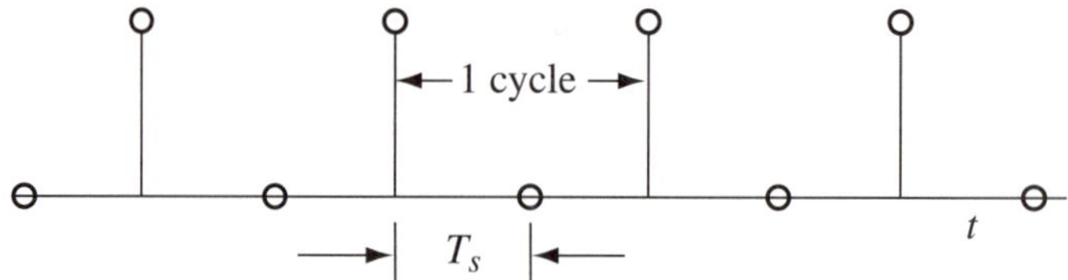

Figure 7.2 The highest sample variation rate detectable in a sampled signal is $f = 1/2T_s = f_s/2$.

That these facts make sense can be seen from Figure 7.2. The highest rate of variation that can exist in a sampled signal is $f_s/2$.

Sampling a signal in the time domain causes its spectrum to become periodic in the frequency range f_s.

Since the sampled function and its spectrum are now both periodic and discrete, they are both ideally suited to computer manipulation. The equation for the Fourier series coefficients of a sampled, periodic signal is called the N-point discrete Fourier transform, and it is usually normalized to $T_s = 1 = 1/f_s$. The Fourier series itself is called the inverse discrete Fourier transform (IDFT).

$$c_m N = F(m) = \sum_{n=0}^{N-1} f(n) e^{-j\frac{2\pi mn}{N}} \qquad \text{DFT} \tag{7.3}$$

$$f(n) = \frac{1}{N} \sum_{m=0}^{N-1} F(m) e^{j\frac{2\pi mn}{N}} \qquad \text{IDFT} \tag{7.4}$$

Our original goal was a computer program that would provide a good approximation to the Fourier coefficients of a continuous-time signal. Clearly, the c_m found from Equation 7.3 are not quite the same as those found from the exact equation at high harmonics. The spectrum of the original signal was not periodic, while the spectrum of the sampled signal is. However, there is a fundamental uncertainty in when things happen with a sampled waveform. If a sampled square wave goes from a sample of $+1$ to a next sample of -1, when does the transition actually occur? We only know it occurs sometime within an interval T_s. So the inaccuracy is only what is inherent in approximating a continuous signal with samples, and we will obtain good results if we use a large number of sampling points over the period. However, the DFT gives the exact spectrum for a discrete signal, no matter how few points are used.

EXAMPLE 7.1

Find the DFT and IDFT for the waveform of Figure 7.3.

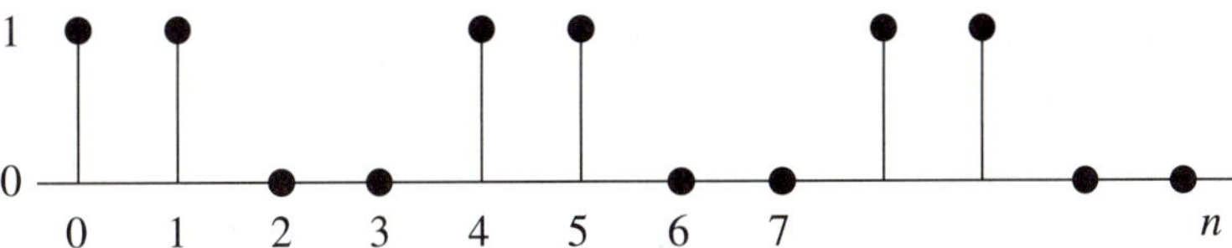

Figure 7.3

Solution

The waveform is periodic in four samples, so we need a four-point DFT:

$$F(m) = \sum_{n=0}^{3} f(n)e^{-j\frac{2\pi mn}{4}} = 1 + 1e^{-j\pi m/2} + 0e^{-j\pi m} + 0e^{-j3\pi m/2}$$

$$F(m) = 1 + e^{-j\pi m/2}$$

So $F(0) = 2$, $F(1) = 1 - j$, $F(2) = 0$, and $F(3) = 1 + j$. The inverse transform consequently is

$$f(n) = \frac{1}{4}\sum_{m=0}^{3} F(m)e^{j\frac{\pi mn}{2}} = \frac{1}{4}[2 + (1 - j)e^{j\pi n/2} + (1 + j)e^{j3\pi n/2}]$$

and after simplifying, we get

$$f(n) = \frac{1}{4}[2 + (1 - j)e^{j\pi n/2} + (1 + j)e^{-j\pi n/2}]$$

$$f(n) = \frac{1}{2}\left(1 + \cos\left(\frac{n\pi}{2}\right) + \sin\left(\frac{n\pi}{2}\right)\right)$$

giving $f(0) = 1, f(1) = 1, f(2) = 0, f(3) = 0$, and $f(4) = 1$. If we want the Fourier series coefficients, they are $c_m = F(m)/N$.

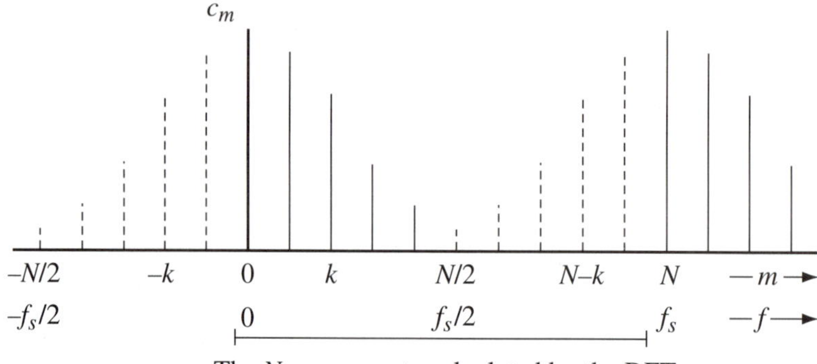

Figure 7.4 Spectrum of the sampled signal based on the DFT equation for even N. The highest unique c_m component present is $c_{N/2}$, and the spectrum is periodic in N samples. Dotted lines are used to indicate a conjugate to the equal-length solid frequency component. (c_0 and $c_{N/2}$ are always real.)

The N spectral components provided by the DFT run consecutively from c_0 to c_{N-1}. We are accustomed to picking up the Fourier series terms as pairs of positive and negative frequencies. The equivalent information is present in the DFT, since the spectrum is now periodic. The c_{-k} term is the same as the c_{N-k} term. Also, the c_{-k} term and the c_k term must still be conjugates. The d-c component is c_0, and the fundamental is composed of the c_1 and c_{-1} terms, as before. If $c_{N/2-1}$ is regarded as the final positive exponential frequency term (N even), the remaining terms are the negative exponential frequencies. Counting d-c as a positive frequency, the DFT gives equal numbers of negative and positive spectral terms. For real $f(n)$, if we know the c_0 through $c_{N/2}$ terms, we know the rest are just the conjugate values (Figure 7.4).

The DFT requires approximately N^2 multiplications in finding the full set of N frequency components from the N time samples. Theorems relating signal spectra show that an N-point DFT (N even) could be found by sorting every other sample of the original signal into two signals of $N/2$ terms each. The spectra of these new signals are then found and the results combined to obtain the original signal's spectrum. This operation would require only two $(N/2)^2$ multiplications to calculate both smaller signals' spectra, and $N/2$ multiplications to recombine them, for a total of $(N + 1)N/2$ multiplications. For large N, this is a savings of nearly 50%.

Now, if N happens to be a power of 2, the two new signals will still each have an even number of terms, and this process can be repeated until we have reduced the original signal to $N/2$ two-point signals whose spectra need to be found and combined. The result is that only about $N \log_2 N$ multiplications are required. The algorithm that implements this process is called a *fast Fourier transform (FFT);* its savings in multiplications is indicated in Table 7.1. A variety of FFT algorithms

Table 7.1 Comparison of relative computation time (number of multiplications) for an *N*-point DFT and for the Cooley–Tukey FFT.

N	N log_2 N	N^2
64	384	4,096
128	896	16,384
256	2048	65,536
512	4608	262,144
1024	10240	1,048,576
2048	22528	4,194,304
4096	49152	16,777,216
8192	106496	67,108,864

exist; the one just described was the first, and it is known as the Cooley–Tukey decimation-in-time FFT. Handling power-of-2 samples is natural for digital devices, so there is little reason not to take advantage of the speed of the FFT over the DFT. Even if it is not important in a particular application, it is best to work with N samples that are powers of 2 in preparation for the day when the speed is required.

7.2 SAMPLED SINEWAVES

The sinusoid is still one of our fundamental signals, and its spectrum is the simplest available from a real function. By considering the consequences of sampling it, we will discover the uniform sampling theorem. After completing this section you will be able to:

- State the uniform sampling theorem.
- Discuss aliasing and how to prevent it.
- Find the spectrum of a sampled sinusoid.

By sampling the sinusoid $v(t) = V_m \cos(\omega t + \phi)$ every T_s seconds so that samples are obtained at instants $t = nT_s$, we generate a discrete-time signal of the form

$$v(nT_s) = V_m \cos(\omega T_s n + \phi) \tag{7.5}$$

where the time variable is now represented by the sample number, n.

Euler's identity may again be used to express the cosine in terms of the simpler exponential functions:

$$v(nT_s) = V_m \cos(\omega T_s n + \phi)$$
$$= V_m\left(\frac{e^{j(\omega T_s n+\phi)} + e^{-j(\omega T_s n+\phi)}}{2}\right) = \left[\frac{V_m e^{j\phi}}{2}\right]e^{j\omega T_s n} + \left[\frac{V_m e^{-j\phi}}{2}\right]e^{-j\omega T_s n}$$

The bracketed term for the positive exponential is the same Euler phasor found for continuous-time signals. We may consequently define the number sequence generated by sampling the cosine as

$$v(nT_s) = 2\,\mathrm{Re}(\vec{V}e^{j\omega T_s n}) \qquad \textbf{(7.6)}$$

where $\vec{V}$ is the Euler phasor.

So far, the description of the discrete-time signal seems nearly the same as that for the continuous-time signal. The key information on amplitude and phase is contained in the Euler phasor for both signal types. There is, however, one very significant difference. Given ω, the continuous-time signal specifies a unique sinusoid

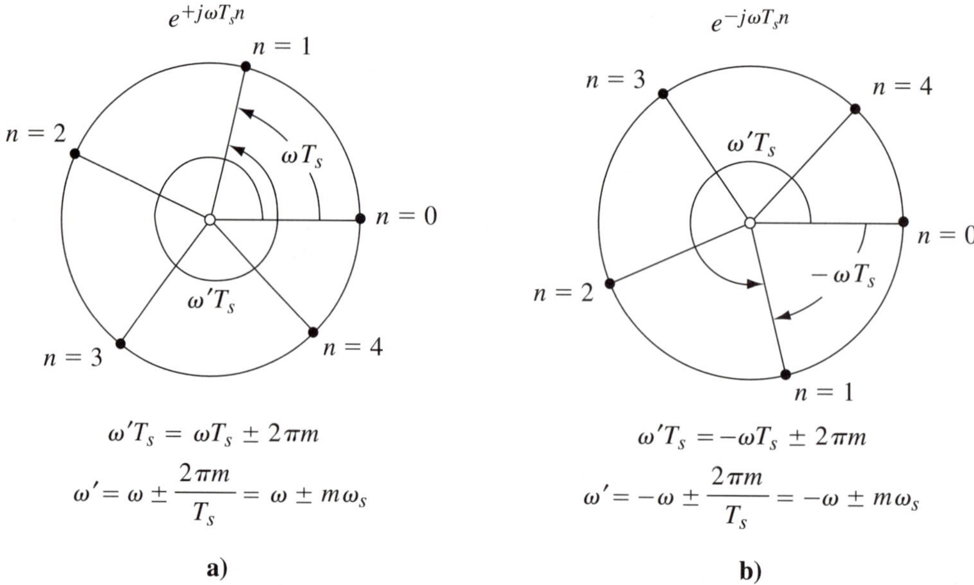

Figure 7.5 (a) The first few sampling points specified by the $e^{j\omega T_s n}$ function, and the frequencies, ω', of other exponentials that would create the same samples. **(b)** A similar situation applies to the $e^{-j\omega T_s n}$ function shown here.

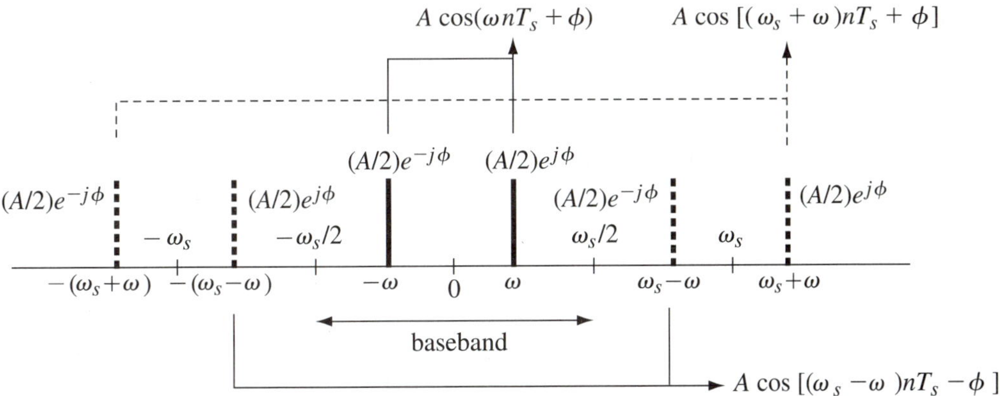

Figure 7.6 Spectrum of a sampled cosine whose frequency is ω, showing the Euler phasors of the individual exponentials. The baseband spectrum is replicated around integer multiples of the sampling frequency, $\pm m\,\omega_s$. The replications are called *images.* Notice that the phases of the cosines alternate because of the Euler phasors of the images.

clearly distinguishable from any other. The discrete-time signal, on the other hand, specifies a set of samples that could have been produced by any of an infinite number of sinusoids. Remembering that $e^{\pm j\omega T_s n} = 1\angle\omega T_s n$, each exponential generates its own unique set of possible frequencies since it is impossible to distinguish between, for instance, $1\angle\omega T_s n$ and $1\angle \pm 2\pi m + \omega T_s n$, where m and n are integers. This situation is pictured in Figure 7.5.

Placing the phasors associated with each exponential along a frequency scale, and remembering that the original negative frequency component takes the conjugate Euler phasor, results in Figure 7.6. Combining the exponentials into cosine functions shows that the same set of real samples would be generated by any cosine with the same Euler phasor and a frequency $\omega' = \pm m\,\omega_s + \omega$, or the conjugate Euler phasor and a frequency of $\omega' = \pm m\,\omega_s - \omega$. Any one of them could have generated the number sequence and caused all the others to be present in the spectrum of the sampled signal. That they all must be present is required for the spectrum to be periodic in an interval ω_s.

EXAMPLE 7.2

The signal $f(t) = \sin(t + 110°)$ is sampled every 3 seconds. Identify at least two other signals that would provide exactly the same samples, and show that you are right.

Solution

The rule for the cosine has been established in Figure 7.6, so it is simplest to convert immediately to the cosine function. The specified signal has a frequency of $\omega = 1$ rad/s, and its Euler phasor is $0.5\underline{/20^\circ} = 0.5\underline{/0.3491}$ radians. Any cosine signal with the same Euler phasor and a frequency $\omega' = \pm m\ \omega_s + \omega$ with produce the same samples. Any cosine with the conjugate Euler phasor and a frequency of $\omega' = \pm m\ \omega_s - \omega$ will also produce the same samples.

The sampling frequency is $\omega_s = 2\pi/T_s = 2\pi/3$ rad/s. Arbitrarily selecting $m = +4$, the three signals we will compare are

$$f(t) = \sin(t + 110^\circ) = \cos(t + 20^\circ)$$
$$g(t) = \cos[(1 + 8\pi/3)t + 20^\circ]$$
$$h(t) = \cos[(-1 + 8\pi/3)t - 20^\circ]$$

where $t = 3n$. We will use MATLAB for the demonstration:

```
>n=0:5;
>f=sin(3*n+1.920);
>g=cos((8*pi+3)*n+0.3491);
>h=cos((8*pi-3)*n-0.3491);
>a=[n' f' g' h']

      n         f         g         h
      0    0.9396    0.9397    0.9397
 1.0000   -0.9785   -0.9785   -0.9785
 2.0000    0.9978    0.9978    0.9978
 3.0000   -0.9971   -0.9971   -0.9971
 4.0000    0.9765    0.9765    0.9765
 5.0000   -0.9363   -0.9363   -0.9363
```

The baseband region consists of frequencies from the range $-\omega_s/2 < \omega < \omega_s/2$. If a sinusoid has a frequency within the baseband, it appears at its proper frequency after sampling. As the input signal frequency is gradually increased, its exponential components move away from each other, as shown in Figure 7.7a. Of course, the same thing happens to all the images as well. When ω reaches $\omega_s/2$, the ω and $\omega_s - \omega$ components fall on top of one another. As ω continues to increase, these components pass one another, and the baseband component (Figure 7.7b) decreases in frequency, although the actual signal frequency is increasing. It is no longer the correct signal that appears in the baseband, but a false one called an *alias*. A time domain presentation of this effect is shown in Figure 7.8.

Aliasing is a sampling rate error. An often-cited example of aliasing in the time domain is the apparent reverse rotation of wheels in movies, most noticeable in the wagon wheels of old westerns. The successive film frames occur too slowly to

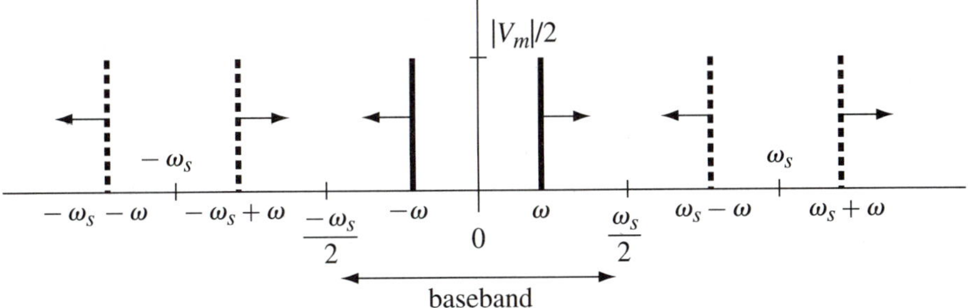

Figure 7.7a The frequency components of the sampled cos ωt move in the direction of the arrows as ω is increased.

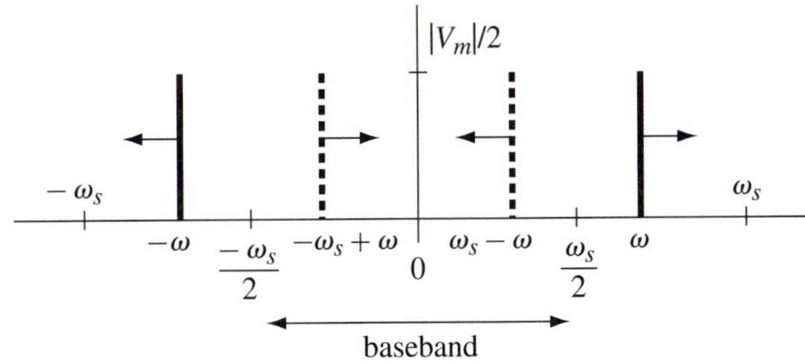

Figure 7.7b The positions indicated are taken by the frequency components of the sampled cos ωt when $\omega_s/2 < \omega < \omega_s$.

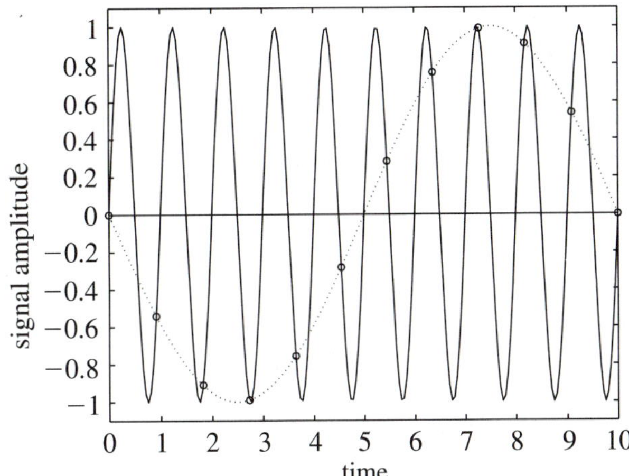

Figure 7.8 If a signal (solid line) is sampled (circles) too slowly, the human eye, as well as plotting routines, see the samples as points along the alias signal (dotted line).

capture the true rotation of the wheels. The same effect is intentionally introduced with strobe lights for rpm measurements. If the strobe flashes once each revolution of a shaft, the shaft visually appears to be stationary, and its speed equals the strobe frequency.

To an observer of the baseband activity on a commercial spectrum analyzer, applying a sinewave of steadily increasing frequency results in a frequency component that reverses direction every time it hits $\omega_s/2$ or 0. It is as if the frequency scale were folded at these points, and $\omega_s/2$ is sometimes referred to as the *folding frequency.*

The only way to avoid aliasing is to prevent signals with frequencies above $\omega_s/2$ from reaching the sampler. An anti-aliasing filter is used for this purpose. If n sinusoids enter the sampler, there will be n sinusoids in the baseband. They will be at their true frequency only if they passed through the anti-aliasing filter.

EXAMPLE 7.3

A signal $x(t) = 6 \cos 5t + 4 \cos 15t + 2 \sin 34t$ is sampled at 12 rad/s. List the frequencies of all baseband sinusoids.

Solution

The baseband region extends from −6 rad/s to 6 rad/s. The 5 rad/s cosine satisfies the sampling theorem and appears as itself.

The 15 rad/s cosine will produce a $\omega_s - \omega = 12 - 15 = -3$ rad/s exponential in the baseband. This will be matched by a component from $-\omega_s + \omega = +3$ rad/s to produce a baseband sinusoid at 3 rad/s.

The 34 rad/s sinewave produces components at $m\omega_s \pm \omega = 12m \pm 34$. The $m = 3$ image will consequently produce a +2 rad/s exponential in the baseband. Another image will produce the −2 rad/s signal to give a 2 rad/s sinusoid. The baseband spectral magnitude is shown in Figures 7.9a and b.

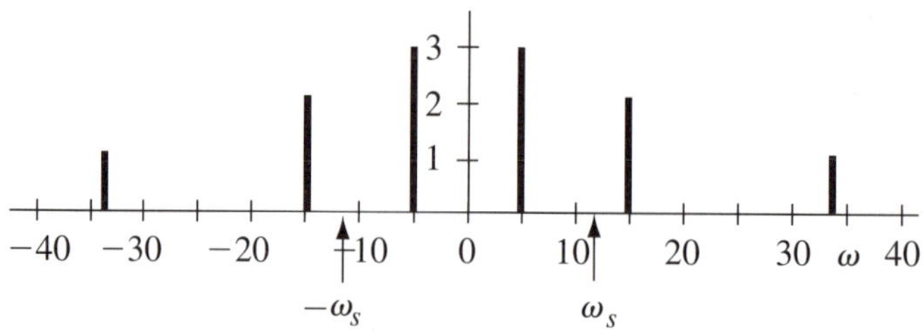

Figure 7.9a Input signal frequency components.

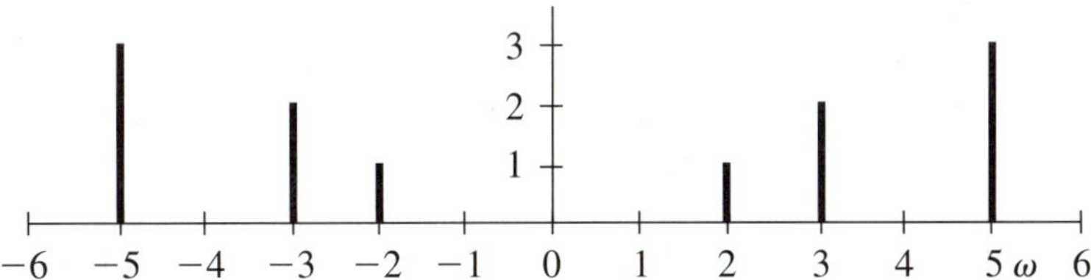

Figure 7.9b Baseband frequency components after sampling.

The observations of this section may now be summarized in the following theorem.

> **Uniform Sampling Theorem**
>
> When a periodic signal is sampled at a rate of f_s samples per second, its spectrum is replicated around integer multiples of the sampling frequency. If f_m is the highest frequency present in the signal and
>
> $$f_s \geq 2f_m \qquad \textbf{(7.7)}$$
>
> the replications will not overlap, and the original signal can be recovered from the samples with a low-pass filter that passes only the baseband frequencies. If $f_s < 2f_m$, the replications will overlap, and the original signal may become mixed with aliased components so as to be unrecoverable.

The lower limit for the sampling frequency is called the *Nyquist rate.* Sampling at the Nyquist rate would require that an ideal low-pass filter be used for the recovery.

7.3 DISCRETE-TIME FOURIER TRANSFORM (DTFT)

The spectrum of aperiodic discrete-time signals is found from the discrete-time Fourier transform. The DTFT has the same relationship to the DFT that the Fourier transform has to the Fourier series. After completing this section you will be able to:

- Find the spectrum of a discrete-time pulse.
- Explain the difference between the DFT and the DTFT.
- Provide an expression for an ideally sampled signal.

Although we have been able to deduce the sampling theorem using sinusoidal signals as a guide, a more general approach will formalize the theorem in a mathematical statement. A great many theorems concerning properties of signals are derived from the Fourier transform using techniques similar to those to be used in this section. Rarely are integrals actually evaluated, but their general properties are deduced from functional notation or by changing the order in which summations and/or integrations are performed. Some mathematical sophistication may be required to appreciate the simplicity of the derivations, but the derivation process is not part of this section's objectives.

The DTFT could be developed by approximating the Fourier transform in the same way we developed the DFT equation. Instead, we will take a different approach, using an expression for an ideal sampler.

An *ideal sampler* is formed from a periodic impulse waveform:

$$\delta_T(t) = \sum_{n=-\infty}^{\infty} \delta(t - nT_s) \qquad \text{Ideal Sampler} \tag{7.8}$$

Being periodic, the ideal sampler may be expanded in a Fourier series. The integral for c_m is especially simple because the only place δ_T is nonzero in the range $-T_s/2 \leqq t \leqq T_s/2$ is at $t = 0$.

$$c_m = \frac{1}{T_s} \int_{-T_s/2}^{T_s/2} \delta_T(t) e^{-jm\omega_s t}\, dt = \frac{1}{T_s} \int_{0-}^{0+} \delta(t) dt = \frac{1}{T_s}$$

Then the Fourier series representation of the ideal sampler is

$$\delta_T(t) = \frac{1}{T_s} \sum_{m=-\infty}^{\infty} e^{jm\omega_s t} \qquad \text{Ideal Sampler} \tag{7.9}$$

These two ways of expressing the ideal sampler waveform lead to two important equations. Suppose $f(t)$ is a continuous-time signal whose Fourier transform is (Eq. 6.6)

$$F(\omega) = \int_{-\infty}^{\infty} f(t) e^{-j\omega t}\, dt$$

We will obtain a discrete-time signal, $g(t)$, by sampling $f(t)$ with the ideal sampler. This process is shown in Figure 7.10.

$$g(t) = f(t)\delta_T(t) = f(t) \sum_{n=-\infty}^{\infty} \delta(t - nT_s) \tag{7.10}$$

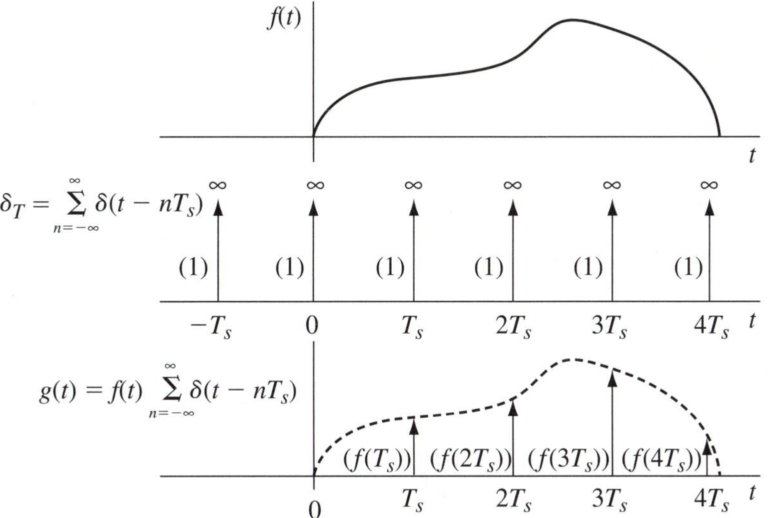

Figure 7.10 An ideal sampler consists of a unit impulse wavetrain. The impulses have zero width, so they pick up the value of f at an instant. The ideally sampled signal consists of an impulse train where the area of each impulse records the original function's value. Although the impulses are always of infinite amplitude, they are drawn to have their length represent their weighting. This representation of discrete data with impulses is an alternative to stem plots.

Using the Fourier series representation of the ideal sampler, the Fourier transform of $g(t)$ becomes

$$G(\omega) = \int_{-\infty}^{\infty} g(t)e^{-j\omega t}\,dt = \int_{-\infty}^{\infty} f(t)\left[\frac{1}{T_s}\sum_{m=-\infty}^{\infty} e^{jm\omega_s t}\right]e^{-j\omega t}\,dt$$

Changing the order of integration and summation, and comparing the integral to the definition of the Fourier transform, gives the following result:

$$G(\omega) = \frac{1}{T_s}\sum_{m=-\infty}^{\infty}\left\{\int_{-\infty}^{\infty} f(t)e^{-j(\omega - m\omega_s)t}\,dt\right\} = \frac{1}{T_s}\sum_{m=-\infty}^{\infty} F(\omega - m\omega_s)$$

This result is a mathematical statement of the sampling theorem: The spectrum of the sampled signal consists of replications of the original signal's spectrum centered on integer multiples of the sampling frequency, and with an amplitude change of $1/T_s$. Figure 7.11 shows the spectrum of the original signal and of the sampled signal for two sampling conditions.

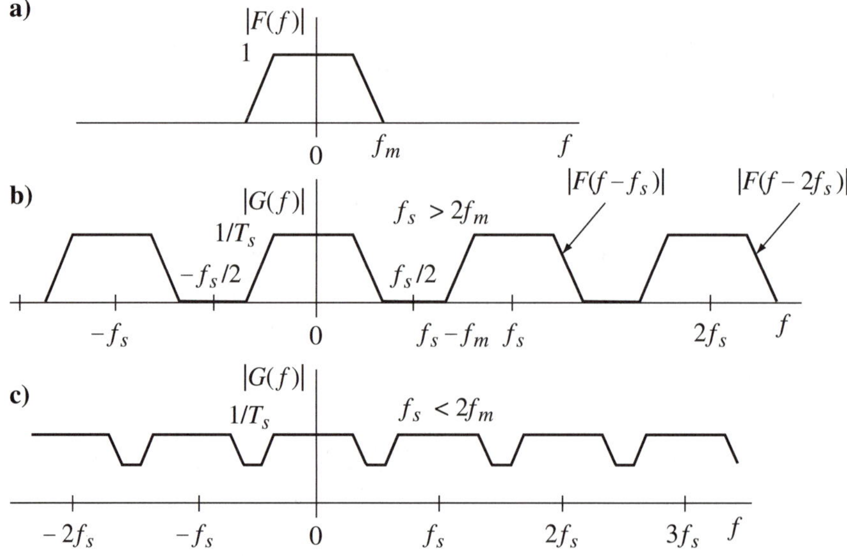

Figure 7.11 (a) Continuous-time and **(b)** discrete-time spectrum when sampling is slightly above the Nyquist rate. Sampling causes the original spectrum to be replicated around integer multiples of the sampling frequency. **(c)** Sampling at less than the Nyquist rate causes aliasing, and the original signal cannot be recovered.

$$G(\omega) = \frac{1}{T_s} \sum_{m=-\infty}^{\infty} F(\omega - m\omega_s) \qquad \text{Sampling Theorem} \tag{7.11}$$

If the other expression for an ideal sampler (Eq. 7.8) is used with $f(t)$ and is inserted into the Fourier transform, we have

$$G(\omega) = \int_{-\infty}^{\infty} g(t)e^{-j\omega t}\, dt = \int_{-\infty}^{\infty} f(t) \sum_{n=-\infty}^{\infty} \delta(t - nT_s)e^{-j\omega t}\, dt$$

Interchanging the order of summation and integration again makes the integration easy because of the properties of the delta function:

$$G(\omega) = \sum_{n=-\infty}^{\infty} \int_{t=-\infty}^{\infty} \delta(t - nT_s)f(t)e^{-j\omega t}\, dt = \sum_{n=-\infty}^{\infty} f(nT_s)e^{-j\omega T_s n}$$

The result is called the *discrete-time Fourier transform (DTFT)*. It is a continuous function of ω, and it is periodic in ω_s since $G(\omega + \omega_s) = G(\omega)$. These properties of $G(\omega)$ suggest a change of notation that explicitly indicates a continuous spectrum periodic in $f_s = 1/T_s$, and reestablishes the symbolic correspondence between the samples of the original time signal and its spectrum. This change is

$$G(\omega) \rightarrow F(e^{j\omega T_s})$$

While somewhat cumbersome, this notation immediately identifies F as the spectrum of a sampled signal $f(nT_s)$, and we do not need to say so in words. With this notation change, our result becomes:

$$F(e^{j\omega T_s}) = \sum_{n=-\infty}^{\infty} f(nT_s)e^{-j\omega T_s n} \qquad \text{DTFT} \tag{7.12}$$

The inverse discrete-time Fourier transform (IDTFT) may be obtained by multiplying both sides of Equation 7.12 by $e^{j\omega k T_s}$ and integrating over ω_s. The result is

$$f(nT_s) = \frac{1}{\omega_s}\int_{\omega_s} F(e^{j\omega T_s})e^{jnT_s\omega}\,d\omega \qquad \text{IDTFT} \tag{7.13}$$

Equation 7.11 expresses the spectrum of the sampled signal in terms of the spectrum of the original continuous-time signal, while Equation 7.12 shows how to find the spectrum of a discrete-time signal from the value of its samples. These equations had identical origins and must provide identical results; they differ only in their point of view.

If f turns on at $n = 0$, and also vanishes for $n > N - 1$, the DTFT and the DFT are identical except that the N point DFT only provides samples of $F(e^{j\omega T_s})$ at $\omega = m\omega_s/N$. These samples will be close enough together to allow a plot of F if zero fill is used to make the period large after the pulse ends (See Figure 6.7).

Closed-form expressions for the DTFT spectra of many common signals may be found from properties of geometric progressions. If an N-term sum is designated S_N, then

$$S_N = \sum_{n=0}^{N-1} x^n = 1 + x + x^2 + \cdots + x^{N-2} + x^{N-1}$$

So

$$xS_N = x + x^2 + \cdots + x^{N-2} + x^{N-1} + x^N$$

$$\therefore (1 - x)S_N = 1 - x^N$$

$$S_N = \sum_{n=0}^{N-1} x^n = \frac{1 - x^N}{1 - x} \tag{7.14}$$

And if $\lim_{N\to\infty} x^N = 0$, then

$$S = \sum_{n=0}^{\infty} x^n = \frac{1}{1-x} \qquad \textbf{(7.15)}$$

EXAMPLE 7.4

Determine the DTFT for the discrete-time unit impulse, $\delta(n)$, defined as follows

$$\delta(n) = \begin{cases} 0 & n \neq 0 \\ 1 & n = 0 \end{cases} \quad \text{discrete unit impulse} \qquad \textbf{(7.16)}$$

Solution

Using Equation 7.12, its spectrum is

$$\Delta(e^{j\omega}) = \sum_{n=-\infty}^{\infty} \delta(n)e^{-j\omega n} = 1 \qquad \textbf{(7.17)}$$

where the sum collapses to a single term because of the definition of the discrete unit impulse. The result shows that the discrete impulse contains all frequencies in equal amounts.

EXAMPLE 7.5

Determine the spectrum of $f(nT_s)$ in Figure 7.12.

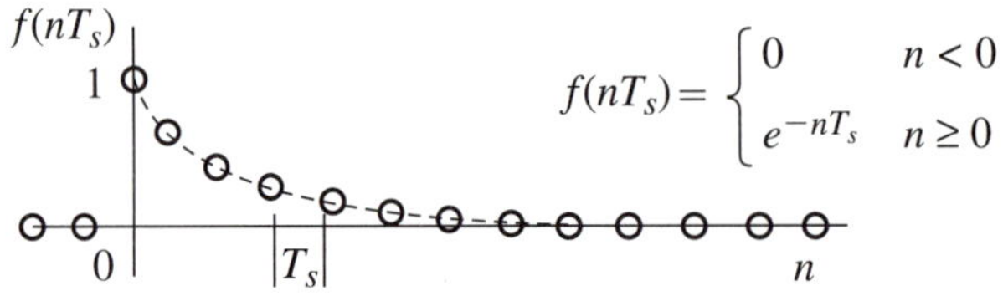

$$f(nT_s) = \begin{cases} 0 & n < 0 \\ e^{-nT_s} & n \geq 0 \end{cases}$$

Figure 7.12

Solution

The spectrum of $f(nT_s)$ is found from Equations 7.12 and 7.15:

$$F(e^{j\omega T_s}) = \sum_{n=0}^{\infty} e^{-nT_s}e^{-jn\omega T_s} = \sum_{n=0}^{\infty} (e^{-(1+j\omega)T_s})^n = \frac{1}{1 - e^{-(1+j\omega)T_s}}$$

In the study of discrete-time systems, a generalization of the DTFT known as the z transform is often used to characterize signals. The z transform can be derived by finding the Laplace transform of an ideally sampled signal.

7.4 MATLAB LESSON 7

The DFT and IDFT are fully compatible with computer operations and are often used to approximate the Fourier transform, Fourier series, and the DTFT. This lesson will concentrate on providing MATLAB functions that use the DFT/IDFT for these purposes. After completing this section you will be able to:

- Use and interpret the FFT and IFFT MATLAB commands.
- Do Fourier series problems without integrations or indeterminates.
- Call application programs of your own design.

While we can calculate the DFT by hand for simple examples, most applications require a large N and are impractical without an actual computer program.

MATLAB EXAMPLES

fft **ifft** **fftshift**

The MATLAB DFT program is called the **fft** command as $X = fft(x)$, where x has been described in an N-element vector. If N is a power of 2, a fast Fourier transform algorithm will be used. If N is not a power of 2, a much slower DFT routine will result. $X = fft(x,512)$ specifies a 512-point FFT. If x contains less than 512 points, zero fill will be provided to expand the time record in the *fft* calculation. If x has more than 512 points, it will be truncated at 512 for the *fft* calculation. These variations apply equally to the **ifft** command.

The elements of the c_m vector produced by the DFT from a program statement like

```
>cm = fft(f,N)/N;
```

are, in order,

$$c_m = [\underbrace{c_0 \quad c_1 \quad c_2 \ \cdots \ c_{(N/2-1)}}_{N/2 \text{ terms}} \quad \underbrace{c_{-N/2} \ \cdots \ c_{-2} \quad c_{-1}}_{N/2 \text{ terms}}]$$

Because the spectrum of the DFT is periodic, this is as good a way to show one full cycle as any. If we are thinking in terms of the Fourier series, however, we would prefer to see a spectrum symmetrical about d-c. This is almost achieved with the command >dm = fftshift(cm); which produces a d_m vector whose elements are

$$d_m = [\underbrace{c_{-N/2} \cdots c_{-2} \;\; c_{-1}}_{N/2 \text{ terms}} \;\; \underbrace{c_0 \;\; c_1 \;\; c_2 \cdots c_{(N/2-1)}}_{N/2 \text{ terms}}]$$

Note that the d-c component is dm element number $N/2 + 1$. Applying the **fftshift** command to the d_m vector shifts it back to the c_m vector.

In many applications we may wish to obtain a signal's spectrum, modify it, and see the consequences back in the time domain. If so, keep in mind that **fft** finds $F = Nc_m$ and that **ifft** also needs F, not c_m. In addition, **ifft** expects the frequency components supplied to it to be in the order of the elements of the c_m vector, not the d_m vector. With these cautions, let us prepare a few MATLAB functions.

In Chapter 6, we wrote a function called FS.m that could plot partial Fourier sums. Before we could use that function we had to: (1) obtain an equation for $x(t)$; (2) integrate to find the expression for c_m; and (3) evaluate c_m and resolve any indeterminate conditions. Now we will create a function called *Partsum* to accomplish the same thing, provided only that we can describe $x(n)$ over one period in a 1024-element vector. Our description of x starts with the $t = 0$ value and ends just before the $t = T$ value.

File Partsum.m

```
function [y,t]=Partsum(x,M,win,T)
% Partsum calculates the partial Fourier series of x with period T
%         including out to the Mth harmonic
%                 win=0 specifies a rectangular window (default value)
%                  win=1 specifies a hamming window
%                   win=2 specifies a hanning window
%                    win=3 specifies a triangular window
% x must be described over T starting from t = 0 in a 1024-point vector
% The command format is [partial_sum,time]=Partsum(x,M,win,T)
% Used without output arguments, Partsum plots the results

if nargin==2               % provide default T and w for two input arguments
T=1;w=1;
end

if nargin==3               % provide default T for three input arguments
T=1;
end

if nargin>=3               % determine the desired window
     if win==3
     w=triang(2*M+1);
     elseif win==2
```

```
        w = hanning(2*M+1);
        elseif win==1
        w = hamming(2*M+1);
        else
        w = 1;
        end
end
                                             % begin program
cm = fft(x,1024);                            % calculate the Fourier
                                             % coefficients
ccm = ffshift(cm);                           % center on d-c (see
                                             % Figure 7.13)
m = -M:M;
dd = ccm(513+m).*w';                         % window the desired harmonics
ddm=[zeros(1,512-M) dd zeros(1,511-M)];      % zero fill the rest
dm=ffshift(ddm);                             % put back to proper order
yy=real(ifft(dm));                           % take the inverse transform
tt=(0:1023)*T/1024;                          % create the time vector
                                             % end program
if nargout==0                                % determine the output desired
     plot(tt,real(yy))
```

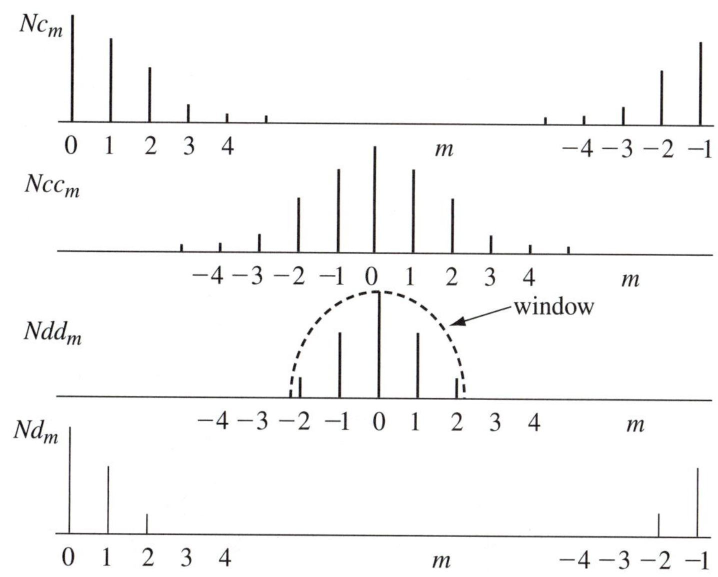

Figure 7.13 Operations performed in *Partsum.m* are pictured. Starting with the DFT spectrum (c_m), the spectrum centered on d-c is obtained (cc_m), multiplied by a rectangular window to terminate the series after M harmonics. The surviving harmonics then have the desired window function applied (dd_m) and are shifted back to the proper order (d_m) for use by the IDFT. (*Partsum* does not divide the **fft** results by N because it would just have to undo it when using the **ifft** command.)

```
        xlabel('t seconds')
        ylabel('fp(t)')
        grid
elseif nargout==1
        y=real(yy);
elseif nargout==2
        y=real(yy);t=tt;
end
```

The actual program portion of *Partsum* is quite small. By forcing the user to enter all four input parameters, allowing only a rectangular or Hamming window, and delivering only a graph as the output, the number of programming lines could easily be cut in half. Figure 7.13 shows the essential programming steps.

EXAMPLE 7.6

Find the Fourier partial sum for the sinusoidal pulse of Figure 7.14. Include out to the twentieth harmonic, and use an appropriate window.

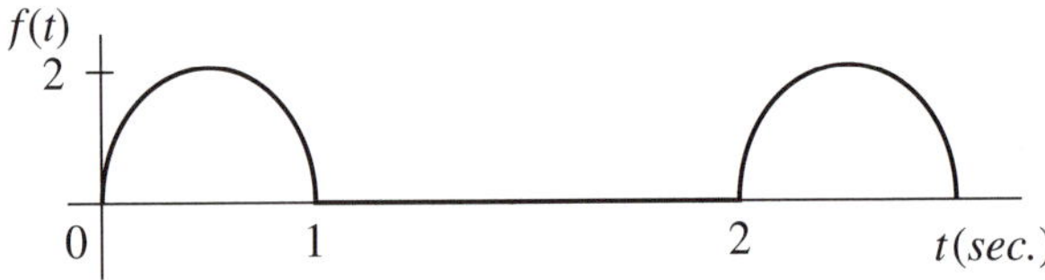

Figure 7.14

Solution

This solution assumes that *Partsum.m* has been put on a floppy and that the floppy has been added to the MATLAB search path (use the *cd a:* directive for Microsoft Windows systems). All that remains is to describe f over a full period in a 1024-point vector. Since f is continuous, a rectangular window should be used.

The pulse takes 1/3 of the period, or 1024/3 = 341.33 points. We need to generate half a sinewave in 341 points. Colon notation is the easiest way to keep track of the number of points used:

```
>f=[2*sin(pi*(0:340)/340) zeros(1,683)];   % defines f
>plot(f)                                   % make sure it is what you wanted
>size(f)                                   % make sure it is 1024 points
>Partsum(f,20,0,3)                         % call the function
>title('20 Harmonic Partial Sum with Rectangular Window')
```

The result is shown in Figure 7.15.

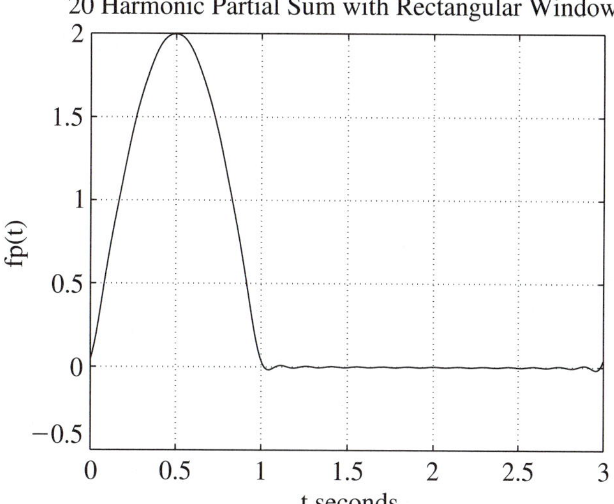

Figure 7.15 This shows the result of using the function *Partsum* to plot the 20 harmonic Fourier partial sum for the waveform of Example 7.6.

As in this example, decisions must usually be made about how best to approximate the time function with 1024 points. Minor variations are possible in the number of points used for a certain fraction of the cycle or whether the normalized colon runner should end at 1 or just short of 1. These one-data-point variations should produce negligible differences when using a 1024-point DFT.

Another useful application of the DFT is to plot the amplitude or power spectrum of a signal. For a Fourier series or the DFT, the spectrum is discrete and a stem plot is most appropriate. For a Fourier transform or a DTFT, the spectrum is continuous and a regular plot is best. Once again we will describe the time domain signal in a 1024-point vector.

File: Dispect.m

```
function [mm,c]=Dispect(x)
% Dispect stem plots the DIscrete SPECTrum of a periodic x
% over the frequency range containing 98% or more of the signal power
% x must be described in a 1024-element vector over one full cycle.
% Command format is [harmonic,spectral_amplitude]=Dispect(x)
cm=fft(x,1024)/1024;
ccm=fftshift(cm);
```

```
totpow=x*x'/1024;          % determine the total power in the waveform
P=(abs(cm(1)))^2; n=1;     % calculate the power in the d-c term
while P < 0.98*totpow
     n=n+1;       %determine the frequency range that includes 98% of P
     P = P + 2*(abs(cm(n)))^2;
end
nn=-n:n;
ddm=ccm(513+nn);

if nargout==0       % plot Fourier series amplitude spectrum
     stem(nn,abs(ddm))
     xlabel('m (harmonic number)')
     ylabel('spectrum amplitude')
     grid

else
c=ddm; mm=nn;
end
```

EXAMPLE 7.7

Plot the amplitude spectrum for the periodic waveform of Figure 7.16a.

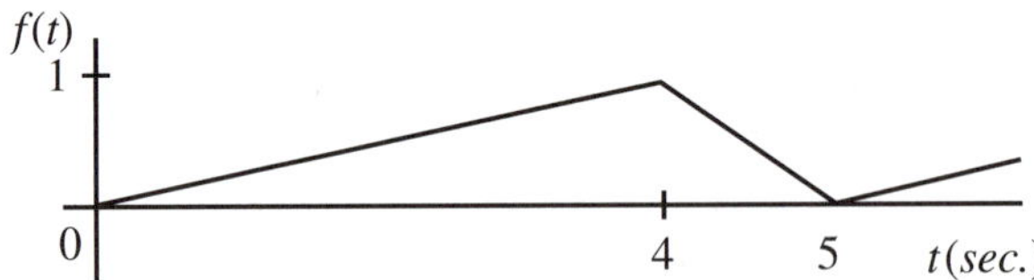

Figure 7.16a

Solution

Assuming *Dispect.m* is in the MATLAB search path, we only need to describe *f.* There are 1024 points/5 seconds = 204.8 samples/s.

```
>f=[(0:818)/819 (205:-1:1)/205];
>plot(f)       % check the description of f
>size(f)       % is it 1024 points?
>Dispect(f)
>title('Spectrum a Triangular Wave (4:1 rise/fall ratio)')
```

The result is Figure 7.16b.

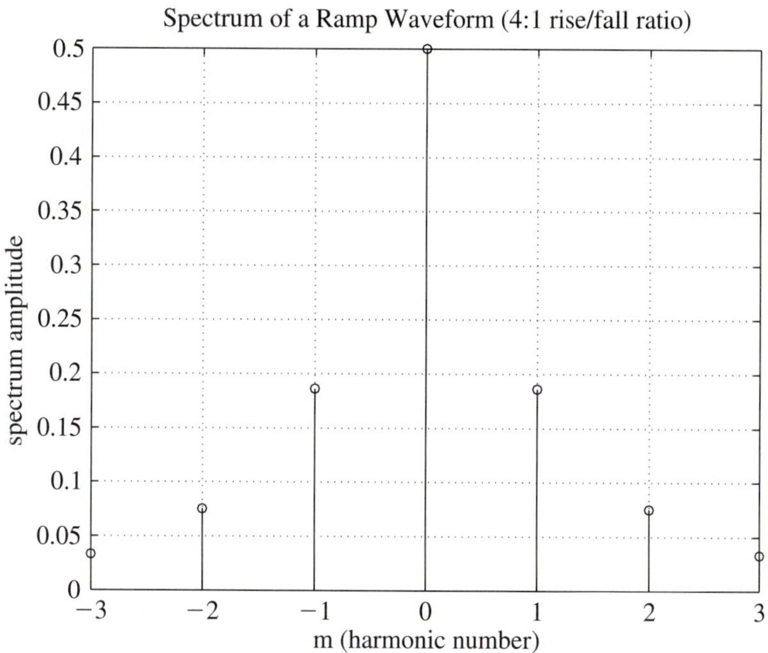

Figure 7.16b A graph of the spectrum produced by the function *Dispect* for the ramp of Example 7.7.

For a power spectrum plot, we can use *Dispect* to do most of the work.

File: Powspect.m

```
function []=Powspect(x)
% Powspect stem plots the POWer SPECTrum of a periodic signal x
% over the frequency range containing 98% or more of the signal power
% x must be described in a 1024-element vector over one full cycle.
% Command format is Powspect(x)
[m,c]=Dispect(x);             % get spectral amplitudes and harmonic numbers
P=x*x'/1024;                  % calculate total power in signal
c=100*abs(c).^2/P;            % calculate % of total power in each component
stem(m,c)
ylabel('% of total power')
xlabel('m harmonic number')
grid
```

EXAMPLE 7.8

Plot the power spectrum of the periodic signal of Figure 7.17a.

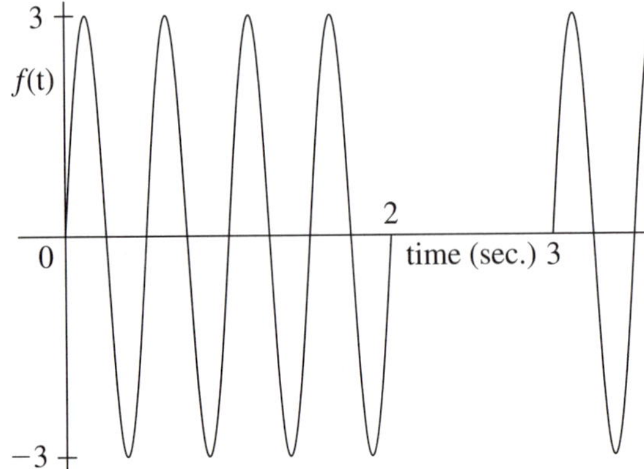

Figure 7.17a

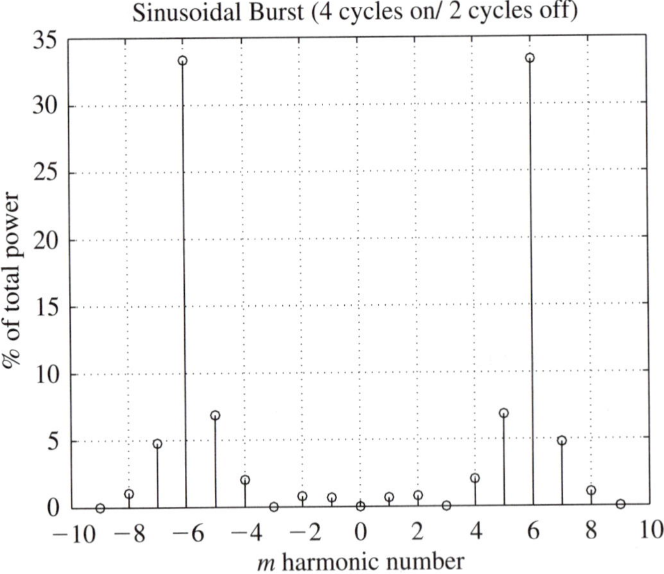

Figure 7.17b The power spectrum plot produced by the function *Powspect* for the waveform of Example 7.8.

Solution

There are 341.33 points per second, so the four-cycle sinusoidal burst must cover $4(2\pi)$ radians in 683 points. Remember that phase is unimportant in determining a waveform's power, so pick an origin that makes the waveform easiest to describe. Dispect.m must be in the search path when Powspect.m is compiled.

```
>f=[3*sin((0:682)*8*pi/683) zeros(1,341)];
>plot(f)
>size(f)
>Powspect(f)
>title('Sinusoidal Burst (4 cycles on/2 cycles off)')
```

The plot is shown in Figure 7.17b.

We can also plot the DTFT of a pulse. There are two new factors involved in this process. The spectrum of a DTFT is continuous, so we need to use the **plot** function instead of the **stem,** and we need to have enough spectral points to obtain a smooth curve. We will accomplish the latter by limiting the description of the pulse to 128 points and zero-filling to obtain 1024 samples. In addition, T_s must be known to calibrate the frequency axis. This was unnecessary with periodic functions, since frequencies could be replaced by harmonic number. Most of the workload is taken by *Dispect;DTFT* just adds the proper plotting routine and adjusts scaling factors.

File: DTFT.m

```
function [] = DTFT(x,fs)
% dtft plots the spectrum of a pulse x described in 128 samples
% taken at a sampling rate of fs.
% Command format is dtft(x,fs)
[m,y]=Dispect(x);
yy=1024*y; f=m*fs/1024;
plot(f,abs(yy))
xlabel('frequency (Hz)')
ylabel('spectral density')
grid
```

EXAMPLE 7.9

Use the DFT to estimate the spectrum of the discrete-time pulse $f(n) = e^{-n/5}\,u(n)$ shown in Figure 7.18.

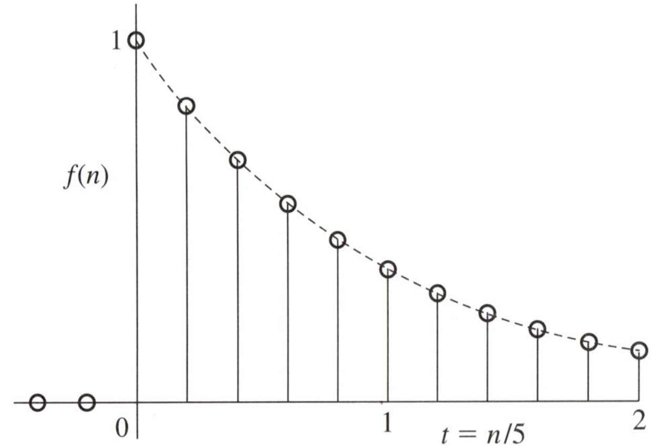

Figure 7.18

Solution

This pulse is a special case of Example 7.5 with $T_s = 1/5$. The exact spectrum was determined in that example to be given by

$$F(e^{j\omega T_s}) = \frac{1}{1 - e^{-T_s}e^{-j\omega T_s}} = \frac{1}{1 - e^{-0.2}e^{-j0.2\omega}}$$

The DFT should be able to approximate the spectrum of this pulse quite well. The 128*th* term will only be $e^{-128/5} = 7.62 \times 10^{-12}$ which should be close enough to zero to call it a pulse. We will use the program *DTFT* to obtain the graph, and then add a few calculated points of the exact spectrum, shown as circles, for comparison. The results are shown in Figure 7.19.

```
>n=0;127                    % need 128 samples
>f=exp(-n/5);               % describe the pulse
>dtft(f,5)                  % call dtft
% compare to the exact result
>m=-10:10;
>freq=2.5*m/10;                  % -fs/2 ≤ freq ≤ fs/2
>F=1./(1-exp(-.2)*exp(-j*2*pi*freq/5));
>hold
>plot(freq,abs(F),'o')
>title('f = exp(-n/5)')
```

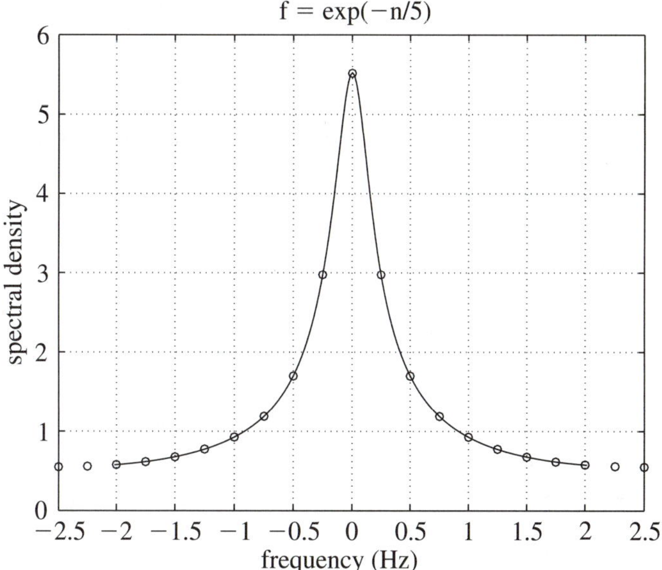

Figure 7.19 The spectral density plot produced by the function *DTFT* for the waveform of Example 7.9. The circles show exact values that were superimposed on the *DTFT* plot.

One obvious application of the DFT is to estimate the spectrum of an incoming signal. This is a fairly involved topic, because the DFT will calculate the spectrum of a fictitious signal that is periodic in the samples collected. This is not quite the same as the spectrum of the actual signal.

One technique used in spectral estimation is to impose a periodicity on the samples by windowing them. The same types of windows we have used in the past are used for this purpose, but the window is applied to the time samples instead of to the frequency components. The windows reduce the size of the samples, and the spectrum must be multiplied by a constant to correct for this. The correction factor is 2 for triangular and Hanning windows and 1.853 for the Hamming window. These factors can be determined by inverting the average value of a windowed signal of all 1's:

```
>f=ones(1,1024);
>g=hamming(1024);
>b=f*g/1024        % sums the weighted samples and divides by N
b = 0.5396
>factor = 1/b
factor = 1.8534
```

We can easily demonstrate the effect of windowing the time samples on the spectrum observed. We start by sampling a signal of known frequencies and see the results of using the DFT to obtain its spectrum. Things to consider are: (1) Are the amplitudes of the signals correct? (2) Do there appear to be signal components present that were not in the input signal? (3) Is equal treatment given to all sinusoids?

A simplistic and incomplete answer to these questions is that if the time samples are not windowed, the spectrum will show rippling. This rippling causes errors in the calculated spectrum amplitudes and may suggest that there are signal components where none are actually present. It may also obscure small spectral components. With a rectangular window, a sinewave that completes exactly an integer number of cycles in the data record will be shown as a single frequency, while any other sinewave will be shown with a frequency spread. This second sinewave will also have no calculation made exactly at its true frequency, so it would be difficult to estimate its actual amplitude even if rippling were not present. A combination of an appropriate window and padding of the data record with zero fill can help alleviate both of these problems and give a better estimate of the actual signal spectrum. The resolution available in a DFT is f_s/K, where K is the number of actual signal samples, not including the zero fill.

EXAMPLE 7.10

Graph and discuss the accuracy of the spectrum for the signal $f(t) = 2\sin[2\pi(2000/256)t] + 3\sin(2\pi 10t)$ if sampled at 100 Hz, using a 256-point DFT with both a rectangular and a Hamming window.

Solution

The sampling rate is high enough to satisfy the sampling theorem. The results are shown in Figure 7.20. The programming follows.

```
>n = 0:255; t = n/100;
>f=2*sin(2*pi*2000*t/256)+3*sin(2*pi*10*t);
>w=hamming(256); F=fft(f)/128;F(1)=F(1)/2;     % Do not double d-c component
>g=f.*w'; G=1.853*fft(g)/128; G(1)=G(1)/2;     % Correct for Hamming window
>freq=(0:255)*100/256;                         % Set up frequency scale
>stem(freq,abs(F)); axis([0 15 0 3]) grid
>plot(n,f); axis([0 255 -5 5]) etc.            % Obtain graphs. Use axis to
                                                 set scales
```

With a rectangular window, the signal at 7.8125 Hz causes no rippling, because it completes exactly 20 cycles in the 256 samples. Its amplitude is slightly low due to the rippling caused by the 10-Hz component. The 10-Hz-component amplitude is very low, because no calculation is made exactly at that frequency. Significant frequency components seem to be present at other frequencies.

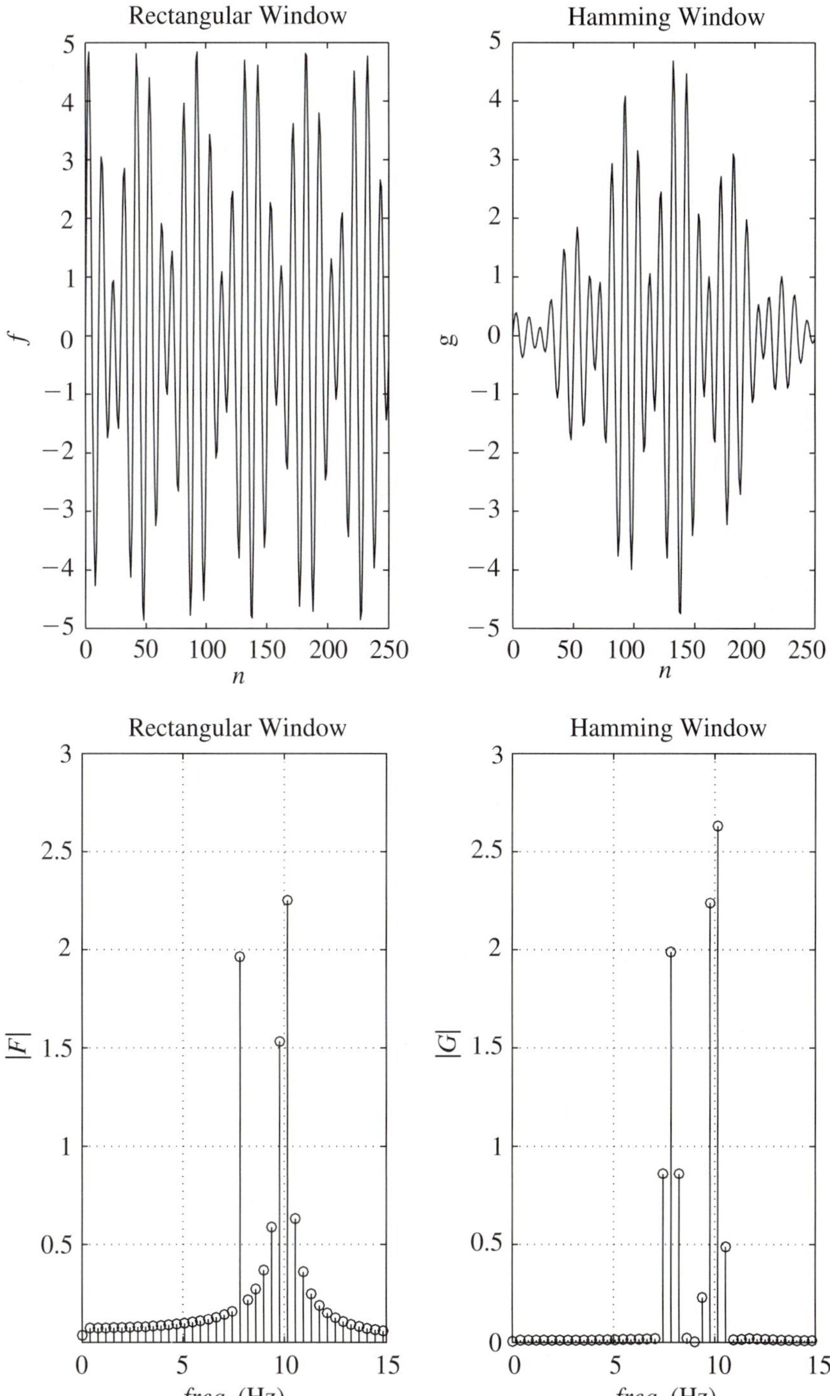

Figure 7.20 256 samples of a waveform consisting of a 7.8 Hz sinewave of amplitude 2 and a 10 Hz sinewave of amplitude 3 are collected to produce signal *f*. Signal *g* results from using a Hamming window on the samples of *f*. A comparison of their DFT spectra shows that the windowed signal gives a better estimate of the original signal's actual spectrum.

With the Hamming window, each sinusoid has a width to its spectral line, so all the sinusoids are given more equal treatment. The 7.8125-Hz signal amplitude is correct. The 10-Hz component is closer to the correct amplitude. Except for the width of the spectral lines, no other frequencies are apparent.

CHAPTER SUMMARY

Discrete-time signals arise through sampling continuous-time signals or are created with a computer program. For a set of samples to correctly represent the original signal, they must be taken at a sampling rate $f_s > 2f_m$, where f_m is the highest frequency present in the signal. If this condition is not met, false signals called aliases will appear in the baseband. A sampled signal has a frequency spectrum that is periodic in the interval f_s.

The spectrum of a periodic discrete-time signal is given by the N-point DFT. It may be derived by applying the ideal sampler to pick up N samples over one period of a signal represented by the Fourier series. The spectral density of aperiodic discrete-time signals is found from the DTFT. It may be derived by ideally sampling a signal represented by the Fourier transform.

Only the DFT is fully compatible with a computer, since both its signal and its spectrum are discrete and periodic and can be completely represented with N points. The DFT gives the exact spectrum for a periodic discrete-time signal of any number of points. The FFT is a superefficient algorithm for calculating the DFT, but it requires that N be a power of 2.

MATLAB programs have been developed in this chapter to use the DFT to:

1. Provide a Fourier series partial sum without requiring integrations or resolution of indeterminate values.
2. Graph the amplitude spectrum or the power spectrum of the exponentials in a periodic signal.
3. Graph the amplitude spectrum of an aperiodic, discrete-time pulse.

In addition, the general problem of spectrum analysis has been discussed and the use of a time domain window demonstrated.

PROBLEMS

Section 7.1

1. Find the four-point DFT and IDFT for the waveform of Figure P7.1.

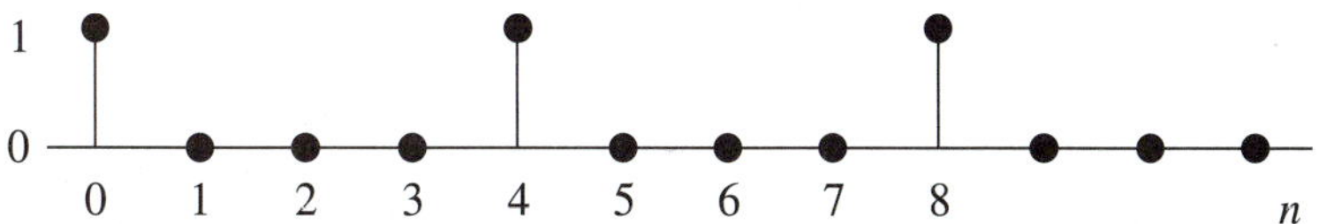

Figure P7.1

2. Given that N is even, show that the DFT component $c_{N/2}$ is always real.
3. Find the four-point DFT for the waveform of Figure P7.3.

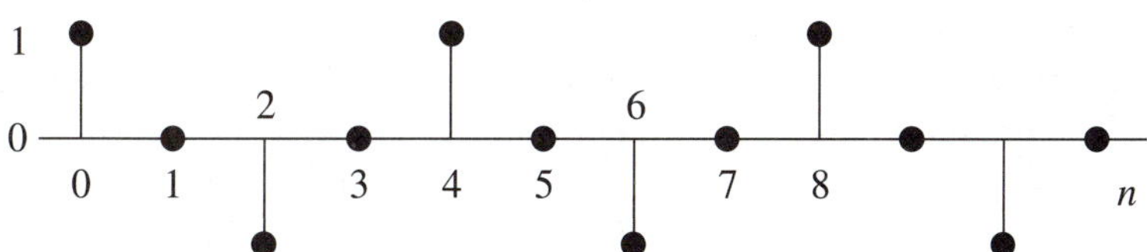

Figure P7.3

4. Determine the one-point DFT, two-point DFT, three-point DFT, and four-point DFT for the waveform of Figure P7.4.

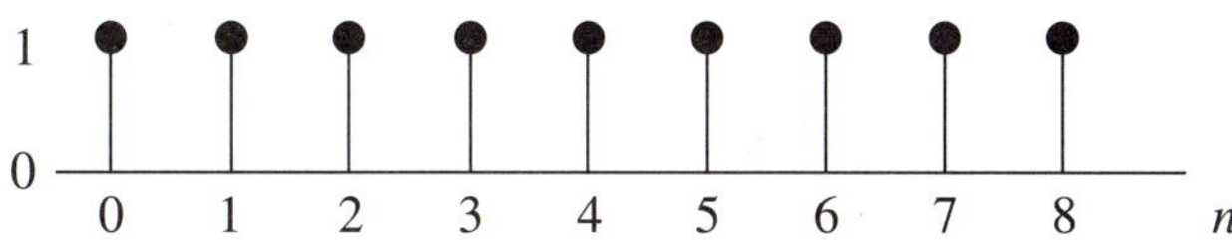

Figure P7.4

5. A 37-point DFT is calculated from samples collected every 40 μs. What frequency does c_{30} represent in the original waveform?
6. The sampling process shown in Figure P7.6 could result in any of the following sample sets:

$f_1(n) = [1\ 1\ 0\ 0]$ $\quad f_2(n) = [1\ 0\ 0\ 1]$ $\quad f_3(n) = [0\ 0\ 1\ 1]$ $\quad f_4(n) = [0\ 1\ 1\ 0]$

Find and compare the spectral amplitudes for any two sets of samples.

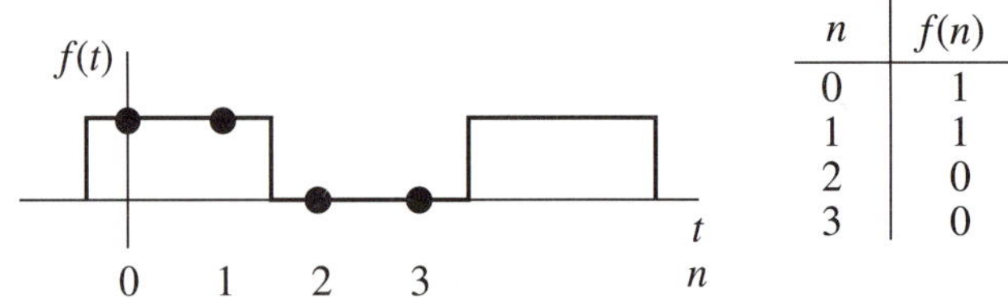

Figure P7.6

Section 7.2

7. The signal $x(t) = \cos(200t + 56°)$ is sampled every millisecond.
   a. Calculate the fifth sample of $x(t)$.
   b. Cite another signal with an angle of 56° that will produce the same set of samples, and calculate the value of its fifth sample.
   c. Cite a signal with an angle of −56° that will produce the same set of samples, and calculate the value of its fifth sample.
8. The signal $x(t) = \cos(440t + 21°)$ is sampled every 0.5 milliseconds.
   a. Calculate the 12th sample of $x(t)$.
   b. Indicate another signal with an angle of 21° that will produce the same set of samples, and calculate the value of its 12th sample.
   c. Indicate a signal with an angle of −21° that will produce the same set of samples, and calculate the value of its 12th sample.
9. The signal $x(t) = 2 \sin 2t + 3 \sin 8t + \sin 16t$ is sampled every 1.2 seconds. List the frequencies of the baseband sinusoids that result.
10. A sampling rate of 8 kHz is used on a recorded tone containing equal-amplitude sinusoids at frequencies of 500 Hz, 2 kHz, and 6 kHz. Determine the baseband frequencies, and discuss the possible amplitudes of these baseband signals.
11. A signal that takes up a frequency range of 8–12 kHz is sampled at 14 kHz. Sketch the resulting spectrum from 0 to 14 kHz. Explain how the signal could still be recovered despite being sampled at below the Nyquist rate.

Section 7.3

12. Find a closed-form expression for the DTFT of $x(n)$, assuming $T_s = 1$.

$$x(n) = \begin{cases} 1 & 0 \le n \le 9 \\ 0 & \text{elsewhere} \end{cases}$$

Sketch its magnitude over the range $0 \le f \le 0.5$.

13. Use the DTFT to find the spectrum of the following signal:

$$x(n) = \begin{cases} 1 & n = 0, 8 \\ 0 & \text{elsewhere} \end{cases}$$

Sketch it over a range $0 \le f \le 0.5$ (assume $T_s = 1$).

14. Use the DTFT to obtain expressions for the spectra of the two signals indicated, and sketch them over the range $0 \le f \le 0.5$ (assume $T_s = 1$). Which signal do you expect to have the larger high-frequency content?

a. $x(n) = \begin{cases} 1 & -2 \le n \le 2 \\ 0 & \text{elsewhere} \end{cases}$ b. $x(n) = \begin{cases} 1 & -1 \le n \le 1 \\ 0.5 & n = \pm 2 \\ 0 & \text{elsewhere} \end{cases}$

15. Find an expression for the spectrum of each of the following two signals, and sketch their magnitudes. Assume $T_s = 1$.

a. $x(n) = \begin{cases} (0.9)^n & n \ge 0 \\ 0 & \text{elsewhere} \end{cases}$ b. $x(n) = \begin{cases} (-0.9)^n & n \ge 0 \\ 0 & \text{elsewhere} \end{cases}$

16. The signal $x(n)$ has a total energy W of 1 joule.

$$x(n) = \begin{cases} Ka^n & n \ge 0 \\ 0 & \text{elsewhere} \end{cases} \qquad W = \sum_{n=0}^{\infty} x(n)^2$$

a. Determine the requirement on K to provide this amount of energy.
b. Obtain an equation for the spectrum of x at $f_s/2$. From it determine whether there is more high-frequency energy in a signal with $a = 0.2$ or in a signal with $a = 0.8$.

Section 7.4

17. For each of the waveforms of Figure P7.17, show how you would describe $f(t)$ to a MATLAB 1024-point DFT program.

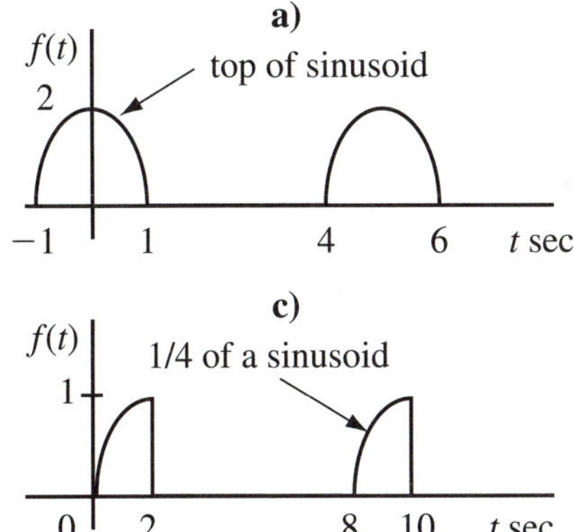

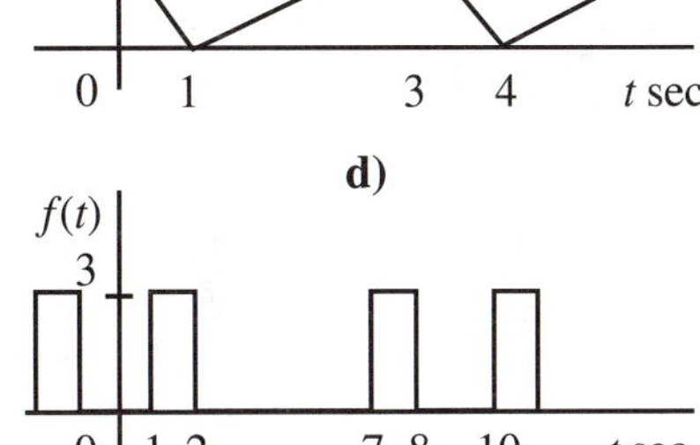

Figure P7.17

18. Prepare two graphs on a single sheet. The top graph is to be a 10-harmonic Fourier partial sum of the waveform in Figure P7.18 using a rectangular window. The bottom graph is to be a 10-harmonic partial sum using a Hamming window.

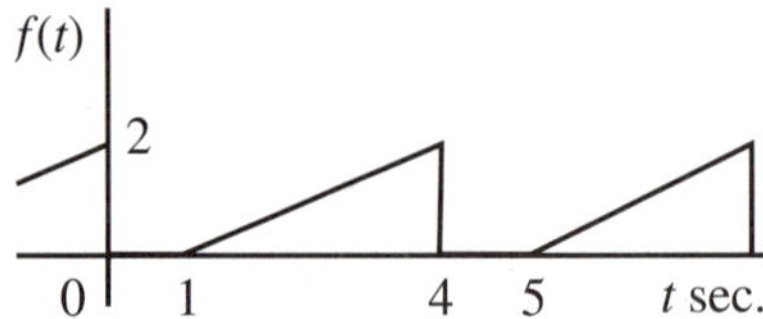

Figure P7.18

19. Prepare two graphs on a single sheet. The top graph is to be a 15-harmonic Fourier partial sum of the waveform in Figure P7.19 using a Hanning window. The bottom graph is to be the amplitude spectrum of the same waveform as windowed.

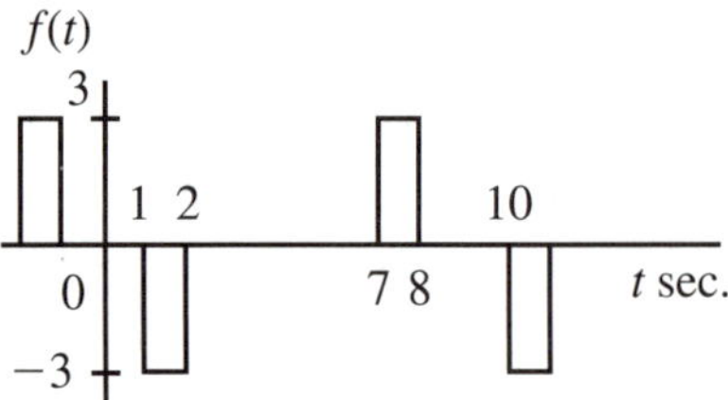

Figure P7.19

20. A type of signal used in network analyzers is called a *chirp*. An 8-second chirp is given by $x(t) = \sin[\omega(t)t]$, where $\omega(t) = 2\pi(1 + t)$ for $0 \leq t \leq 8$. Assume the chirp is repeated every 8 seconds. Obtain the power spectrum of this signal if it is sampled at 128 Hz.

21. For the sinusoidal burst of Figure P7.21, prepare two graphs on a single sheet. The first graph is to be a 20-harmonic partial sum using a window of your choice. The second graph is to be the power spectrum of the signal.

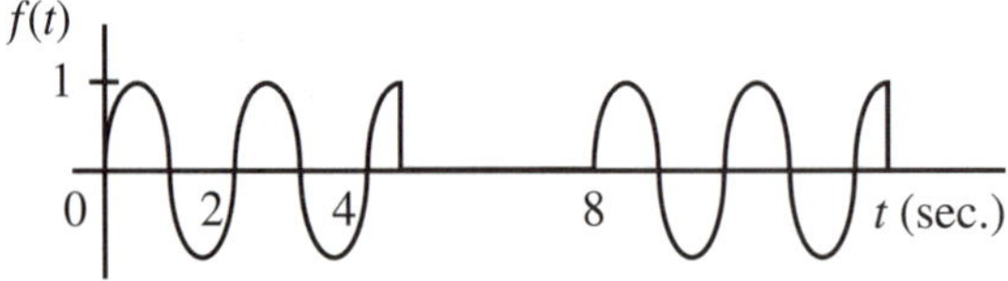

Figure P7.21

Advanced Problems

22. Modify the Powspect function to allow for an output listing of the percentage of total power by sinewave harmonic instead of only a graphical output. Demonstrate the first five terms in its listing for the waveform of Figure P7.17d.
23. Plot on the same graph the spectral density amplitude for $x(t) = \sin 0.4\pi t$ if sampled at 10 Hz and (a) x is a pulse on for one cycle, (b) x is a pulse on for two cycles. Based on the results, what would you predict the spectral density would look like if the pulse were on for three cycles?
24. The signal $x(t) = \sin 4\pi t + 2 \sin 8\pi t + 3 \sin 12\pi t$ is sampled at 9 Hz.
    a. Observe the spectrum obtained using a 1024-point DFT.
    b. Observe the spectrum obtained if 256 samples are collected and the rest are zero-filled to give a 1024-point DFT.
    c. Observe the spectrum obtained if 256 samples are Hanning windowed before the zero fill is added.

    Compare the various spectra for amplitude accuracy, frequency resolution, and ability to detect small signals.

8

DISCRETE-TIME SYSTEMS

OUTLINE

OBJECTIVES

1. Define the key signal for discrete-time systems.
2. Define a normalized frequency variable for discrete-time systems.
3. Describe a linear difference equation with constant coefficients.
4. Find the transfer function of a difference equation.
5. Determine the stability of a discrete-time system.
6. Identify basic MATLAB commands for discrete-time systems.

INTRODUCTION

Now that we know how to sample a signal properly, we want to begin real-time digital signal processing. Each input voltage sample will have its amplitude converted to a numerical value, and the number will be delivered to a computer. The computer will quickly calculate a corresponding output number and deliver it to a port, where it will be converted back to a voltage amplitude. This processing cycle repeats continuously. The basic hardware required for this process is discussed in Appendix A. Since we do not have hardware in this text, we will have to simulate the process by filtering files or MATLAB vectors containing the numerical input data.

The development of discrete-time systems will parallel our development of continuous-time systems. The program the computer runs is a linear difference equation. These are similar to differential equations and are easily solved for sampled exponential signals. Relationships between input and output discrete-time phasors will result in transfer functions. From the transfer functions we will be able to sketch the system's pole-zero plot or determine its frequency response to sampled sinusoids. The transfer function also will reveal the system's natural response and, therefore, tell us whether or not the system is stable.

8.1 THE *z* DOMAIN

In working with linear continuous-time systems, we chose to specialize in exponentially increasing or decreasing sinusoids and to invoke linearity to further simplify the signals to $v(t) = \vec{V}e^{st}$. This resulted in a generalization of frequency to $s = \sigma + j\omega$ and defined an s domain. Linear discrete-time systems will build on this foundation. After completing this section you will be able to:

- Define the key signal for discrete-time systems.
- Relate the s domain and the z domain.

The key signal for continuous-time systems is $v(t) = \vec{V}e^{st}$. If we look at this signal only at discrete values of time given by $t = nT_s$, where n is an integer and T_s is the sampling interval, the fundamental discrete-time signal would be

$$v(t) = v(nT_s) = \vec{V}e^{snT_s} = \vec{V}(e^{sT_s})^n \tag{8.1}$$

Some notation changes are appropriate to reflect the fact that we will be processing numbers rather than voltages or currents. There are applications in switched capacitor circuits where the signal is indeed carried along in the form of a voltage (or charge) on capacitors. In most applications, however, the signals of interest are simply a sequence of numbers whose original units are no longer of importance. The number sequence may be created by sampling a continuous-time voltage or current, but it also may be generated by a computer program or delivered by a compact disk or other storage device. We will consequently adopt a generic $\{x\}$ to denote the input number sequence, and use $\{y\}$ to represent the output sequence.

Since s is a complex constant, e^{sT_s} is just another complex constant. We may as well simplify our notation and define

$$z \equiv e^{sT_s} \tag{8.2}$$

which makes our key discrete-time signal

$$x(n) = \vec{X}z^n \tag{8.3}$$

This equation provides the transition between the time or *sample domain* and the frequency or *z domain.*

The notation changes have not affected the nature of the phasor $\vec{X}$. It still represents the initial ($t = 0$) amplitude and phase information of a sinusoidal signal whose amplitude may be changing exponentially. It could be the Euler phasor or any other phasor agreed to by defining a phasor transformation.

Consider the nature of the key discrete-time signal for several common situations:

1. For a d-c signal, $s = 0$ and $z = e^{sT_s} = 1$. The key signal is $x(n) = \vec{X}(1)^n = \vec{X}$, where the phasor is a real number.
2. For a simple exponential, $s = \sigma$, and $z = e^{\sigma T_s}$. The key signal is $x(n) = \vec{X}(e^{\sigma T_s})^n$. Again the phasor will be real. If $\sigma > 0$, then $z > 1$, and the signal grows exponentially with n. If $\sigma < 0$, then $z < 1$ and the signal decays exponentially with n.
3. The signal of greatest interest is still the constant-amplitude sinusoid, for which $s = j\omega$ and $z = e^{j\omega T_s} = 1\angle\omega T_s$. The key signal becomes $x(n) = \vec{X}(e^{j\omega T_s})^n$, which is interpreted through a phasor transformation.

EXAMPLE 8.1

The signal $x(t) = 12e^{-t}\cos(4t + 20°)$ is sampled at $t = n\pi/5$ seconds. Determine the Euler phasor, z, and the key signal $x(n)$.

Solution

The Euler phasor is defined as $\vec{X} = 6\angle 20°$. Since $s = -1 + j4$,

$$z = e^{sT_s} = e^{(-1+j4)\pi/5} = e^{-\pi/5}e^{j0.80\pi} = 0.5335e^{j0.80\pi}$$

Converting to proper units, 20° = 0.3491 radians, so

$$\vec{X} = 6\angle 20° = 6e^{j0.3491}$$

and

$$x(n) = 6e^{j0.3491}(e^{-\pi/5}e^{j0.80\pi})^n \qquad \text{or} \qquad x(n) = 6(0.5335)^n\, e^{j(0.80\pi n + 0.3491)}$$

The signal is a sequence of samples that will be interpreted the same way regardless of whether they were obtained using an appropriate sampling rate or not.

8.2 NORMALIZED FREQUENCY

In sampled systems, the signal frequency is only unique relative to the sampling frequency. Furthermore, system frequency response must be periodic, since sampled signal spectra are periodic in the range f_s. These facts suggest working with a normalized frequency variable so that the results are independent of the sampling frequency used. After completing this section you will be able to:

- Define a normalized frequency variable for discrete-time systems.
- Express a sinusoidal signal in terms of the normalized frequency.

The signal frequency, ω, and the sampling interval, T_s, always appear together as $e^{j\omega T_s}$ in the expressions for sampled waveforms. This also gives the sampled signal a frequency spectrum that is periodic in the interval $\omega T_s = 2\pi$ or $\omega = \omega_s$. These factors suggest shifting to a normalized frequency variable and using a standard range for the frequency domain information. Two normalized frequency variables are suggested, as indicated in Equation 8.4. (Also see Figure 8.1.) Both are used in the literature. We will prefer the v variable, which measures the frequency relative to the folding frequency of $f_s/2$.

$$\omega T_s = 2\pi f T_s = 2\pi\underbrace{(f/f_s)}_{\Omega} = \pi\underbrace{(2f/f_s)}_{v} \tag{8.4}$$

As with the DFT, the frequency range used for signal spectra may be taken as $0 \leq v < 2$, since it represents one full period of the spectrum, or it may be shifted to show the baseband region of $-1 \leq v < 1$. Usually the frequency response of a discrete-time system is given for the phasor of the positive exponent only, $z = e^{+j\omega T_s} = e^{j\pi v}$, just as was done for continuous-time systems. This phasor exists over the baseband range $0 \leq v < 1$.

Because we always work with baseband frequencies, there is a tendency to think v is confined to that range. Actually, v can take on any value.

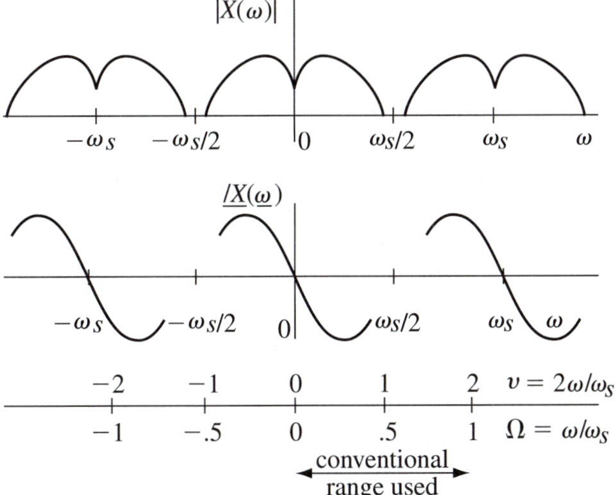

Figure 8.1 Spectrum of a sampled real $x(t)$ and the normalized frequency variables v and Ω found in the literature. The spectrum is conventionally given for $0 \leq v \leq 2$ or for half that range. The frequency response of discrete-time systems is given over the range $0 \leq v \leq 1$.

EXAMPLE 8.2

Find the values of z and v associated with the sampled sinewave

$$x(n) = 6 \sin(3n + 30°)$$

and recommend a phasor for the key signal.

Solution

A phasor transformation with the sine as the base function is

$$X_m e^{-\sigma T_s n} \sin(\omega T_s n + \theta) \quad \leftrightarrow \quad \vec{X} z^n = [X_m e^{j\theta}] z^n$$

The phasor is $6e^{j30°}$ using this transformation.

The sample variable comes from inserting $t = nT_s$, making

$$\omega t = \omega T_s n = 2\pi f T_s n = \pi(2f/f_s)n = \pi v n$$

By comparison, $\pi v = 3$. It is worth noting that the sampled sinusoid always takes the general form $\cos(\pi v n + \phi)$, making v obvious by inspection.

By definition,

$$z \equiv e^{sT_s} = e^{j\omega T_s} = e^{j\pi v} = e^{j3}$$

and the key signal is

$$x(n) = \tilde{X}e^{j\pi v n} = [6e^{j30°}]e^{j3n}.$$

8.3 THE DIFFERENCE EQUATION

In continuous-time systems, a circuit ultimately defined the time domain relationship between input and output through a differential equation. In discrete-time systems, a computer program defines that relationship directly through a difference equation. After completing this section you will be able to:

- Describe a linear difference equation with constant coefficients.
- Classify the difference equation according to its impulse response.
- Solve a linear difference equation by iteration.
- Provide an overview of the processing cycle.

Of the many possible prescriptions that might be conceived for creating the output sequence, only *linear* difference equations with *constant coefficients,* as defined by Equation 8.5, will be considered. This equation describes how the present output $y(n)$ is to be constructed from present and past inputs and past outputs. The functional notation $y(n - i)$ indicates the value of $y(n)$ that occurred i samples earlier. Such a system is both *causal and time-invariant.*

$$y(n) = \sum_{i=0}^{M} a_i x(n - i) - \sum_{i=1}^{N} b_i y(n - i) \tag{8.5}$$

Systems in which the b_i are all zero are called Mth-order *finite impulse response* (FIR) systems. They are inherently stable, and are especially interesting because they have no counterpart in the continuous-time world except for the trivial case of $M = 0$ (a resistive circuit).

If at least one of the b_i is nonzero, Equation 8.5 describes an Nth-order *infinite impulse response* (IIR) system. These are similar to continuous-time systems, and it will be found that many well established designs of continuous-time systems may be adapted for use in IIR filters. Such systems involve feedback, because past outputs influence the current output. Unfortunately, with feedback comes the potential for instability.

Difference equations have one significant advantage over differential equations. Any difference equation is solvable by iteration. The process simply involves creating a table, putting in the starting conditions, and filling in the successive lines in the table from the known input signal $x(n)$.

EXAMPLE 8.3

Find the first five output values for the difference equation subjected to a discrete unit step input:

$$y(n) = x(n) - x(n-2) + \frac{1}{2}y(n-1) \qquad \text{if } x(n) = \begin{cases} 0 & n < 0 \\ 1 & n \geq 0 \end{cases}$$

Solution

Since the input is zero prior to $n = 0$, we insist that the output $y(n) = 0$ for $n < 0$. This is the property of *causality.* The iteration table is shown below. It is helpful to summarize the effects of the input terms in an $f(n)$ column. In this case, only two columns remain to be combined to determine each $y(n)$. The values in the first row should be self-evident. Values in succeeding rows depend only on the input signal and values found in previous rows of the table.

n	−x(n − 2) +	0x(n − 1) +	x(n) =	f(n)	+ 0.5y(n − 1) =	y(n)
0	0	0	1	1	0	1
1	0	0	1	1	+0.5(1)	+3/2
2	−1	0	1	0	+0.5(3/2)	+3/4
3	−1	0	1	0	+0.5(3/4)	+3/8
4	−1	0	1	0	+0.5(3/8)	+3/16

Notice, in this case, that the solution might be expressed as $y(n) = 0$ if $n < 0$, $y(0) = 1$, and $y(n) = 3(1.2)^n$ if $n \geqslant 1$.

Although the iteration procedure is fundamental and straightforward, it does not provide much insight into the system characteristics. In addition, if the iterative procedure is done by hand calculation, an error in any row will propagate through all subsequent rows. The iterative approach also does not generally lead to a closed-form solution for $y(n)$, but it is sometimes possible to deduce a closed-form solution from the results. This was the case in Example 8.3 for $n \geq 1$. Finally, most difference equations involve many more terms than our example, making the iteration procedure impractical. Since the difference equation is a program that a processor will run, you can write the program and let the processor do the iterations.

It may be helpful to consider how the processing takes place and what attributes are needed by the processor. In real-time digital signal processing, the processor will begin a sampling cycle by directing that the input signal be sampled. The sample will be presented to an analog-to-digital (A/D) converter, which turns the signal amplitude into a binary number. The A/D unit will deliver its latest conversion, $x(n)$, directly to an input port or to an internal processor register, and it will raise a flag to signal that the new sample is ready. Sometimes an interrupt is initiated by the flag. The processor will collect the latest input sample and begin calculating the output specified by the difference equation.

The processor will need to have placed its most recent inputs and outputs in storage and have a fast and easy way to recover them. The linear difference equation requires that the latest input, along with past inputs and outputs, be multiplied by fixed constants and summed to calculate the next output, $y(n)$. This *multiply and accumulate* operation is fundamental to digital signal processing, and the processing chips are optimized to accomplish it efficiently. Once the latest output has been calculated, it is sent to its next destination. This often is an output port for digital-to-analog (D/A) conversion back to a continuous-time signal. The processor waits for the sampling interval to expire and then calls for the next sample to be collected. This process repeats forever.

In some applications real-time processing is not necessary. Photos from Mars, for instance, may be processed enough to show that the camera system is working properly, but fine details and precise coloring can await later processing if the signal is stored on tape. Any input number sequence may be stored in memory and processed like any other file. A general-purpose computer can be used instead of costly high-speed processors. Since this also does not require special sampling hardware, we will eventually use this technique to demonstrate some digital filtering.

8.4 THE TRANSFER FUNCTION

The linear difference equation is easily solved for the key signal $x(n) = \tilde{X}z^n$. The result is a transfer function relating the output phasor to the input phasor as a function of z. After completing this section you will be able to:

- Find the transfer function of a difference equation.
- State the delay theorem.
- Find the frequency response of a discrete-time system.

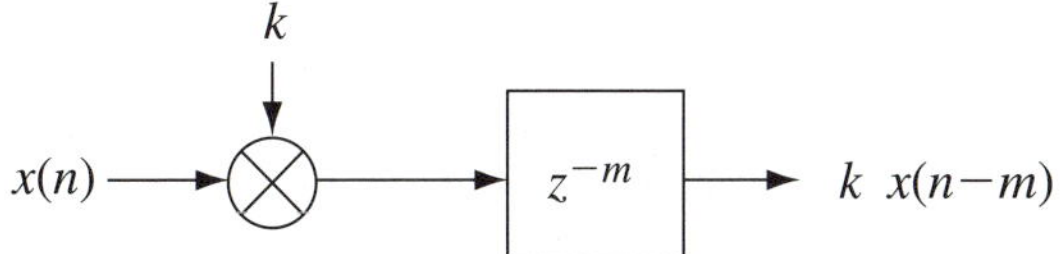

Figure 8.2 The delay theorem may be characterized in a simple block diagram. Combining a delay with multiplication by a constant, discrete-time signals are modified in magnitude and phase, exactly as impedances modified continuous-time signals.

In linear continuous-time systems, a voltage $v(t) = \vec{V}e^{st}$ was applied to each circuit element, and a current $i(t) = \vec{I}e^{st}$ resulted. This led to the definition of *impedance* and turned differential equations into algebraic equations. The equivalent concept needed for linear difference equations is the delay theorem. For our key signal, if $x(n) = \vec{X}z^n$, then $x(n - m) = \vec{X}z^{(n-m)} = z^{-m}\vec{X}z^n = z^{-m}x(n)$.

Delay Theorem

$$x(n - m) = z^{-m}x(n) \tag{8.6}$$

A block diagram representation of the delay theorem, incorporating an amplitude change, is shown in Figure 8.2. The multiplication by k may occur before or after the delay.

We hypothesize that a input signal of the form $x(n) = \vec{X}z^n$ will result in a forced response of the form $y(n) = \vec{Y}z^n$ in a linear system. Then for the class of signals having this form, a transfer function may *always* be found by starting with the difference equation

$$y(n) = \sum_{i=0}^{M} a_i x(n - i) - \sum_{i=1}^{N} b_i y(n - i)$$

substituting $x(n) = \vec{X}z^n$ and $y(n) = \vec{Y}z^n$ and applying the delay theorem

$$\vec{Y}z^n = \sum_{i=0}^{M} a_i z^{-i}\vec{X}z^n - \sum_{i=1}^{N} b_i z^{-i}\vec{Y}z^n$$

The z^n terms cancel, leaving

$$\vec{Y}\left(1 + \sum_{i=1}^{N} b_i z^{-i}\right) = \sum_{i=0}^{M} a_i z^{-i}\vec{X}$$

Finally, we see that the output and input phasors are related by a ratio of polynomials in z:

$$\frac{\vec{Y}}{\vec{X}} = \frac{\sum_{i=0}^{M} a_i z^{-i}}{1 + \sum_{i=1}^{N} b_i z^{-i}} \tag{8.7}$$

To find the frequency response of the discrete-time system, we simply substitute $z = e^{j\pi v}$.

 **EXAMPLE 8.4**

A signal $x(n) = 12\cos(\pi n/4 + 30°)$ is input to a processor running the program

$$y(n) = x(n) - x(n-2) + \frac{1}{2}y(n-1)$$

Find the output sinusoid.

Solution

Making the standard substitutions, we have

$$\vec{Y}z^n = \vec{X}z^n - z^{-2}\vec{X}z^n + \frac{1}{2}z^{-1}\vec{Y}z^n$$

Canceling z^n terms and regrouping gives

$$\vec{Y}\left(1 - \frac{1}{2}z^{-1}\right) = \vec{X}(1 - z^{-2})$$

For the input signal given, we define a phasor transformation

$$12\cos\left(\frac{n\pi}{4} + 30°\right) \leftrightarrow 12\angle 30° \qquad \left(v = \frac{1}{4}\right)$$

and substitute for z;

$$z = e^{j\pi/4} = 1\angle 45^\circ = \frac{1}{\sqrt{2}}(1 + j1)$$

The result is

$$\vec{Y}\left(1 - \frac{1}{2}e^{-j\pi/4}\right) = (12\angle 30^\circ)(1 - e^{-j\pi/2})$$

$$\vec{Y}\left(1 - \frac{1}{2\sqrt{2}} + j\frac{1}{2\sqrt{2}}\right) = (12\angle 30^\circ)(1 + j)$$

which we continue to systematically reduce to a polar-form number

$$\vec{Y} = \frac{(12\angle 30^\circ)(\sqrt{2}\angle 45^\circ)}{(0.6465 + j0.3536)} = \frac{16.97\angle 75^\circ}{0.7369\angle 28.68^\circ} = 23.03\angle 46.32^\circ$$

Finally, we reverse the phasor transformation to obtain

$$y(n) = 23.03\cos\left(\frac{\pi n}{4} + 46.32^\circ\right)$$

Phasor calculations in the z domain are the same as those in the s domain, but they involve an extra layer of calculations to replace the exponential terms with Euler's identity.

Example 8.4 is equivalent to the type of problem initially faced in an a-c circuits course; it helps reinforce the phasor concept. By now, however, we recognize that the system's transfer function gives all the important information and does not require us to state a phasor transformation or even specify details about the input signal.

The transfer function takes the form of a ratio of polynomials in z, and factoring the polynomials will identify the poles and zeros of the system in the z plane. Furthermore, by letting $z = e^{j\pi v} = 1\angle \pi v$ we obtain the frequency response of the system. Notice that as v varies from 0 to 2, it takes us once around the unit circle of the z plane and through the landscape created by the poles and zeros. As v continues to increase, it just keeps making additional trips through the same landscape, showing again that the frequency response of the discrete-time system is periodic in the range $0 \le v \le 2$. Since the frequency response is unique only in the range $0 \le v \le 1$, it is particularly easy to have a computer make the calculations.

EXAMPLE 8.5

Find the frequency response of the difference equation

$$y(n) = x(n) - x(n-2) + \frac{1}{2}y(n-1)$$

Solution

To find the frequency response, we need the transfer function form of the difference equation. Substituting $x(n) = \vec{X}z^n$, $y(n) = \vec{Y}z^n$, and using the delay theorem gives

$$y(n) - \frac{1}{2}y(n-1) = x(n) - x(n-2)$$

$$\vec{Y}\left(1 - \frac{1}{2}z^{-1}\right) = \vec{X}(1 - z^{-2})$$

$$H(z) = \frac{\vec{Y}}{\vec{X}}(z) = \frac{(1 - z^{-2})}{\left(1 - \frac{1}{2}z^{-1}\right)} = \frac{z^2 - 1}{z(z - 0.5)}$$

The pole-zero diagram for the system is shown in Figure 8.3a. It shows that the frequency response will start and end on zeros. The pole at the origin affects only the phase of the response, since it is equidistant from all points on the unit circle. The pole at $z = 0.5$ should produce a peaking in the frequency response somewhere between $v = 0$ and $v = 0.5$.

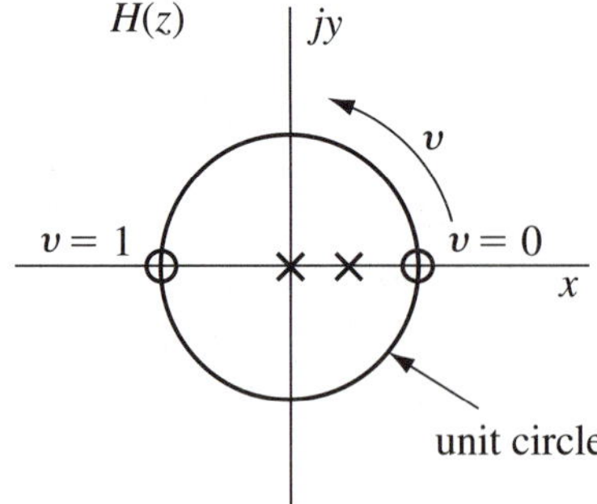

Figure 8.3a The pole-zero diagram for the system of Example 8.5.

If we want an expression for the frequency response in terms of v, we would try to simplify:

$$H(e^{j\pi v}) = \frac{e^{j2\pi v} - 1}{e^{j\pi v}(e^{j\pi v} - 0.5)}$$

No special rules have been developed for sketching these types of functions, but we can always calculate the frequency response. We only need to provide the $H(z)$ polynomials. To a computer, calculating the frequency response is not much different in the s or z domain (Figure 8.3b).

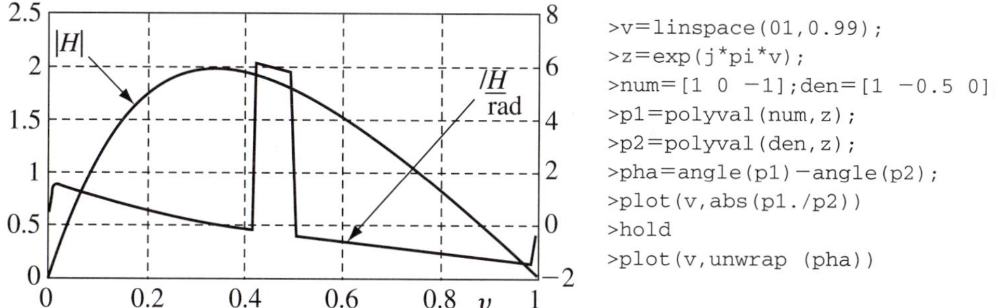

Figure 8.3b The frequency response is shown for the system of Example 8.5. The programming provided will produce an unwrapped phase version of this response.

If we have a transfer function to implement or a difference equation whose frequency response is in question, the relationship between them is always the same; it is summarized here for convenience.

Theorem

The Difference Equation/Transfer Function Relationship

The *n* domain | **The *z* domain**

$$y(n) + \sum_{i=1}^{N} b_i\, y(n-i) = \sum_{i=0}^{M} a_i\, x(n-i) \Leftrightarrow \frac{\vec{Y}}{\vec{X}} = \frac{\sum_{i=0}^{M} a_i\, z^{-i}}{1 + \sum_{i=1}^{N} b_i\, z^{-i}}$$

As with a-c circuits, it is considered poor practice to mix domains. The n and z variables should never appear together except in the definition that relates the two domains, $x(n) = \vec{X}z^n$, or temporarily as the algebraic transition is being made from one domain to the other.

8.5 STABILITY IN THE z DOMAIN

The frequency response of a difference equation is meaningless if it has a natural response that is unstable. The numerical values generated by an unstable difference equation increase steadily until they exceed the capacity of the accumulator and registers. Any forcing function present is simply overwhelmed by the runaway natural response. After completing this section you will be able to:

- Determine the stability of a discrete-time system.
- Describe the natural response of a linear difference equation.

There are basically two ways to check the stability of a program. In the time domain, the simplest test is to apply a discrete unit impulse to the difference equation and see if the output dies out. The discrete unit impulse was defined as

$$\delta(n) = \begin{cases} 1 & n = 0 \\ 0 & n \neq 0 \end{cases} \tag{7.16}$$

and since it forces the equation only when $n = 0$, it activates the system's natural response. During this test, it is unnecessary to retain delayed input terms in the difference equation, since only the feedback of the output terms affects stability.

EXAMPLE 8.6

Determine if the following difference equation is stable:

$$y(n) = x(n) - 2x(n - 2) - y(n - 1) + \frac{3}{4}y(n - 2)$$

Solution

The iteration table is set up with only the $x(n)$ input term. The input terms are irrelevant in establishing stability. The test signal is $x(n) = \delta(n)$. It is necessary to continue the iterations either until a mathematical expression for the output is obtained or until some pattern is established that can be shown to continue indefinitely. In complicated difference equations, this may require an extensive iteration table.

n	x(n)	$+\frac{3}{4}y(n-2)$	$-y(n-1)$	=	y(n)
0	1	3/4(0)	0		1
1	0	3/4(0)	−(1)		−1
2	0	3/4(1)	−(−1)		7/4
3	0	3/4(−1)	−(7/4)		−10/4
4	0	3/4(7/4)	−(−10/4)		61/16

In this case we see that for $n > 1$, the $y(n-2)$ and $y(n-1)$ terms are always the same sign and that the $y(n-1)$ term is always as large as the last output, so the equation is unstable. It has an oscillatory and ever-increasing natural response.

Even though we have used the simplest possible input and eliminated any delayed input terms, using an iteration table is still very prone to numerical and interpretation errors. A better approach is available in the z domain. The denominator of a transfer function is the *characteristic equation* of the system, and its poles represent the natural z values of the system. We will designate these poles as z_p. Turning on any input creates infinitesimal amounts of signal at all z values. Those that receive infinite gain show up as the system's natural response. (If multiple poles occur, terms like $n^k(z_p)^n$ may also be generated.)

If the natural response terms take the form $y_n(n) = \vec{Y}z_p^n$, then the natural response terms will grow or decay with sample number according to the *magnitude* of the pole value,

$$|y_n(n)| \propto |z_p|^n$$

Reinforcing this concept is the fact that the locations of the z plane poles are related to s plane locations through the defining relationship

$$z = e^{sT_s} = e^{(\sigma+jw)T_s} = e^{\sigma T_s}e^{j\omega T_s} = e^{\sigma T_s}\angle \pi v = |z_p|\angle \pi v \tag{8.8}$$

from which it is seen that $|z| > 1$ corresponds to $\sigma > 0$ the (RHP). Allowed pole locations for stable systems are shown in Figure 8.4.

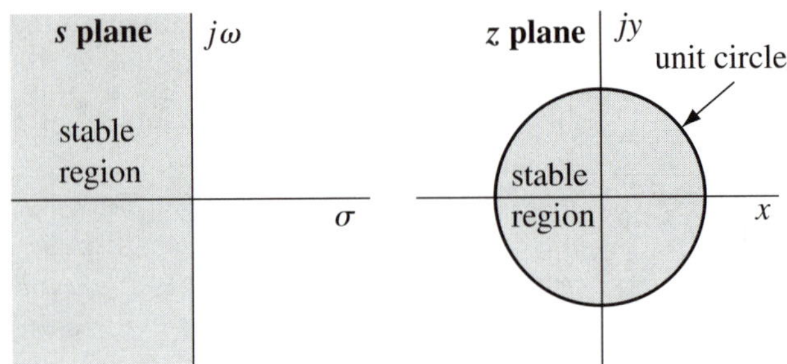

Figure 8.4 For stability, roots of the characteristic equations for continuous-time systems and discrete-time systems must lie in the shaded regions of the s and z planes, respectively.

To summarize:

If $|z_p| > 1$: The natural response will grow without bound. Such a system is *unstable* and it will not be useful.

If $|z_p| = 1$: The natural response neither grows nor decays and is classified as *conditionally stable.* This condition corresponds to $\sigma = 0$ in the s plane. (If such roots are repeated, however, the response will grow without bound.)

If $|z_p| < 1$: The natural response is a transient response, and the system is *stable.* This condition corresponds to $\sigma < 0$ in the s plane.

It is perhaps obvious, but if the difference equation does not include any delayed output terms, the characteristic equation reduces to unity. If the transfer function is expressed in terms of polynomials in positive powers of z, all the poles are at $z = 0$, and there is no possibility of an unstable system. This is characteristic of FIR systems.

EXAMPLE 8.7

Determine the natural response of a system whose difference equation is

$$y(n) = x(n) - 2x(n - 2) - y(n - 1) + \frac{3}{4}y(n - 2)$$

Solution

We obtain the transfer function from the difference equation in the usual way.

$$y(n) = x(n) - 2x(n-2) - y(n-1) + \frac{3}{4}y(n-2)$$

$$\vec{Y}\left(1 + z^{-1} - \frac{3}{4}z^{-2}\right) = \vec{X}\left(1 - 2z^{-2}\right)$$

$$\frac{\vec{Y}}{\vec{X}} = \frac{1 - 2z^{-2}}{1 + z^{-1} - \frac{3}{4}z^{-2}} = \frac{z^2 - 2}{z^2 + z - \frac{3}{4}} = \frac{z^2 - 2}{\left(z - \frac{1}{2}\right)\left(z + \frac{3}{2}\right)}$$

The poles indicate that the natural response will be

$$y_n(n) = K_1\left(\frac{1}{2}\right)^n + K_2\left(\frac{-3}{2}\right)^n$$

where K_1 and K_2 are arbitrary constants. The pole at $z = -3/2$ represents a growing oscillation that makes the system unstable.

8.6 MATLAB LESSON 8

Many of the function calls provided for continuous-time systems have their counterparts in discrete-time systems. Of course, all of the polynomial-related commands are still usable. After completing this section you will be able to:

- Identify basic MATLAB commands for discrete-time systems.
- Test a discrete-time system with step or impulse inputs.
- Find the frequency response of a discrete-time system.
- Find the output sample sequence, given the system and its input sequence.

Most commands for discrete-time systems expect the system polynomials to be in negative powers of z, such as $1 - z^{-1} + 2z^{-2} + 3z^{-3} + \cdots$, where the leftmost entry must be the z^0 term. If the system numerator and denominator polynomials are of the same degree, it does not matter whether you interpret the polynomials as positive powers of z or negative powers of z. If you seem to be getting strange results, use the help facility to see which way the polynomials are expected. If it still is not clear, add a leading or trailing zero to the polynomial vector. If the leading zero does not change the result, positive powers of z are expected; if a trailing zero does not change the result, negative powers of z are expected.

MATLAB EXAMPLES

dimpulse dstep

The simplest way to test the stability of a discrete-time system in the time domain is to use **dimpulse** to generate the iteration table. The discrete step function is another standard test signal. Positive powers of z are expected.

```
>num=[1 0 -1];            % from Example 8.5, num = z^2 - 1 ?or?
                            (1 - z^-2)
>den=[1 -.5];             % den = z - 5 ?or? 1 - .5z^-1
>dimpulse(num,den)        % system is noncausal as specified
>den=[1 -.5 0];           % den = z^2 - .5z + 0 ?or? 1 - .5z^-1
                            (no change)
>y=dimpulse(num,den);     % must need positive powers of z!
```

When MATLAB solves a difference equation, it delays the application of the input by a number of samples equal to the difference in the degrees of the denominator and numerator polynomials. Usually, the point of application of the impulse or step is obvious from the output results.

```
>de=[1 -.5 0 0 0 0 1];
>y=dimpulse(num,de,8)     % ask for the first eight outputs

>tt=0:5;
>y=dstep(num,den,tt);
>stem(tt,y)               % compares to the Example 8.3 iteration
                            table

>den2=[1 1 -3/4];         % finding poles of the transfer function
                            from Example 8.7
>r=roots(den2);           % roots expects positive powers of z, as
                            it always has
>abs(r)                   % any pole with a magnitude over 1
                            indicates instability
```

freqz

freqz expects polynomials in negative powers of z. Since it evaluates $H(z)$ along the unit circle ($z = e^{j\pi v}$), interpreting the polynomials incorrectly amounts to multiplying H by some power of z, which affects the phase but not the magnitude of the result.

```
>freqz(num,den)           % plots frequency response curves in the z
                            domain

>h=freqz(num,den);
>v=(0:511)/512;           % the default freqz uses a 512-point FFT
>plot(v,abs(h))           % allows other plotting options
>plot(v,180*angle(h)/pi)

>w=linspace(0,pi);        % calculation frequencies given in the
                            normalized Ω variable
```

```
>h=freqz(num,den,w);
>plot(w,abs(h))
>help freqz                      % other options
```

filter

The **filter** command performs the difference equation on a specified input number sequence. The output is the (file) filtered input. The **dimpulse** and **dstep** functions are often-used special cases. Polynomials in negative powers of z are expected.

```
>x=[0 ones(1,10) zeros(1,30)];  % prepare a 10-sample-long square input
                                  pulse
>num=1;                          % make up a simple H(z)
>den=[1 -.2 .2];
>freqz(num,den)                  % see what the frequency response looks
                                  like
>y=filter(num,den,x);
>stem(y)                         % see what the output looks like in the n
                                  domain
>grid
```

EXAMPLE 8.8

Plot the output $y(n)$ given that $x(n) = \sin(0.05\pi n)$ is passed through a filter with a transfer function of

$$H(z) = \frac{z^2 - 1}{z(z - 0.5)}$$

Solution

Using 100 samples will give 2.5 cycles of the input, which should be enough to show steady-state conditions. The transfer function is the same one plotted in Figure 8.3b.

```
>n=0:99;                   % prepare a 100-sample input signal
>x=sin(.05*pi*n);          % it will be a sinewave with v = 0.05
>num=[1 0 -1];             % define H(z)
>den=[1 -.5 0];            % H ≈ 0.6 /+1.3 rad @ v = 0.05 (Fig. 8.3b)
y=filter(num,den,x);
>plot(n,y);
>grid
>hold
>plot(n,x,'--')
```

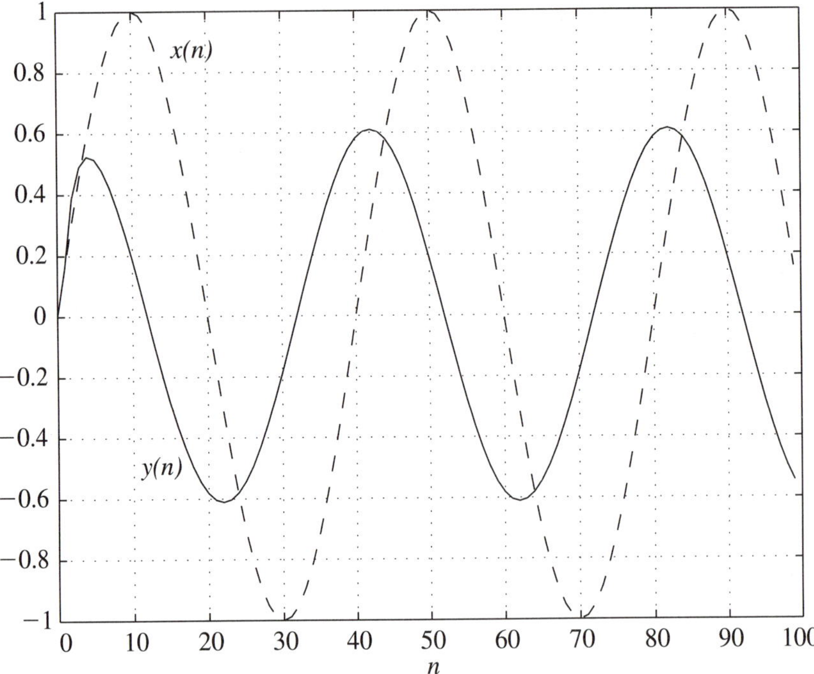

Figure 8.5 Input and output samples for Example 8.8. A start-up transient is evident in the first 10 samples.

The result (Figure 8.5) shows that the output is about 0.6 times the amplitude of the input and that it leads the input by about eight samples or $0.05\pi(8) = 1.26$ radians.

CHAPTER SUMMARY

The behavior of a discrete-time system is described in the form of a linear difference equation. This equation is also the program a digital processor will execute. It describes how the current output is calculated from current and past inputs and possibly past outputs. If past outputs are included, the system has an infinite impulse response similar to continuous-time systems but may be unstable. If no past outputs are included, the system has a finite impulse response and is always stable. In principle, the output of a linear difference equation can be determined from the input using an iteration table.

The exponential signal takes the form z^n in discrete-time systems. Assuming signals of this form and applying the delay theorem, we are able to convert a linear difference equation into a transfer function in the z domain. We can represent the system with a pole-zero diagram, identify the poles with the system's natural response, and observe that any pole outside the unit circle represents an unstable natural response.

Evaluating the transfer function along the unit circle gives the frequency response of the system. If a system is found that has a desirable frequency response, we can transform back to the sample domain to find the difference equation that gives that response.

The development of discrete-time systems has closely paralleled that of continuous-time systems. In most respects, discrete-time systems are simpler, but phasor and frequency response calculations for them are more involved. In the absence of simple frequency response estimation techniques, we are even more reliant on computer tools for evaluating discrete-time systems. The computer, of course, is the main reason we are interested in discrete-time systems anyway.

PROBLEMS

Section 8.1

1. Express each of the following signals by their phasor representation $\vec{X}z^n$, where $\vec{X}$ is the Euler phasor. Identify z in each case.

$$Ae^{\sigma T_s n} \cos(\omega T_s n + \theta) \leftrightarrow \vec{X} z^n$$

a. $2e^{0.1n}$ b. $4 \cos(\pi n - 45°)$
c. $12e^{-n} \cos(0.2n + 14°)$ d. $6(1.2)^n \sin(3n - 25°)$

2. Find the signal whose phasor representation is given. Assume the phasor transformation is:

$$Ae^{\sigma T_s n} \cos(\omega T_s n + \theta) \leftrightarrow [Ae^{j\theta}]z^n$$

a. $12(-j)^n$ b. $2(0.2)^n$
c. $6\angle 30°(1 + j)^n$ d. $-j(2\angle 30°)^n$

Section 8.2

3. A sampled signal is being nulled out by a difference equation implementing the equivalent of a notch filter. The signal is at a frequency of 2.5 kHz and is being sampled at 12 kHz. What signal frequency would be nulled if the sampling rate were reduced to 9 kHz.?

4. A digital filter is designed for cutoff at $v = 0.15$. What is the frequency of the signal where cutoff occurs if the sampling rate is 8 kHz?

Section 8.3

5. Use iteration to find the first five terms of the difference equation if $x(n)$ is the unit step function.

a. $y(n) - \frac{1}{2}y(n-1) = x(n) - \frac{1}{2}x(n-1)$

b. $y(n) + \frac{1}{2}y(n-2) = x(n) - \frac{1}{2}x(n-1)$

6. Use an iteration table to find the first five output samples from the system whose difference equation is $y(n) = x(n) - 0.2y(n-1) + 0.1y(n-2)$ and $x(n)$ is a unit impulse.

Section 8.4

7. Find $H(z) = \vec{Y}/\vec{X}$ for the following difference equations.
   a. $y(n) = x(n) - 0.2y(n-1) + 0.1y(n-2)$
   b. $y(n) = x(n) - 0.5x(n-2) + 0.5y(n-2) - 0.5y(n-4)$
8. Find the transfer function represented by the following difference equations.
   a. $y(n) = x(n) + x(n-1) - 0.75y(n-1) + 0.5y(n-3)$
   b. $y(n) = 0.25x(n) - 0.5x(n-1) + x(n-2) - 0.5x(n-3) + 0.25y(n-4)$
9. Given that the input to the system of Problem 7a is $x(n) = 12\cos 0.2\pi n$, find the forced response for $y(n)$.
10. The input to a system with $H(z) = 1 - 0.5z^{-1} + 0.5z^{-2}$ is $x(n) = 2\cos 0.2\pi n$. Determine the forced response of the output.
11. The input to a system with $H(z) = 1 - z^{-4}$ is $x(n) = \cos 0.3\pi n$. Determine the forced response of the system.
12. Find the gain at d-c, $v = 0.5$, and $v = 1$ for the following transfer functions.

    a. $H(z) = \dfrac{z^2 + 1}{z^2 + 0.2z + 0.8}$

    b. $H(z) = 1 - z^{-1} + z^{-2}$
13. Determine the difference equation that implements each of the following transfer functions.

    a. $H(z) = \dfrac{z^2 - z + 2}{z^3 - 0.25z}$

    b. $H(z) = \dfrac{z^3 - 1}{z^3 - 0.2z^2 + 0.2z - 1}$
14. Determine the difference equation for each of the following systems.

    a. $H(z) = \dfrac{z^4 + 4z^3 - 2z^2 + z - 1}{z^4}$

    b. $H(z) = \dfrac{z^4 - 1}{z^4 + 0.5}$

15. What problems, if any, can you see in implementing the following transfer function?

$$H(z) = \frac{z^4 - 1}{z^3 + 0.5z - 0.25}$$

Section 8.5

16. Test the stability of the system in Problem 7b using an iteration table. Follow the output until the result is clearly established.
17. Test the stability of the system in Problem 8b using an iteration table.
18. Test the stability of each system in Problem 14 by finding the location of its poles in the z plane.
19. Test the indicated difference equations for stability by locating the poles of each transfer function.
    a. $y(n) = 0.14x(n) + 0.14x(n - 1) + 1.02y(n - 1)$
    b. $y(n) = 0.5x(n) - 0.3x(n - 2) - 2y(n - 1) - y(n - 2)$

Section 8.6

20. Use the **dstep** command to find the first 12 outputs from the systems of Problem 5.
21. Plot the magnitude of the gain versus frequency for the difference equations indicated. Use a linear scale for $|H|$.
    a. $y(n) = 0.1373x(n) + 0.1373x(n - 1) + 0.7254y(n - 1)$
    b. $y(n) = 0.0301x(n) - 0.0301x(n - 2) + 1.8439y(n - 1) - 0.9398y(n - 2)$
    c. $y(n) = 0.9288x(n) - 1.7682x(n - 1) + 0.9288x(n - 2) + 1.7682y(n - 1) - 0.8576y(n - 2)$
    d. $y(n) = 0.8627x(n) - 0.8627x(n - 1) + 0.7254y(n - 1)$
22. Test the stability of the systems in Problem 19 using the **dimpulse** command.
23. An input signal $x(t) = \sin 250\pi t + \cos 1000\pi t + \cos 3000\pi t$ is sampled at $t = n/6000$. It is filtered with a "comb filter" whose difference equation is $2y(n) = x(n) + x(n - 6)$. Use about 100 samples.
    a. Plot $x(n)$ vs. n.
    b. Plot the magnitude of the filter transfer function for $0 \leqslant v < 1$.
    c. Plot $y(n)$ vs. n.
    d. Explain the results of (c) based on the transfer function and the known input.
24. Repeat Problem 23 for $x(t) = \cos 1000\pi t - \sin 3000\pi t + \cos 11500\pi t$. Explain the results.

25. Create a 300-sample signal consisting of $x = x1 + x2$, where $x1$ is a low-frequency square wave and $x2$ represents high-frequency noise.

$$x1 = [\text{ones}(1{,}75) \quad -1{*}\text{ones}(1{,}75) \quad \text{ones}(1{,}75) \quad -1{*}\text{ones}(1{,}75)]$$
$$x2 = 2{*}\cos(0.7{*}\pi{*}n) + 3{*}\cos(0.9{*}\pi{*}n)$$

  a. Plot x vs. n.
  b. Filter the signal with the low-pass filter difference equation of Problem 21a and plot the output vs. n.
  c. Put the output of the first pass through the filter again. Plot the new output waveform. (A sinusoid is a poor test signal for a filter, because the foregoing process can be repeated over and over and the sinusoid will eventually emerge free of noise. Any other waveform, such as the square wave of this problem, is degraded each time it passes through the filter.)

9

IIR FILTER DESIGN

OUTLINE

OBJECTIVES

1. Use the bilinear transform to create digital filters from analog filters.
2. Use computer tools to find difference equation coefficients.
3. Properly implement IIR designs.

INTRODUCTION

The feedback in IIR difference equations produces very efficient transfer functions. They require fewer multiply and accumulate operations to accomplish as much as an equivalent FIR system. They also approximate the feedback inherent in continuous-time systems. This suggests the possibility that IIR systems could be used to provide digital equivalents to the classic analog filters. It would be very desirable to be able to make use of all the expertise that went into the development of those filters.

There are basically two methods of achieving this goal: impulse invariance designs and bilinear transform designs. The impulse invariance approach finds the impulse response of the desired analog filter, samples it, and creates a digital filter with the same impulse response samples. Because sampling is involved, aliasing will result unless a high-order prototype filter is the basis for the design so that its impulse response has little high-frequency content. For the same reason, high-pass or stopband filters cannot be designed with this method. The impulse invariance approach involves a somewhat unfamiliar mathematical procedure, is limited in the types and orders of filters it can be successfully used on, and offers no advantages over bilinear transform designs. The bilinear transform approach will be covered exclusively.

The feedback in IIR filters can also produce problems. Register overflows during additions and processor word size limits on the accuracy to which filter coefficients can be set can have effects that are aggravated by the feedback. These issues must be considered when implementing an IIR filter.

9.1 BILINEAR TRANSFORMATION

The bilinear transformation is well-known to mathematicians and other theorists. While it is unlikely we would have stumbled upon it as a tool for creating digital filters, it is easy to show that it has exactly the kinds of features we need. After completing this section you will be able to:

- Use the bilinear transform to create digital filters from analog filters.
- Set filter cutoff frequencies precisely.

The bilinear transformation relates two complex variables s and z as:

$$s = C\frac{z - 1}{z + 1} \tag{9.1}$$

where C is a real constant. It has unusual properties that have made it of interest in various fields of study. Students of transmission lines may notice a similarity between the bilinear transformation and the equation of the reflection coefficient. The Smith chart is based on the bilinear transform.

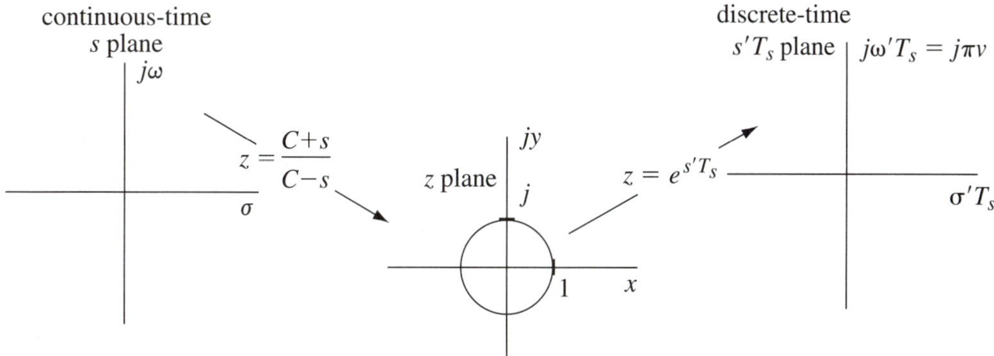

Figure 9.1 The bilinear transformation maps the poles and zeros of continuous-time systems into the *z* plane in such a way that their frequency response along the *jω* axis becomes compressed on to the unit circle of the *z* plane. Each full revolution around the unit circle gives one full cycle of the periodic frequency response of the discrete-time system.

There are two *s* planes involved in discussing the bilinear transform, and this can be a point of confusion unless it is clear what we are attempting to accomplish with the transformation. We have already defined a *z* plane that is related to the *s* plane of a *sampled* signal as $z = e^{sT_s}$. We will temporarily denote this "digital," or discrete-time *s,* plane with primes. We also have, as our starting point, a desirable transfer function defined in the "analog," or continuous-time, *s* plane. The bilinear transformation is being used to move the poles and zeros of the desired $H(s)$ into the *z* plane in the hope that doing so will result in a digital filter similar to the original analog filter. This process is suggested in Figure 9.1.

Solving Equation 9.1 for *z,* we obtain

$$z = \frac{C + s}{C - s} = \frac{(C + \sigma) + j\omega}{(C - \sigma) - j\omega}$$

$$|z| = \left|\frac{(C + \sigma) + j\omega}{(C - \sigma) - j\omega}\right| = \sqrt{\frac{(C + \sigma)^2 + \omega^2}{(C - \sigma)^2 + \omega^2}} < 1 \qquad \text{if } \sigma < 0 \tag{9.2}$$

As a consequence of Equation 9.2, poles or zeros in the left half of the analog *s* plane map into poles or zeros within the unit circle of the *z* plane. Digital systems whose poles are within the unit circle of the *z* plane are stable. In other words, a stable $H(s)$ becomes a stable $H(z)$ using the bilinear transformation. Since stability is mandatory for a system to be useful, the bilinear transform passes the first crucial test.

Checking the analog frequency response conditions, $s = j\omega$, shows that the entire $j\omega$ axis is mapped into the unit circle of the *z* plane.

$$z = \frac{C+s}{C-s} = \frac{C+j\omega}{C-j\omega} = 1\angle 2\arctan\left(\frac{\omega}{C}\right) \tag{9.3}$$

The frequency response of the digital system is also taken along the unit circle of the z plane, and at any given position on the unit circle these frequencies are related by

$$e^{j\pi v} = e^{j2\arctan(\omega/C)} \qquad \frac{\pi}{2}v = \arctan\left(\frac{\omega}{C}\right)$$

$$\omega = C\tan\frac{\pi v}{2} \tag{9.4}$$

The analog frequency and the digital baseband frequencies are directly related but in a highly nonlinear fashion, since an infinite range of $0 \leq \omega \leq +\infty$ is compressed into half the circumference of the unit circle, $0 \leq v \leq 1$. If we start with a known $H(s)$ and apply the bilinear transform to obtain an $H(z)$, each value of $z = e^{j\pi v}$ results in a numerical value for $H(z)$ that would also be obtained from plugging one particular $j\omega$ value in the original $H(s)$. The fact that they are directly related means a low-pass analog filter will transform into a low-pass digital filter, a bandpass filter into bandpass filter, and so on. Further, an analog Chebyshev filter will produce an equiripple digital filter passband, but the maximums and minimums will not have the same frequency spacing in the two systems. The constant C can be used to line up the responses at one particular frequency, such as cutoff.

Alternatively, for $v \ll 1$, $\tan\theta \approx \theta$, and the analog and digital systems can be made to agree in a general sense at low frequencies:

$$2\pi f = \frac{C\pi}{2}\left(\frac{2f'}{f_s}\right)$$

$$f = \left(\frac{C}{2f_s}\right)f' = C\left(\frac{T_s}{2}\right)f'$$

$$f = f' \qquad \text{if } C = \frac{2}{T_s} \tag{9.5}$$

$C = 2/T_s$ is routinely used in control system applications or when sampling significantly above the Nyquist rate. Filter designers otherwise prefer to use C to precisely identify some salient feature, such as the cutoff frequency or center frequency, of the digital filter.

EXAMPLE 9.1

Find $H(z)$ to provide a second-order 2-dB Chebyshev–based low-pass digital filter with $\upsilon_{co} = 0.3$ using the bilinear transformation.

Solution

The prototype filter has the following transfer function (Table 5.3):

$$H_p(p) = \frac{0.6538}{p^2 + 0.8038p + 0.8231}$$

C is found from equating cutoff frequencies:

$$\omega_{co} = 1 = C \tan\frac{\pi\upsilon_{co}}{2} = C \tan 0.15\pi = 0.5095C$$

So $C = 1.9626$, and, transforming directly from the prototype,

$$p = 1.9626\frac{z-1}{z+1}$$

Then

$$H(z) = \frac{0.6538}{\left[1.9626\frac{z-1}{z+1}\right]^2 + 0.8038\left[1.9626\frac{z-1}{z+1}\right] + 0.8231}$$
$$= \frac{0.6538(z+1)^2}{3.8518(z-1)^2 + 1.5775(z-1)(z+1) + 0.8231(z+1)^2}$$

Multiplying and grouping powers of z gives

$$H(z) = \frac{0.6538(z+1)^2}{6.2524z^2 - 6.0574z + 3.0974} = \frac{0.1046(z+1)^2}{z^2 - 0.9688z + 0.4954}$$

The resulting response is shown in Figure 9.2. Note that the decibel ripple has been preserved and the cutoff is at the desired value. Since the prototype filter has zero gain at infinite frequency, the digital filter will have zero gain at $\upsilon = 1$. This is another nice feature of using the bilinear transformation.

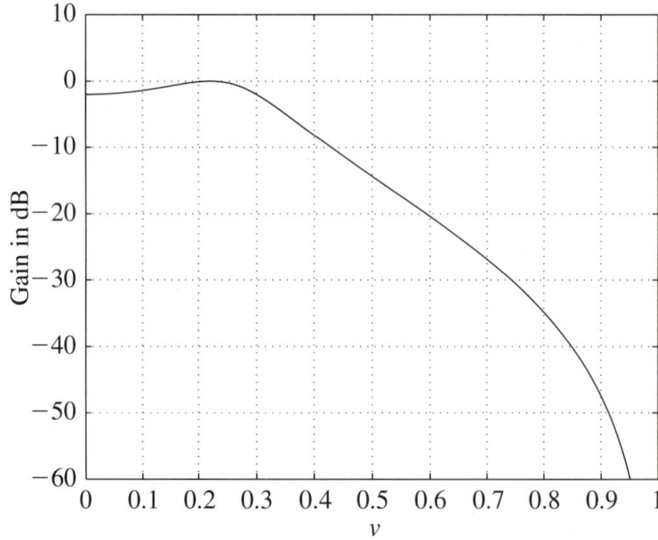

Figure 9.2 Frequency response of a bilinear transformed 2-dB Chebyshev low-pass filter with $v_{co} = 0.3$.

Implementing a bandpass digital filter using the bilinear transform involves a bit more planning. Referring to Figure 9.3, we want to be able to specify the digital filter cutoff frequencies v_H and v_L. From the frequency warping formula (Eq. 9.4) we can determine the required analog bandpass filter specifications, ω_c and B:

$$B = \omega_H - \omega_L = C\left[\tan\frac{\pi v_H}{2} - \tan\frac{\pi v_L}{2}\right] \tag{9.6}$$

$$\omega_c^2 = \omega_L\omega_H = C^2 \tan\frac{\pi v_L}{2}\tan\frac{\pi v_H}{2} \tag{9.7}$$

We will arbitrarily set ω_c to unity and calculate the required value for C. This will give the simplest bandpass transformation, but it does not affect the final result in any way. A three-step process will be used:

1. Select the desired analog prototype.
2. Transform it to a bandpass filter using

$$p = \frac{s^2 + 1}{Bs}$$

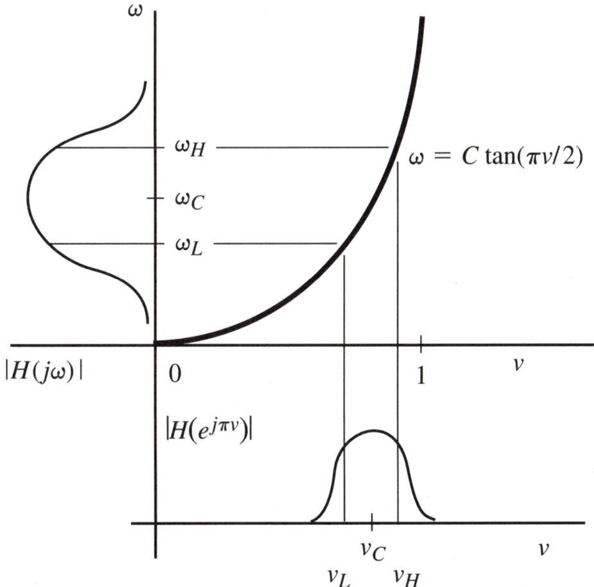

Figure 9.3 Frequency warping in the bilinear transformation as applied to bandpass filter design.

3. Transform the analog bandpass filter to a digital filter using

$$s = C\frac{z - 1}{z + 1}$$

EXAMPLE 9.2

Determine $H(z)$ for a second-order Butterworth–based bandpass filter with $v_H = 0.35$ and $v_L = 0.25$. Use the bilinear transform.

Solution

Using Equation 9.6 we get

$$1 = C^2 \tan\frac{0.25\pi}{2}\tan\frac{0.35\pi}{2} = C^2(0.4142)(0.6128)$$

$$C = 1.9849$$

Then the required B is

$$B = 1.9849(0.6128 - 0.4142) = 0.3942$$

The Butterworth prototype is converted to a bandpass filter:

$$H_p(p) = \frac{1}{p^2 + \sqrt{2}p + 1}$$

$$H_{BP}(s) = \frac{1}{\left[\dfrac{s^2+1}{0.3942s}\right]^2 + \sqrt{2}\left[\dfrac{s^2+1}{0.3942s}\right] + 1}$$

$$H_{BP}(s) = \frac{0.1554s^2}{s^4 + 0.5575s^3 + 2.1554s^2 + 0.5575s + 1}$$

Now we apply the bilinear transform:

$$H_{BP}(z) = \frac{0.1554\left[1.9849\dfrac{z-1}{z+1}\right]^2}{\left[1.9849\dfrac{z-1}{z+1}\right]^4 + 0.5575\left[1.9849\dfrac{z-1}{z+1}\right]^3 + 2.1554\left[1.9849\dfrac{z-1}{z+1}\right]^2 + 0.5575\left[1.9849\dfrac{z-1}{z+1}\right] + 1}$$

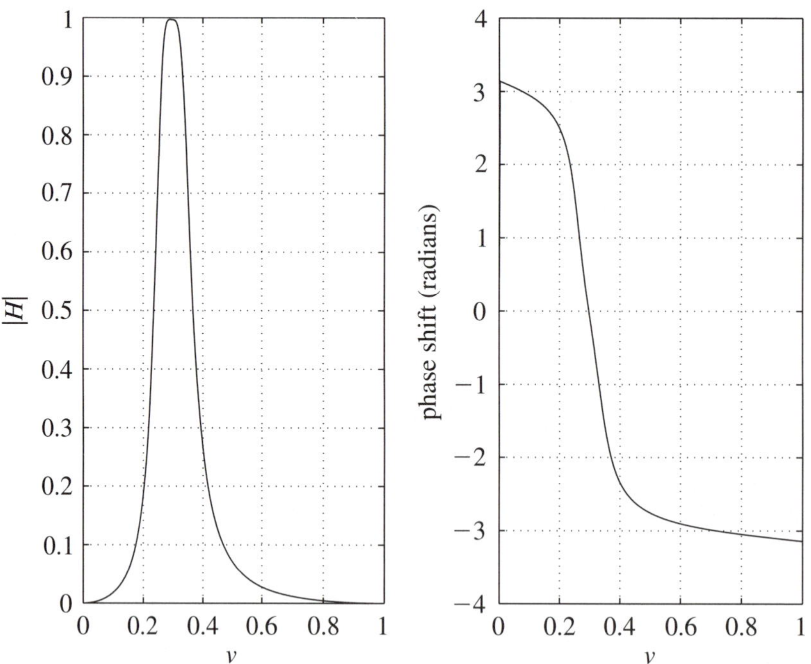

Figure 9.4 Second-order bandpass filter derived from the bilinear transformation.

A "little" algebra reduces the result to

$$H(z) = \frac{0.0201(1 - 2z^{-2} + z^{-4})}{1 - 2.1192z^{-1} + 2.6952z^{-2} - 1.6923z^{-3} + 0.6414z^{-4}}$$

and a frequency plot (Figure 9.4) confirms that the desired response has been achieved.

The task of simplifying the transfer functions of digital filters to their final form is tedious, and the numbers involved are never "nice," so either the process will be computerized or a good deal of key punching on the calculator will be necessary. The bilinear transform has the advantage that the process is at least routine and is identical to the type of procedure used to make a particular analog filter from a prototype analog filter. Stopband and high-pass filters may also be created using the procedures demonstrated in this section.

9.2 MATLAB LESSON 9

The IIR design techniques involve numeric and algebraic calculations that are cumbersome even for low-order filters. Fortunately, MATLAB tools exist that make these techniques practical. After completing this section you will be able to:

- Use computer tools to find difference equation coefficients.
- Try out the Yule–Walker computer-aided IIR design program.

The commands of this chapter expect all z domain polynomials to be in negative powers of z and create z domain polynomials of that same form. Polynomials from the s domain remain in their usual form.

MATLAB EXAMPLES

bilinear

Using a desired $H(s)$ as the starting point, the bilinear derived $H(z)$ may be found. C may be set to $2/T_s$ or used to match an analog frequency to a digital frequency.

```
>anum=[1 0 4]; aden=[1 1 4];                   % notch filter, notch at
                                                 wo = 2, fo = 1/π
>freqs(anum,aden)                              % may as well take a
                                                 look
>[dnum,dden]=bilinear(anum,aden,8/pi,1/pi);    % fs = 8/π, fo = 1/π
                                                 (vo = 1/4)
>freqz(dnum,dden)                              % notch is at vo = 0.25
>[dnum,dden]=bilinear(anum,aden,8/pi);         % fs = 8/π, C = 2/Ts
>freqz(dnum,dden)                              % notch is at vo ≈ 0.236

% bilinear can also work directly with poles, zeros, and a multiplying
  constant
>[az,ap,ak]=cheb1ap(6,1);                      % Chebyshev 6th-order
                                                 1-dB prototype
>[dz,dp,dk]=bilinear(az,ap,ak,2/pi, 1/(2*pi)); % fs = 2/π, fco =
                                                 1/(2π), vco = 2fco/
                                                 fs = 0.5
>dnum=dk*poly(dz); dden=poly(dp);
>freqz(dnum,dden)
>dnum                                          % trying getting this by
                                                 hand!
```

butter cheby1 cheby2 ellip

Bilinear transformed versions of the standard filters are available directly in single commands. For example, the sixth-order Chebyshev 1-dB digital filter polynomials could have been found from the following:

```
>[dnum,dden]=cheby1(6,1,.5);      % 6th-order Chebyshev, 1-dB ripple,
                                    vco = 0.5
>freqz(dnum,dden)

>[dnum,dden]=butter(8, .4);       % 8th-order Butterworth, vco = 0.4
>freqz(dnum,dden)

>w=[.14 .34];
>[n,d]=ellip(5, 1, 30,w);   % 5th-order Cauer bandpass filter, 1-dB passband
                              ripple,
                            % 30-dB min. stopband atten., passing from
                              0.14 ≤ v ≤ 0.34
>freqz(n,d)                 % help ellip for other options
```

yulewalk

The *Yule–Walker* program designs IIR filters based on the desired magnitude response for its transfer function. The user specifies a set of frequencies starting at $v = 0$ and running to $v = 1$ in one vector and the desired magnitude response at each of these frequencies in a second vector. The user also sets the order of the filter. The Yule–Walker program then seeks numerator and denominator polynomials that provide an optimum fit to the points specified within the order constraint.

Multiple passbands and passbands of differing gain are possible with Yule–Walker designs. Common sense suggests that few points should be specified if only low-order polynomials are allowed. Experience with the classical filters also suggests that rapid changes from pass to stop regions will require either high-order filters or lots of rippling in the pass- and stop-bands.

Designing a digital filter based on one of the classical analog filters and using the Yule–Walker program are fundamentally different processes. With the classical approaches, the designer basically knows what the filter characteristics are going to be from the start. With Yule–Walker, it is expected that many trial specifications may need to made and the resulting filter characteristics observed before a final design is decided on. In cases where the standard filters are viable options, it may be very difficult to find a Yule–Walker design that is preferable. In comparing potential filter designs, the number of multiply and accumulate operations should be the same.

The Yule–Walker program provides a very useful alternative design approach for IIR digital filters. The polynomials that give an "optimum" match are those that produce the least error at the specified response points. This does not mean it is an optimum design or even that it produces a good filter. The user of the program is responsible for achieving that. The user must also realize that it is possible to specify very poor filters and to achieve them with great accuracy using computer-aided design.

EXAMPLE 9.3

Use the Yule–Walker program to design a digital low-pass filter with $v_{co} = 0.22$.

Solution

One of the difficulties in using Yule–Walker is setting the cutoff frequency. Since the program will try to match the filter response to the points specified in the design, specifying the -3-dB point is one possibility, provided no other points are specified nearby. On the other hand, the -3-dB point may not be the proper cutoff criterion if rippling occurs in the passband. The result for one set of specifications is shown for three different filter orders in Figure 9.5a.

As the order of the filter increases, rippling appears in the passband. The response never does pass right through the -3-dB point specified. Looking at the third-order filter on a decibel scale (Figure 9.5b) shows it to be very similar to the Chebyshev type 2 filters, which we have not particularly favored in the past. Both designs require seven multiplications. Further exploration would be needed to see if other v and m vectors would yield better results.

```
>v=[ 0 .1 .2 .22 .3 .7 1];
>m=[ 1 1 1 .707 0 0 0];
>[n,d]=yulewalk(3,v,m);
>h=freqz(n,d);
>vv=(0:511)/512;
>plot(v,m,'o')
>hold
Curent plot held
>plot(vv, abs(h))
>grid
```

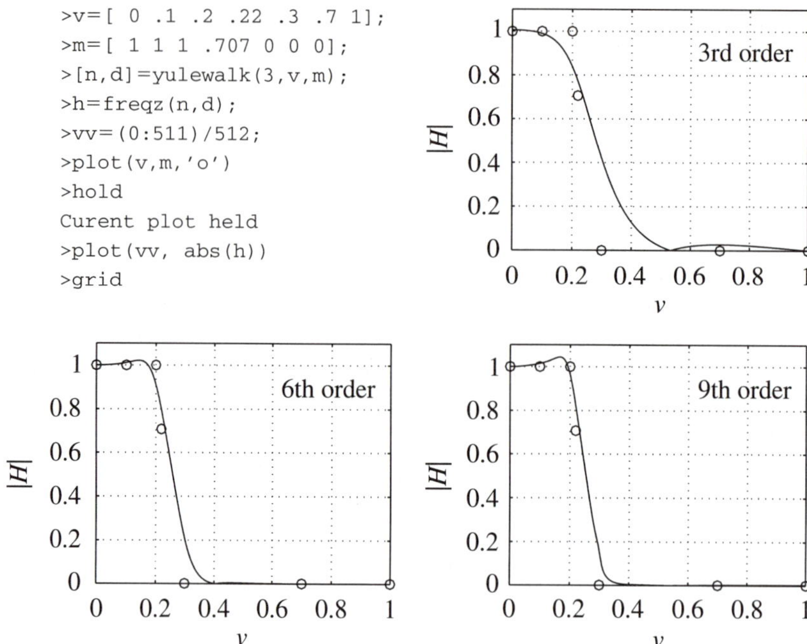

Figure 9.5a Yule–Walker filter designs of varying order. Circles show filter specification points.

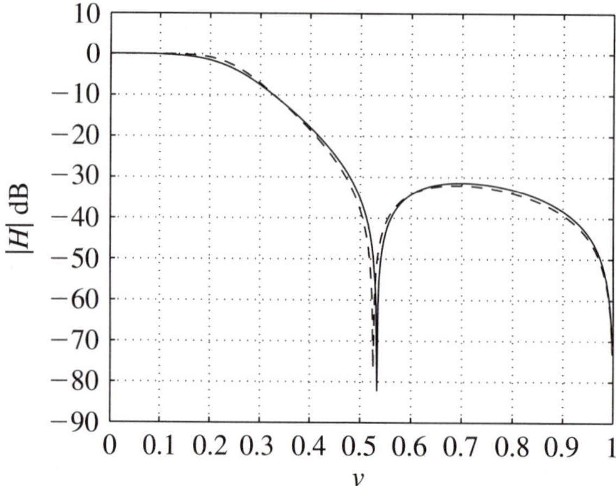

Figure 9.5b Comparison of the 3rd order Yule–Walker design (solid line) specified in Figure 9.5a with a 3rd order Chebyshev type 2 filter (dashed line) having a minimum stopband attenuation of 32 dB @ $v = 0.48$.

9.3 IIR IMPLEMENTATION

Practical implementation of any digital filter is limited by things like the processor word size, register overflow during additions, and round-off or truncation errors. These factors can become especially troublesome in IIR filters due to their inherent feedback. After completing this section you will be able to:

- Properly implement IIR designs.
- Minimize effects of coefficient error with cascaded sections.
- Recommend actions to take in the event of accumulator overflow.
- Recognize and avoid limit cycles.

All of our theoretical analysis has assumed linear systems and the ability to specify filter coefficients as accurately as we wish. Real processing systems are not quite that ideal. A direct implementation of an IIR system is shown in Figure 9.6a. It shows how a hardware implementation consisting of shift registers, constant multipliers, and summers could be configured for a direct calculation of the difference equation. Direct implementation form 2 is shown in Figure 9.6b. It cuts the number of shift registers that are required in half by cleverly implementing the numerator, $N(z)$, and denominator, $D(z)$, of the transfer function separately and cascading them.

$$H(z) = [1/D(z)][N(z)]$$

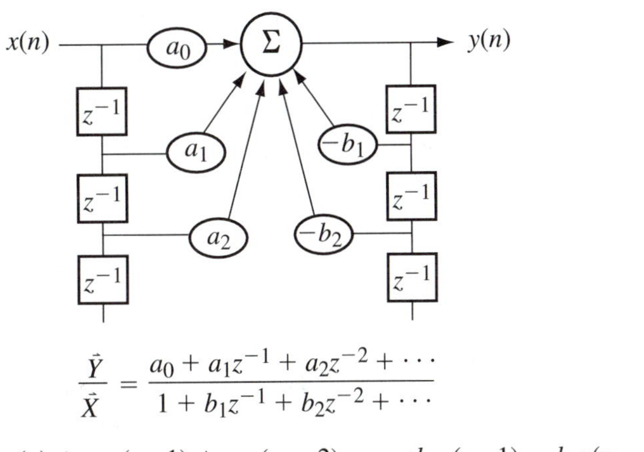

Figure 9.6a Direct implementation form 1 for a difference equation.

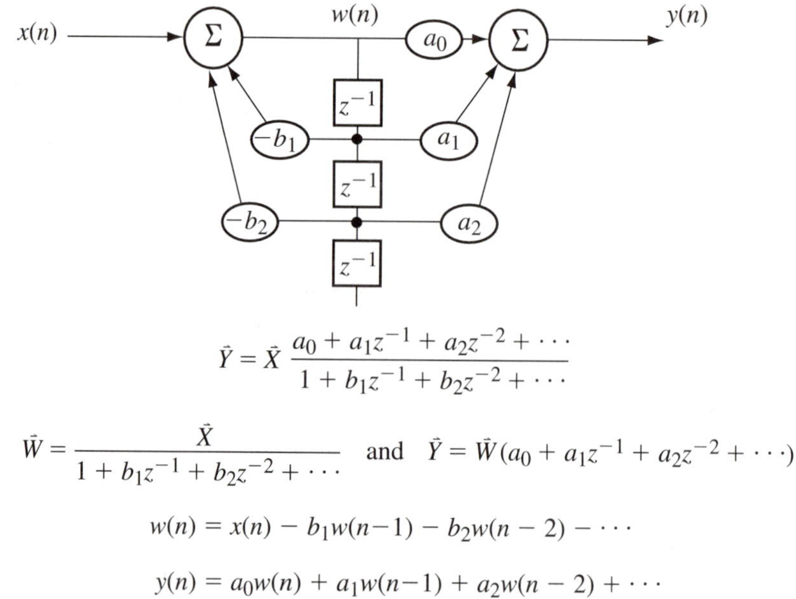

Figure 9.6b Direct implementation form 2 for a difference equation.

Neither form of direct implementation of a transfer function works well if the accuracy with which the filter coefficients can be specified is limited excessively by the processor word size. That, of course, is not a surprise. The trouble is that the feedback inherent to IIR filters compounds the inaccuracy, and can also lead to instability. To demonstrate this problem suppose we wish to implement a fourth-order Chebyshev–based low-pass filter (1-dB ripple, $\upsilon_{co} = 0.20$) obtained from the bilinear transformation:

$$H(z) = \frac{0.0018z^4 + 0.0073z^3 + 0.0110z^2 + 0.0073z + 0.0018}{z^4 - 3.0543z^3 + 3.8290z^2 - 2.2925z + 0.5507} \quad \textbf{(9.8)}$$

The impulse response for this filter is shown in Figure 9.7a. An indication of the problem can be obtained by thinking in terms of applying a d-c signal ($x(n) = 1$ for all n) to the difference equation or zero frequency ($z = e^{j0} = 1$) to the transfer function. The low-pass filter transfer function has a d-c gain of unity. The numerator terms sum to 0.0250. The denominator terms must consequently do the same, but that involves adding or subtracting coefficients with digits in the ones position in order to get a result whose most significant digit is in the 10^{-2} position. The higher the order of the filter, the worse the accuracy problem becomes.

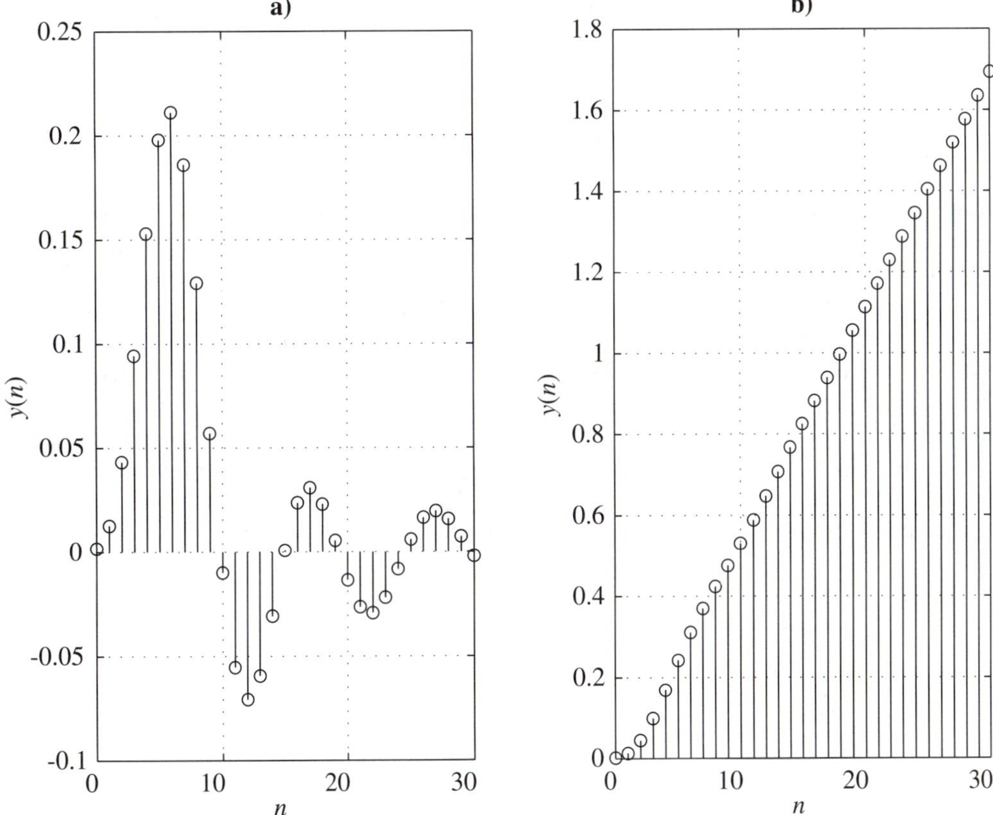

Figure 9.7 Impulse responses for a fourth-order, 1-dB, $v_{co} = 0.2$ bilinear transformed Chebyshev low-pass filter: **(a)** uses full coefficient accuracy and is a stable system; **(b)** limits the coefficients to two significant figures and is unstable.

If the filter coefficients could be specified to only two significant figures, the denominator polynomial of Equation 9.8 would have to be represented as

$$1.0z^4 - 3.1z^3 + 3.8z^2 - 2.3z + 0.6 = (z - 1)^2(z - 0.55 \pm j0.54)$$

which has a repeated pole on the unit circle. The system is unstable and has the impulse response shown in Figure 9.7b. (The calculated roots assume the polynomial coefficients are exact, and therefore the roots have been allowed more significant figures than the coefficients. We also assume that order of magnitude changes between the numerator and denominator coefficients can be handled separately, or we would need at least five significant figures to represent both sets of coefficients in exactly the same way.)

Since the filter is unstable, we will have to provide for a larger word size if we want to implement it. Suppose we can represent the filter coefficients to three significant figures. Then the denominator becomes

$$1.00z^4 - 3.05z^3 + 3.83z^2 - 2.29z + 0.55 = (z - 0.77 \pm j0.53)(z - 0.75 \pm j0.24)$$

and all of the roots are within the unit circle. The system is stable. Its frequency response is compared to that of the ideal filter in Figure 9.8. Although the filter is stable, it is a very poor approximation to the Chebyshev response desired.

The effects of coefficient accuracy can be further reduced by implementing transfer functions as cascaded quadratic sections. To accomplish this the numerator and denominator polynomials of $H(z)$ are factored. Complex roots are always recombined with their conjugates so that all filter coefficients remain real. Any quadratic numerator term can be combined with any quadratic denominator term. In high-order filters, many combinations are possible and some may work better than others. Normally each quadratic section would be implemented using the direct form 2 sections shown in Figure 9.9. Since the extra shift registers are not

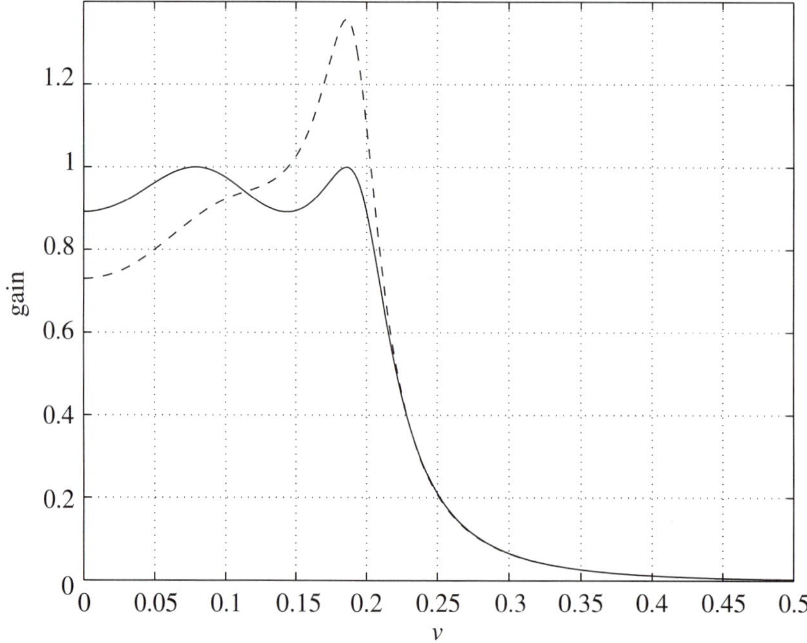

Figure 9.8 Fourth-order Chebyshev–based filter response with full (solid line) and three-significant-figure (dashed line) coefficient accuracy. The poles of each system are within the unit circle.

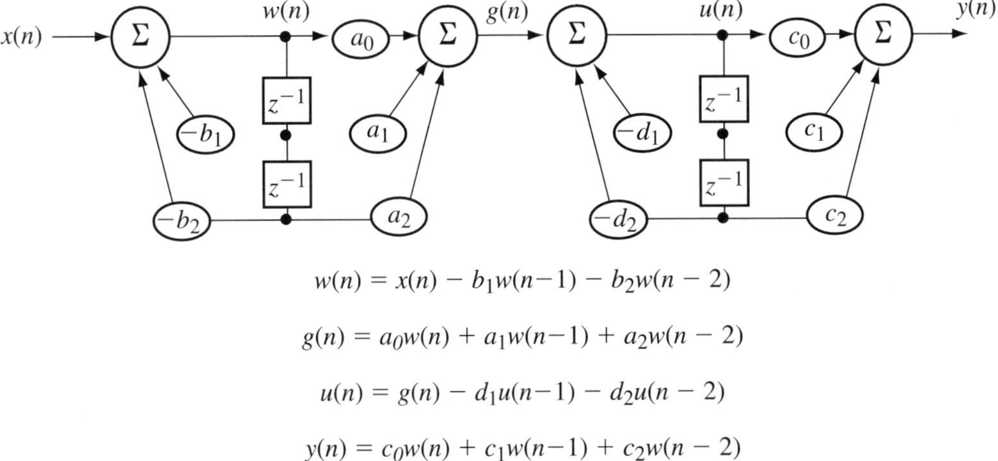

Figure 9.9 Implementation by cascaded quadratic sections reduces the effects of coefficient inaccuracy. The direct form 2 equations are shown.

an issue for our demonstration, a direct form 1 implementation will be used. Our $H(z)$ factors as follows:

$$H(z) = H_1(z)H_2(z) = 0.00183\underbrace{\left(\frac{z^2 + 2z + 1}{z^2 - 1.4996z + 0.8482}\right)}_{H_1(z) = \vec{G}/\vec{X}}\underbrace{\left(\frac{z^2 + 2z + 1}{z^2 - 1.5548z + 0.6493}\right)}_{H_2(z) = \vec{Y}/\vec{G}}$$

Allowing three significant figures for the coefficients of the cascaded sections, the direct form 1 difference equations are:

$$\begin{aligned} g(n) &= 0.00183x(n) + 0.00366x(n-1) + 0.00183x(n-2) + 1.50g(n-1) \\ &\quad - 0.85g(n-2) \\ y(n) &= g(n) + 2g(n-1) + g(n-2) + 1.55y(n-1) - 0.65y(n-2) \end{aligned}$$

The results for the cascaded implementation are shown in Figure 9.10. The result is significantly better than the direct implementation, but it is still not a true Chebyshev equiripple response. Increasing the processor word size to handle four significant figures would probably work nicely for this fourth-order filter.

Two other problems that occur in implementing IIR filters are associated with accumulator operations. Accumulator overflow is a nonlinear effect that may override the predictions of linear systems. If two numbers are added in a digital processor, the result may overflow the accumulator. When that happens, an overflow flag is

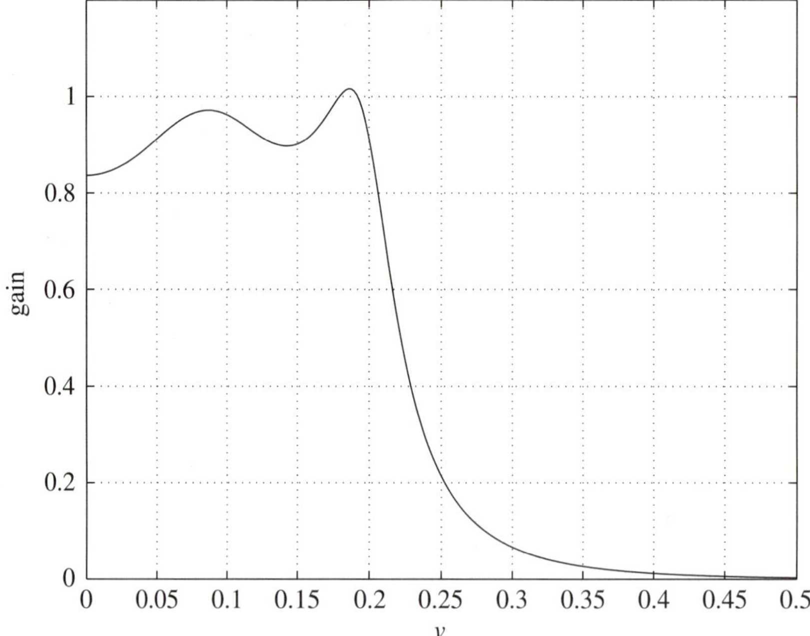

Figure 9.10 Cascade implementation of a fourth-order Chebyshev 1-dB low-pass filter using filter coefficients of three significant figures.

asserted. While the result may still be properly interpreted, there is no way to accommodate a larger number size. If the result is going to a D/A converter, for instance, the converter expects a fixed number of input bits and has no means of receiving or responding to an extra one. The overflow will consequently cause an incorrect value to be delivered to the output. What may have been a large negative number could end up as a large positive number when using a 2's complement number format. Probably the best thing the programmer can do in this situation is to use the flag information to set the output to the maximum positive or negative value, as appropriate. The D/A converter will then show the same kind of saturated output produced by an overdriven amplifier, rather than the sporadic appearances of random values.

Another type of problem occurs due to rounding. In an 8-bit processor, for instance, the 8-bit filter coefficient is multiplied by an 8-bit representation of the signal, producing a 16-bit result. Since we are limiting the results to 8 bits, either we simply throw the lower byte of the multiplication result away (truncate), or we can consult its most significant bit and round the upper byte to the appropriate value. Rounding would give the most accurate result, but it can lead to *limit cycles* or *deadband* effects.

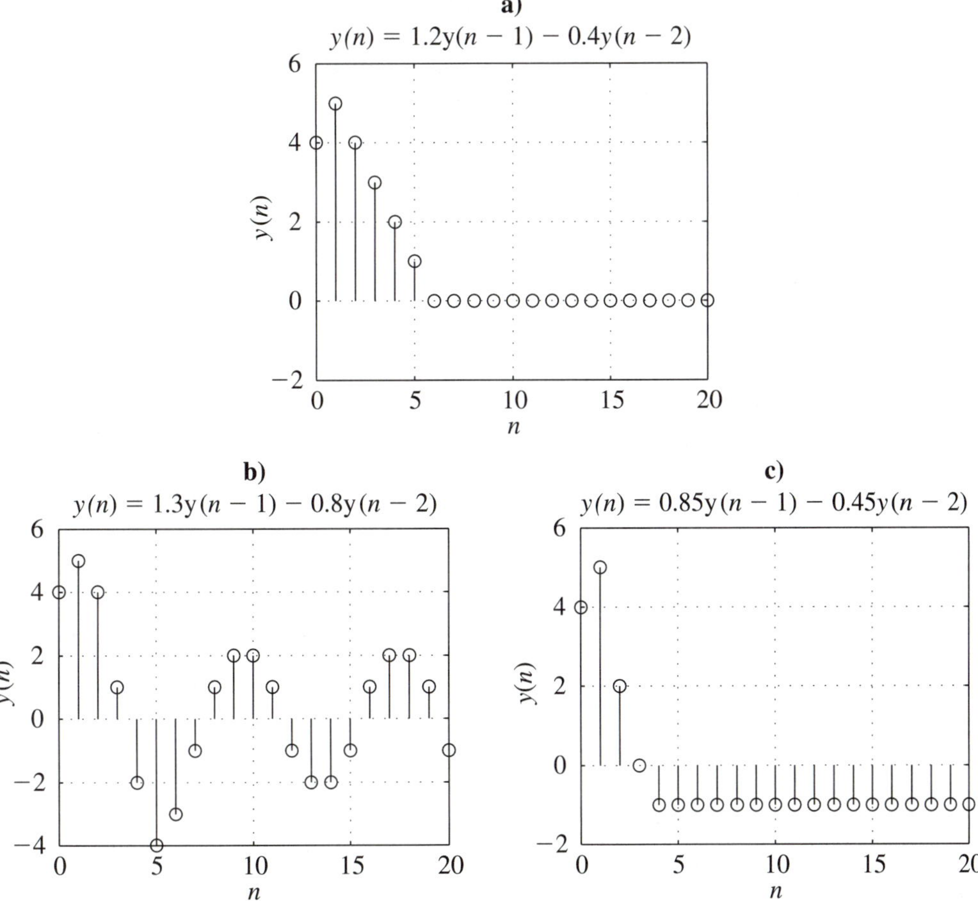

Figure 9.11 Impulse responses for stable second-order systems where round-off is used after each multiply. **(a)** The response settles to zero, as expected in a linear system. **(b)** The response falls into a limit cycle and oscillates forever. **(c)** The response settles at a nonzero value due to deadband loss of any correction signals. Initial values for all responses are based on a four-unit impulse at $n = 0$, which results in a five-unit response at $n = 1$.

To investigate these effects, think of our signal as taking on integer values: −128 to 127 in an 8-bit machine. Neither the order of the filter nor the accuracy of its coefficients is at issue. The three difference equations of Figure 9.11 all have poles within the unit circle of the z plane, and checking their impulse response would demonstrate that they are indeed stable. However, since we want to consider the discrete levels of our signal and use rounding, we will need to write our own program. (An iteration table would be a second choice.) A MATLAB program will be demonstrated, although any standard programming language would do as well. We will supply the program with values for $y(1)$ and $y(2)$. These can be arbitrary integer values, since we can always find a two-sample input pulse that would cause them. In MATLAB, a function file **round.m** will be called to look up the rounded-off result after each multiplication. It rounds to the closest integer value.

File: round.m

```
function [n]=round(y)
% rounds an input value between -127 and 128
% to the nearest integer.
i=1; k=1;
if y<0
     k=-1;                          % save sign of y
end
z=abs(y);
x=linspace(0,128,129)+.5;     % look up table for integers 0-128
while z>x(i)
     i=i+1;
end
n=k*(x(i)-.5);                % return with nearest integer
```

The following program can be entered in the Command Window; but if you want to explore using different filter coefficients, put it in a script .m file.

```
% MATLAB 2nd Order Difference Equation Program
>y(1)=4; y(2)=5;           % 2 initial values are needed to start the
                             program
>for i=3:30                % use as many terms as you want, but start with
                             y(3)
a = 1.3*y(i-1);            % multiply
z = round(a);              % call the rounding subroutine
b = -0.8*y(i-2);           % multiply the next terms
zz=round(b);               % call the rounding subroutine
y(i) = z + zz;
end                        % continue for 30 values
>t=0:29;
>stem(t,y)                 % see the result!
```

Truncation does not create the problems associated with rounding. From a simple-minded point of view, truncation always decreases the value of the output, and therefore it adds a damping factor to the difference equation response. Rounding, on the other hand, may contribute to an overshoot of the output value, which effectively lowers the damping. While these effects are diminished as larger word sizes are used, an IIR design should not be considered complete until it has been simulated and checked for limit cycling.

CHAPTER SUMMARY

The bilinear transformation creates z domain filters from s domain filters through a simple algebraic relationship. It converts stable $H(s)$ transfer functions into stable $H(z)$ transfer functions while retaining identical frequency response characteristics except for a warping of the frequency axis. The frequency axes of each domain can be matched at one frequency or provided a general correspondence at low frequencies. Through the bilinear transform, all of the classical filter types remain available for digital processing as IIR systems.

Practical application of the bilinear transformation for high-order filters would be a nightmare without computer assistance. A computer-aided design technique known as the Yule–Walker program creates IIR filter designs that minimize the error between desired and actual transfer function magnitudes at a set of points selected by the designer.

The feedback inherent in IIR filters will lead to unstable operation unless proper attention is paid to implementation techniques, having adequate processor word size for establishing filter coefficients, or handling of accumulator overflows. A nonlinear oscillation called a limit cycle can result from using accumulator round-off.

PROBLEMS

Section 9.1

1. The standard one-pole low-pass filter is given by $H(s) = 1/(s + 1)$. Find $H(z)$ for an equivalent digital filter using the bilinear transform. Express the results in terms of the cutoff frequency v_{co}.
2. An analog notch filter with its notch at $\omega_o = 1$ rad/s has the transfer function

$$H(s) = \frac{s^2 + 1}{s^2 + 0.1s + 1}$$

   Use the bilinear transformation to find an equivalent digital filter with its notch at $v_o = 0.25$.
3. Use the bilinear transform to design a 0.5-dB second-order Chebyshev–based low-pass digital filter having $v_{co} = 0.35$.
4. Find the bilinear-based digital equivalent for the analog transfer function $H(s) = 10{,}000/(s + 100)^2$. Use $C = 100$ for the bilinear transformation

constant. Sketch the pole-zero plots for $H(s)$ and $H(z)$. Do they have the same number of poles and zeros?

5. Design a second-order Butterworth–based high-pass digital filter having $v_{co} = 0.50$ using the bilinear transform.
6. Obtain an analytical expression for and sketch the $|H(e^{j\pi v})|$ obtained from the difference equation $y(n) = x(n) + ky(n - m)$ for the following values.
   a. $k = 1, m = 4$
   b. $k = -1, m = 4$
   c. $k = 0.95, m = 6$
7. An nth-order Butterworth–based bandpass filter with $v_L = 0.3$ and $v_H = 0.45$ is to be designed using the bilinear transform. Given that the design starts from a Butterworth bandpass filter with a center frequency of 1 rad/s, determine that filter's bandwidth and the C value needed in the bilinear transformation.
8. Use the bilinear transform to design a first-order Butterworth–based bandpass digital filter to have $v_L = 0.5$ and $v_H = 0.7$.
9. Design a first-order bilinear–based stopband digital filter to have $v_L = 0.1$ and $v_H = 0.3$. Use a 0.5-dB Chebyshev prototype.

Section 9.2

10. Design a tenth-order 2-dB Chebyshev–based low-pass digital filter with $v_{co} = 0.20$ using the bilinear transform.
    a. Plot the filter response over a sensible decibel range.
    b. Use the filter to process 250 samples of the waveform

    $$x(n) = \sin 0.20\pi n + 10 \sin 0.25\pi n + 100 \sin 0.55\pi n$$

    Plot the output, and relate it to the filter response.
11. Use the bilinear transform to design a Butterworth–based fifth-order band-pass filter to pass frequencies of $0.15 \le v \le 0.25$.
    a. Determine the filter transfer function, and plot its magnitude and phase response.
    b. Use the filter to process 200 samples of the signal

    $$x(n) = 10 \cos 0.06\pi n + \cos 0.20\pi n + 50 \cos 0.4\pi n$$

    Plot $x(n)$ and output $y(n)$. Discuss the results in terms of the filter's response curve.
12. Given the transfer function

$$H_1(z) = \frac{0.0181 - 0.0543z^{-2} + 0.0543z^{-4} - 0.0181z^{-6}}{1 - 2.9419z^{-1} + 4.7023z^{-2} - 4.6341z^{-3} + 3.0774z^{-4} - 1.2466z^{-5} + 0.2781z^{-6}}$$

   a. Plot the magnitude and phase of $H_1(z)$.
   b. Create a new filter $H_2(z) = H_1(-z)$, and plot its magnitude and phase.
   c. Compare the passbands and group delays of the two filters.

13. Sinusoids of frequency v may be generated digitally by hitting the following difference equations with an impulse. Demonstrate this for $v = 0.02$. What is the difference between the waveforms generated by the two equations?
   a. $y(n) = (\sin \pi v)x(n-1) + (2\cos \pi v)y(n-1) - y(n-2)$
   b. $y(n) = x(n) - (\cos \pi v)x(n-1) + (2\cos \pi v)y(n-1) - y(n-2)$
14. Plot the characteristic of a Butterworth-based bilinear design for a third-order low-pass filter with $v_{co} = 0.2$. Design a third-order Yule–Walker filter to compete with the Butterworth-based design. Show the frequency and magnitude vectors that give what you regard to be the best result, and plot the final Yule–Walker and Butterworth characteristics on the same graph. Discuss the advantages of the two filters in terms of flatness of passband, accuracy of the cutoff frequency, stopband attenuation, and the number of multiply/accumulate operations required.
15. Repeat Problem 14 using a type 1 Chebyshev–based design with a passband ripple of 1 dB as the target characteristic.

Section 9.3

16. Express the given transfer functions as products of first- or second-order functions with real coefficients.
   a. $H(z) = \dfrac{0.3(1 - z^{-3})}{1 + 0.2z^{-1} - 0.6z^{-2} + 0.4z^{-3}}$
   b. $H(z) = \dfrac{1 + 0.2z^{-1} - 0.6z^{-2} + 0.4z^{-3}}{1 - 0.5z^{-3}}$
   c. $H(z) = \dfrac{(1 - 0.4512z^{-1})(1 - 0.2201z^{-1})(1 - 0.512z^{-1} + 0.3131z^{-2})}{1 + 0.1234z^{-2} - 0.4321z^{-4}}$
17. A third-order Chebyshev–based filter design results in

$$H(z) = \frac{0.01(1.15 + 3.44z^{-1} + 3.44z^{-2} + 1.15z^{-3})}{1 - 2.1378z^{-1} + 1.7693z^{-2} - 0.5398z^{-3}}$$

   a. Can the filter be implemented in direct form if its coefficients are limited to two significant figures (the ones and tenths decimal position values)? Show the impulse response of the filter under that implementation.
   b. Provide a cascade implementation of the filter, again limiting the coefficients to two significant figures. Plot $|H|$ vs. v.
18. Use the bilinear transform to design a fourth-order Butterworth–based low-pass filter with $v_{co} = 0.35$.
   a. Can a direct form implementation be obtained for this filter if only two significant figures are allowed in its coefficients?
   b. Show a cascaded form for this filter using two-significant-figure coefficients.
   c. Provide frequency response plots for each of the implementations that is stable.

19. Calculate the first 10 outputs of the difference equations $y(n) = \pm 0.8y(n-1)$ if y is rounded to integer values. Assume $y(0) = 10$.
20. Write a program to plot the first 30 outputs from the difference equation $y(n) = 0.4y(n-1) - 0.4y(n-2) - 0.5y(n-3)$ if each multiplication is rounded to the closest integer value. Use nonzero initial values of your choice in the range $-20 \leq y \leq 20$.

10

FIR FILTER DESIGN

OUTLINE

OBJECTIVES

1. Convert an IIR design into an FIR filter.
2. Create linear phase difference equations.
3. Use a Fourier series and a window function to design FIR filters.
4. Use the DFT for an empirical frequency sampling design of FIR filters.
5. Produce a Parks–McClellan equiripple FIR design.

INTRODUCTION

While IIR filters allow us to benefit from all of the efforts put into the development of classical filter theory, they also tie us to physical laws that need not apply in the discrete-time world. We do not have to provide physical devices, only a linear difference equation.

Finite impulse response filters have both advantages and disadvantages over IIR filters. FIR filters have no feedback and therefore are free from stability problems, including limit cycling, which are issues of some concern with IIR designs. Similarly, while coefficient accuracy will affect the ability to achieve a specific frequency response, it is not as important as when feedback is present. On the other hand, feedback makes for very efficient signal processing in terms of the number of multiply and accumulate operations required.

To those accustomed to the differential equations of continuous-time systems, the idea that filtering could be accomplished using a difference equation containing only input terms seemed far-fetched. In addition, the FIR handbook was a blank page, and early investigators had to find ways to create useful FIR difference equations. Today, a variety of standard design techniques exist for FIR filters. Believe it or not, we already know several of them!

10.1 FIR FROM IIR

This section will show how an FIR system can be created to duplicate the behavior of a stable IIR system. This gives insight into how filtering can be achieved working only with current and past inputs. After completing this section you will be able to:

- Convert an IIR design into an FIR filter.
- Relate impulse response to FIR filter coefficients.

The difference equation for an Mth-order linear FIR system is

$$y(n) = \sum_{i=0}^{M} a_i x(n-i) = a_0 x(n) + a_1 x(n-1) + \cdots + a_M x(n-M) \qquad \textbf{(10.1)}$$

and is implemented with the block diagram of Figure 10.1. The corresponding transfer function is

$$H(z) = \frac{\vec{Y}}{\vec{X}} = \sum_{i=0}^{M} a_i z^{-i} = a_0 + a_1 z^{-1} + \cdots + a_M z^{-M} \qquad \textbf{(10.2)}$$

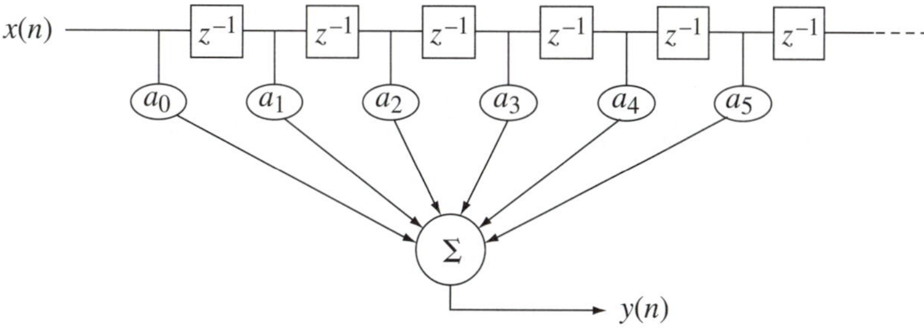

Figure 10.1 FIR implementation block diagram.

and the set of coefficients a_0–a_M make up the impulse response of the system. If that is not immediately evident, make an iteration table and apply a unit impulse. The result is shown in Table 10.1. Even if the a_i are increasing exponentially, the sequence ends after $M+1$ terms. There is just no way for the difference equation to be unstable.

There is a simple method for approximating an IIR filter with an FIR filter. The process involves dividing out the IIR transfer function polynomials to obtain a result in the form of Equation 10.2. The coefficients produced by this division constitute the impulse response of the IIR filter; while they go on forever, we know they must diminish with time if the system is stable. At some point they will become negligible, and it becomes reasonable to represent the infinitely long transient with a finite sequence.

Table 10.1 Unit Impulse Response of an Mth-Order FIR System

$$y(n) = \sum_{i=0}^{M} a_i x(n-i)$$

n	x(n)	$a_0x(n)$	$a_1x(n-1)$	$a_2x(n-2)$	$a_3x(n-3)$	...	$a_Mx(n-M)$	y(n)
0	1	a_0	0	0	0		0	a_0
1	0	0	a_1	0	0		0	a_1
2	0	0	0	a_2	0		0	a_2
3	0	0	0	0	a_3		0	a_3
⋮								⋮
M	0	0	0	0	0	...	a_M	a_M

EXAMPLE 10.1

Find an FIR implementation for a second-order Butterworth-type IIR filter with $v_{co} = 0.2$.

Solution

A second-order bilinear transformed Butterworth filter with the cutoff frequency specified has the transfer function

$$H(z) = \frac{0.0675(z^2 + 2z + 1)}{z^2 - 1.1430z + 0.4128}$$

The synthetic division goes as follows:

$$\begin{array}{r l l l}
 & 0.0675 \quad + 0.2120z^{-1} & + 0.2819z^{-2} \cdots & \\
z^2 - 1.143z + 0.4128 \big) & 0.0675z^2 + 0.1349z & + 0.0675 & \\
 & 0.0675z^2 - 0.0772z & + 0.0279 & \\
\hline
 & \qquad\quad 0.2120z & + 0.0396 & \\
 & \qquad\quad 0.2120z & - 0.2423 + 0.0875z^{-1} & \\
\hline
 & & 0.2819 - 0.08754z^{-1} &
\end{array}$$

Since quite a few terms may be needed, computer assistance is an attractive option.

```
>[n,d]=butter(2,0.2);          % get the Butterworth digital IIR filter
>nn=[n zeros(1,51)];           % get about 50 terms from the synthetic
                                 division
>a=deconv(nn,d);               % generates the first 51 a coefficients
%    Since this is also the impulse response, we could just use
>a=dimpulse(n,d,51);
```

The FIR design is complete. Now we will see how well the Mth-order FIR filter compares with the original IIR filter.

```
>num=a(1:11);               % try a 10th order FIR filter
>h10=freqz(num,1);          % this is the FIR response
>h=freqz(n,d);              % Use the IIR filter as a reference
>v=(0:511)/512;             % set up a frequency variable
>plot(v,20*log10(abs(h)),v,20*log10(abs(h10)))
etc.
```

The results are shown in Figure 10.2.

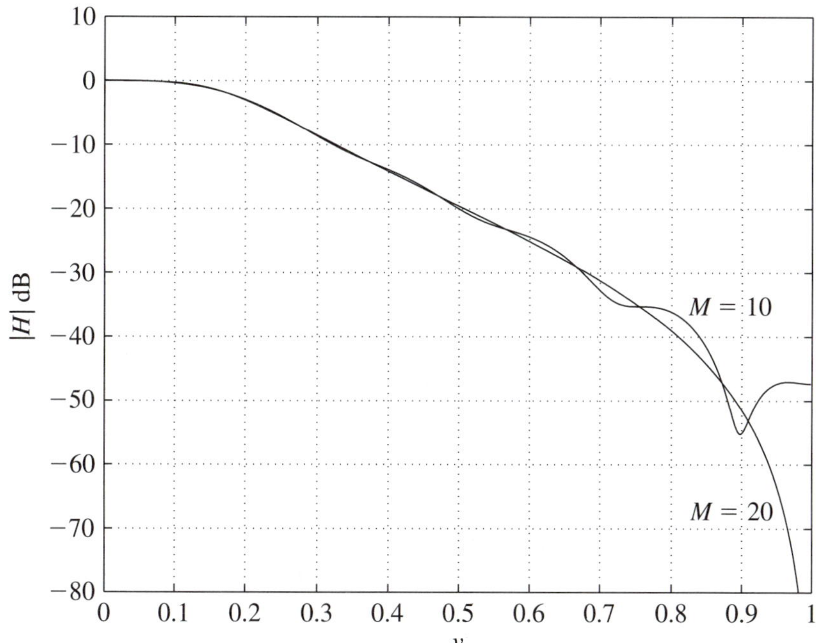

Figure 10.2 Comparison of the IIR second-order Butterworth filter derived from the bilinear transform and the 10th- and 20th-order FIR filters based on having the same impulse response. The 20th-order FIR filter characteristic is indistinguishable from the IIR characteristic over an 80-dB range.

In the case of Example 10.1, a 20th-order FIR filter matches the performance of the IIR filter over a substantial range. A second-order IIR filter requires five multiplications, while the FIR filter requires about four times as many to do the same job over an 80-dB range.

Similar comparisons can easily be made for other filter types (see, for instance, Figure 10.3.) This approach retains the ability to obtain a filter with well-defined passband and cutoff characteristics. It is rarely used, however, probably because it also ties us to the limitations of the classical filters.

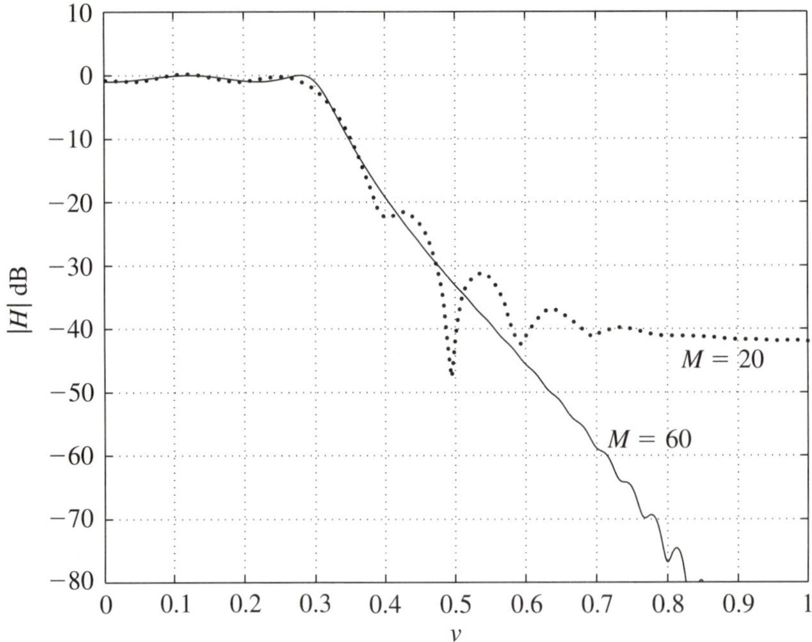

Figure 10.3 Comparison of an IIR fourth-order Chebyshev 1-dB, $\nu_{co} = 0.3$ filter derived from the bilinear transform, and 20th- to 60th-order FIR filters based on having the same impulse response. The $M = 60$ case cannot be distinguished from the exact IIR response in the passband.

10.2 SYMMETRIC DIFFERENCE EQUATIONS

FIR filters can have true linear phase if their coefficients have either even or odd symmetry about the center term (or space). After completing this section you will be able to:

- Create linear phase difference equations.
- Describe the frequency response of a finite averager.
- Describe the difference equation of a comb filter.

A difference equation in which the coefficients obey the relationship $a_i = a_{M-i}$ for all i is called *even symmetric.* Suppose, for instance, that M is an odd number. This gives an even number of difference equation terms,

$$H(z) = a_0 + a_1 z^{-1} + \cdots + a_{M-1} z^{-(M-1)} + a_M z^{-M}$$

and for an even symmetric transfer function we require

$$H(z) = a_0 + a_1 z^{-1} + \cdots + a_1 z^{-(M-1)} + a_0 z^{-M}$$

The frequency response is

$$H(e^{j\pi v}) = a_0 + a_1 e^{-j\pi v} + \cdots + a_1 e^{-j\pi(M-1)v} + a_0 e^{-j\pi M v}$$

$$H(e^{j\pi v}) = e^{-j\pi v \frac{M}{2}}[a_0 e^{j\pi v \frac{M}{2}} + a_1 e^{j\pi v\left(\frac{M}{2}-1\right)} + \cdots + a_1 e^{-j\pi v\left(\frac{M}{2}-1\right)} + a_0 e^{-j\pi v \frac{M}{2}}]$$

$$H(e^{j\pi v}) = e^{-j\pi v \frac{M}{2}}\left[2a_0 \cos \pi \frac{M}{2} v + 2a_1 \cos \pi\left(\frac{M}{2} - 1\right)v \right.$$

$$\left. + 2a_2 \cos \pi\left(\frac{M}{2} - 2\right)v + \cdots\right]$$

$$H(e^{j\pi v}) = 2e^{-j\pi v \frac{M}{2}} \sum_{m=0}^{(M-1)/2} a_m \cos \pi\left(\frac{M}{2} - m\right)v \qquad M \text{ odd} \tag{10.3}$$

The significance of this result is that the cosines are real functions of the variable v. They can only affect the phase by changing sign. If H is a typical filter, the cosine terms will be combining to produce a value of $|H| = 1$ in the passband, so the net sign will not be changing in that region. This makes the exponential entirely responsible for the passband phase, so $\angle H = -(\pi M/2)v$. In the stopband, the cosines will be combining to make $|H| = 0$, so it is likely that net sign changes will occur in the stopband, but no one cares what happens to the phase of rejected signals. The phase delay provided by such a transfer function is the same as its group delay and is given by

$$\angle H = -\tau_p \omega = -\left(\frac{\pi M}{2}\right)v = -\left(\frac{\pi M}{2}\right)\left(\frac{2\omega}{\omega_s}\right) = -\left(\frac{\pi M}{2}\right)\left(\frac{2}{2\pi f_s}\right)\omega = -\left(\frac{M}{2}T_s\right)\omega$$

$$\tau_p = \tau_g = \frac{M}{2}T_s \quad \text{even symmetric equations} \tag{10.4}$$

If M is even, the difference equation contains an odd number of terms. In that case the exponential multiplying the center term may be factored out, giving

$$H(e^{j\pi v}) = e^{-j(M/2)\pi v}\left[a_{M/2} + \sum_{m=0}^{(M/2)-1} 2a_m \cos\left(\frac{M}{2} - m\right)\pi v\right] \qquad M \text{ even}$$

The equation is still even symmetric and has the phase and group delay of Equation 10.4.

Odd symmetric difference equations have coefficients related by $a_{M-m} = -a_m$. They have the same group delay as even symmetric equations but start with a 90° phase shift. The impulse responses for even and odd symmetric FIR systems are shown in Figure 10.4.

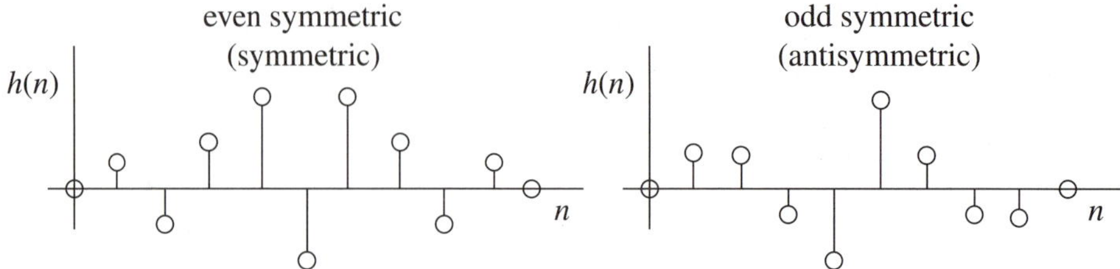

Figure 10.4 Symmetric impulse responses result in linear phase FIR filters.

$$H(e^{j\pi v}) = e^{-j\frac{\pi}{2}(Mv-1)}\left[\sum_{m=0}^{(M-1)/2} 2a_m \sin\left(\frac{M}{2} - m\right)\pi v\right] \qquad M \text{ odd}$$

$$H(e^{j\pi v}) = e^{-j\frac{\pi}{2}(Mv-1)}\left[\sum_{m=0}^{(M/2)-1} 2a_m \sin\left(\frac{M}{2} - m\right)\pi v\right] \qquad M \text{ even}$$

These equations show that we can create even- or odd-ordered FIR transfer functions with linear phase and a magnitude that has either even or odd symmetry. The complex Fourier series has coefficients that meet the symmetry requirements automatically, so it will be a significant FIR filter design tool.

EXAMPLE 10.2

Find the frequency response of the difference equation

$$y(n) = x(n) - 2x(n-1) + 2x(n-3) - x(n-4)$$

Solution

This is an odd symmetric difference equation with even M. The expression for $H(z)$ is

$$H(z) = 1 - 2z^{-1} + 2z^{-3} - z^{-4}$$
$$H(z) = z^{-2}(z^2 - 2z^1 + 2z^{-1} - z^{-2})$$

and the frequency response is found by substituting $z = e^{j\pi v}$:

$$H(e^{j\pi v}) = e^{-j2\pi v}(e^{j2\pi v} - e^{-j2\pi v} - 2e^{j\pi v} + 2e^{-j\pi v})$$
$$H(e^{j\pi v}) = e^{-j2\pi v}(2j \sin 2\pi v - 4j \sin \pi v)$$
$$H(e^{j\pi v}) = e^{-j\pi(2v-1)/2}2(\sin 2\pi v - 2 \sin \pi v)$$

A plot of the result (Figure 10.5) shows a very simple bandpass filter with absolutely linear phase (constant group delay). Usually the filter gain would be normalized to a unity passband.

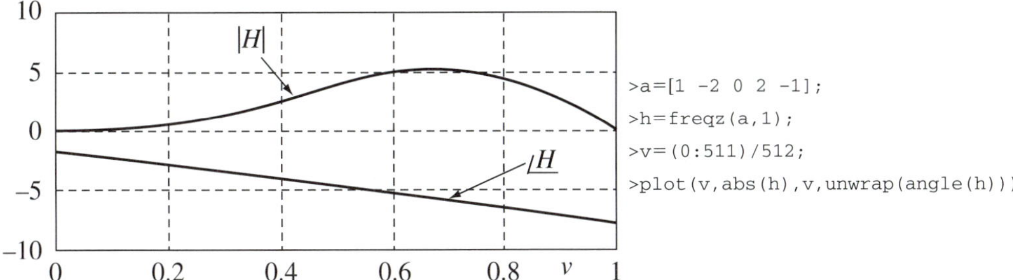

Figure 10.5

The difference equation of Example 10.2 was made simple enough so that the filter gain could be sketched without computer assistance. Usually only high order FIR filters give good results, and computer assistance is essential in obtaining their frequency response. There are two interesting exceptions. The first is the M point finite averager whose difference equation is:

$$y(n) = \frac{1}{M}[x(n) + x(n-1) + x(n-2) + x(n-3) + \cdots + x(n-(M-1))]. \qquad \textbf{(10.6)}$$

The coefficients are all 1 and are therefore even symmetric. This difference equation is particularly easy to implement if M is a power of 2, since no multiplications are required, and the division involves only an arithmetic shift of the summed terms. It provides a running average of the incoming signal. The transfer function of this equation is

$$H(z) = \frac{1}{M}[1 + z^{-1} + z^{-2} + \cdots + z^{-(M-1)}] = \frac{1 - z^{-M}}{M(1 - z)}$$

where we have used Equation 7.14 for the closed-form expression of a geometric series. The frequency response of this difference equation is

$$H(e^{j\pi v}) = \frac{1 - e^{-j\pi M v}}{M(1 - e^{j\pi v})} = \frac{e^{-j\pi M v/2}}{e^{-j\pi v/2}} \frac{(e^{j\pi M v/2} - e^{-j\pi M v/2})}{M(e^{j\pi v/2} - e^{-j\pi v/2})}$$

$$H(e^{j\pi v}) = e^{-j\pi(M-1)v/2} \frac{\sin M\pi v/2}{M \sin \pi v/2} \qquad \textbf{(10.7)}$$

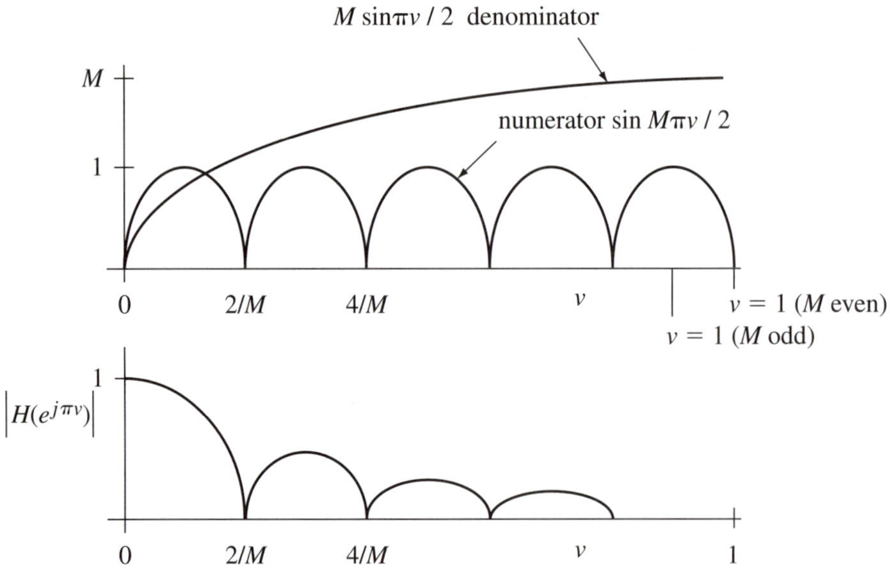

Figure 10.6 Sketching the frequency response of the M-term averager. The result is a poor low-pass filter. Nulls in the filter indicate frequencies at which exactly an even number of samples are taken over the sinewave cycle. These samples will always add to zero.

We intuitively expect that if random noise were present on a signal, then averaging many samples would average out the noise, leaving only the signal. Of course, the signal would have to be changing very slowly in comparison to the noise or it would be averaged out as well. The operation described is that of a low-pass filter; sketching Equation 10.7 will show such a characteristic (Figure 10.6). The finite averager is not a particularly good filter, but it is one of the few for which we can obtain a simple closed-form expression. It is also an easy one to implement on any microprocessor and can be used to demonstrate the frequency response of a real-time DSP system.

Another simple, even symmetric difference equation combines a signal with its echo:

$$y(n) = x(n) + x(n - M) \tag{10.8}$$

The frequency response for the echo equation is

$$H(e^{j\pi v}) = (1 + e^{-jM\pi v}) = 2e^{-jM\pi v/2} \cos \frac{M\pi v}{2} \tag{10.9}$$

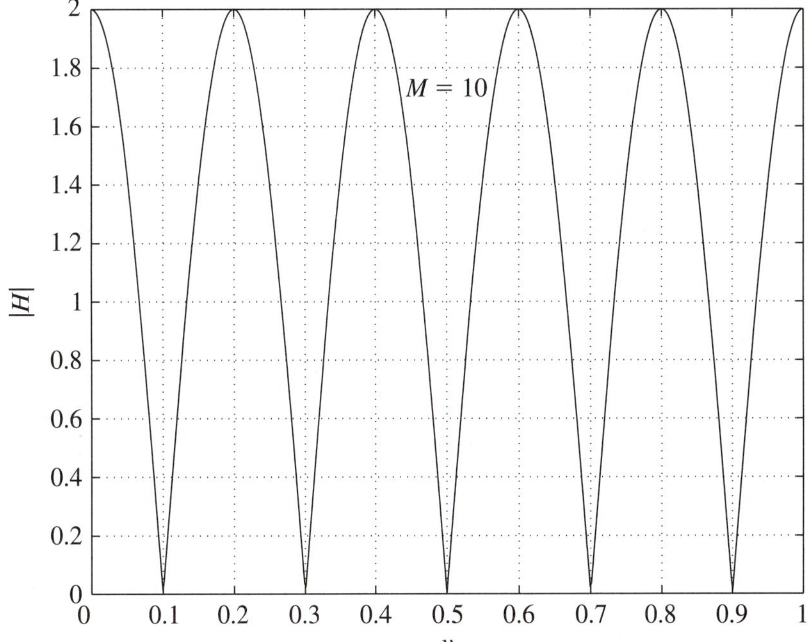

Figure 10.7 Impulse response of an echo delayed MT_s seconds. The result is called a *comb filter.*

which generates the *comb filter* frequency response of Figure 10.7. This response can be explained as follows. If the echo sinusoid is delayed by exactly an odd number of half cycles, it will be 180° out of phase with the input and the two will cancel each other, leading to zero gain. This occurs if

$$MT_s = n(T/2) \qquad (n \text{ odd})$$

or

$$v = n/M \qquad (n \text{ odd}).$$

Similarly, if the echo is delayed an even number of half cycles, the signal and the echo will be in phase and add together, creating a gain of 2 at frequencies of

$$v = n/M \qquad (n \text{ even}).$$

On the other hand, if either $x(n)$ or its echo is inverted, making the difference equation odd symmetric, the frequencies that are passed and stopped are reversed.

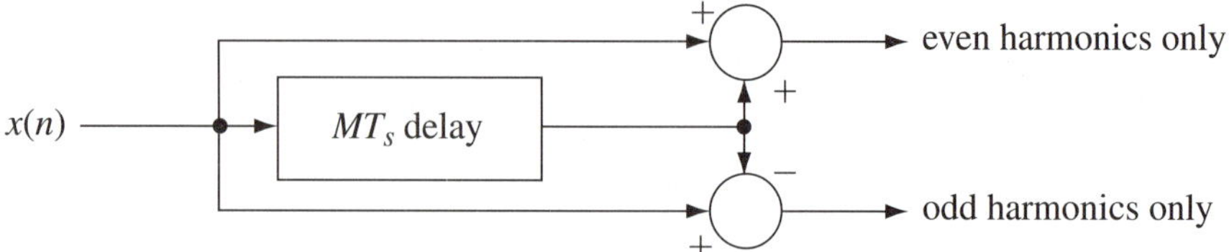

Figure 10.8 Combining a comb filter with an inverter allows a periodic signal to be separated into its even and odd harmonics.

Suppose you have a quasi-periodic signal where the even harmonics carry one piece of information, and the odd harmonics carry another. Using the scheme of Figure 10.8, these independent pieces of information can be recovered without interfering with one another. This is exactly the case with the chrominance and luminance signals in television sets. The circuits that pick them up and separate them once required careful adjustment for good results. Modern sets use a comb filter to accomplish this task with ease.

10.3 WINDOW DESIGNS

The frequency response of a discrete-time system is known to be periodic in v; consequently it may be expanded in a Fourier series. Since we usually want to pass the signal either completely or not at all, the frequency response will be discontinuous, and a window function will be needed to avoid rippling. After completing this section you will be able to:

- Use a Fourier series and a window function to design FIR filters.
- Design and suggest applications for an FIR differentiator.
- Design and suggest applications for a Hilbert transformer.

In the past, the functions we have expanded in a Fourier series have been functions of time, but the Fourier series may be used for representing the periodic function of any variable. It is only necessary to generalize the physical notions of period and fundamental frequency to obtain the required formulas. In terms of the normalized frequency variable, for instance, the period is $v = 2$, and the fundamental "frequency" is $2\pi/\text{“}T\text{”} = \pi$.

If the *desired* transfer function is represented by $H_d(e^{j\pi v})$, the magnitude function may be expanded in a Fourier series as

$$f(t) = \sum_{m=-\infty}^{\infty} c_m e^{jm\omega_o t} \Rightarrow |H_d(e^{j\pi v})| = \sum_{m=-\infty}^{\infty} c_m e^{jm\pi v} \qquad \textbf{(10.10)}$$

where

$$c_m = \frac{1}{T}\int_T f(t)e^{-jm\omega_o t}\,dt \Rightarrow c_m = \frac{1}{2}\int_{-1}^{1} |H(e^{j\pi v})|\,e^{-jm\pi v}\,dv \tag{10.11}$$

Since the Fourier series must be terminated to give a practical FIR filter, we can only *approximate* the desired transfer function magnitude. For an Mth-order FIR filter we would have

$$|H_a(e^{j\pi v})| = \sum_{m=-M/2}^{M/2} c_m e^{jm\pi v} \tag{10.12}$$

Usually we have an $H(z)$ and find its frequency response by substituting $e^{j\pi v}$ for z. Reversing that process suggests an $H(z)$ of the form

$$H_a(z) = \sum_{m=-M/2}^{M/2} c_m z^m \tag{10.13}$$

We cannot implement Equation 10.13 directly, because it is noncausal due to the z^{+m} terms. To make it *causal,* we simply multiply all the terms by $z^{-M/2}$, which delays the outputs by $M/2$ samples, giving

$$H_c(z) = z^{-M/2}H_a(z) = \sum_{m=-M/2}^{M/2} c_m z^{(m-M/2)} = \sum_{i=0}^{M} c_{(M/2-i)} z^{-i}$$

This matches Equation 10.2 provided

$$a_m = c_{(M/2-m)} \qquad \text{for} \qquad 0 \le m \le M \tag{10.14}$$

Whenever we have listed the complex Fourier series coefficients in the past, we started with $c_{-M/2}$ and progressed to $c_{M/2}$. Equation 10.14 shows that this order should be reversed for the difference equation terms (i.e., $a_0 = c_{M/2}$). This matters only for odd symmetric equations, since even symmetric coefficients are the same either way. To reverse the order of the coefficients for the odd symmetric case, just multiply them by -1.

EXAMPLE 10.3

Determine the coefficients of a sixth-order bandpass FIR filter based on a Fourier series expansion. The passband is to be $0.4 \le v < 0.70$.

Solution

Figure 10.9a shows a general transfer function that can be used for low-pass, high-pass, or bandpass designs. For low-pass designs, $v_a = 0$; for high-pass designs, $v_b = 1$.

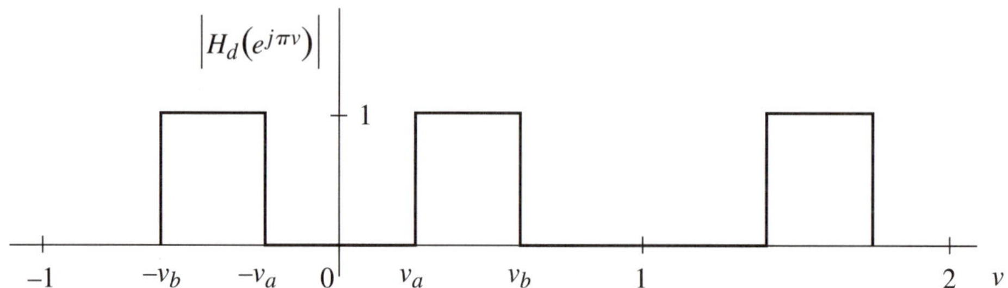

Figure 10.9a A general form of a *desired* filter amplitude response having a single passband.

The Fourier series coefficients are easily found from Equation 10.11 as

$$c_m = \frac{1}{2}\left[\int_{-v_b}^{-v_a} e^{-j\pi m v}\,dv + \int_{v_a}^{v_b} e^{-j\pi m v}\,dv\right] = \int_{v_a}^{v_b} \cos \pi m v\, dv$$

$$c_m = \begin{cases} \dfrac{\sin m\pi v_b - \sin m\pi v_a}{\pi m} & m \neq 0 \\ v_b - v_a & m = 0 \end{cases} \tag{10.15}$$

For the bandpass filter desired in this example, $v_a = 0.4$ and $v_b = 0.7$. Then

$$c_m = \begin{cases} 0.3 & m = 0 \\ \dfrac{\sin 0.7\pi m - \sin 0.4\pi m}{\pi m} & m \neq 0 \end{cases}$$

These coefficients do not have to be flipped, because the general transfer function of Figure 10.9a has even symmetry, which means that $c_m = c_{-m}$.

The coefficients are

$$c_0 = +0.3000, \quad c_{\pm 1} = -0.0452, \quad c_{\pm 2} = -0.2449, \quad c_{\pm 3} = 0.0952$$

and the FIR transfer function is

$$H(z) = 0.0952 - 0.2449z^{-1} - 0.0452z^{-2} + 0.3000z^{-3} - 0.0452z^{-4} - 0.2449z^{-5} + 0.0952z^{-6}$$

Once again a low order filter has been demonstrated to allow the process to be followed step by step. The resulting frequency response is shown in Figure 10.9b.

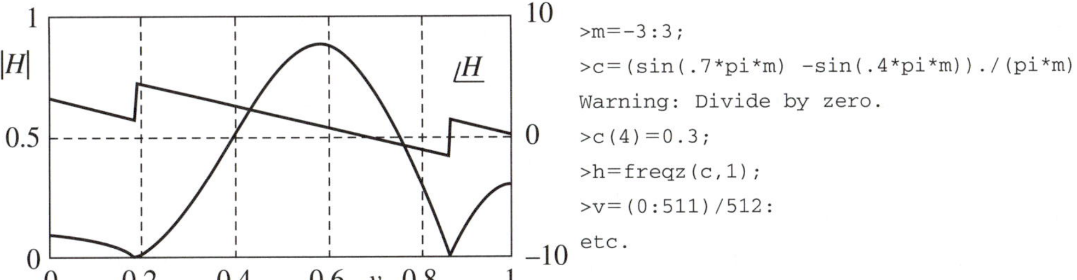

Figure 10.9b A sixth-order bandpass FIR filter characteristic before windowing. The order of the filter is too small to produce a sharp magnitude response, but the phase is certainly linear over the passband.

The desired filter transfer functions are always discontinuous, which means that terminating the Fourier series in a partial sum will result in rippling. Unequal rippling in the passband is unfortunate, but rippling in the stopband ruins the attenuation of the filter. A window function is consequently essential for producing a useful filter, which is why these Fourier-based filters are classified as window designs.

Figure 10.10 compares the response curves for a filter using a variety of standard windows. The rectangular window has the sharpest cutoff, but achieves it by accepting unequal ripple in the passband. It also only guarantees about 22 dB of attenuation in the stopband and is not usually acceptable. The Hamming and Hanning windows slightly sacrifice the rate of fall-off outside the passband to achieve a smoother passband and greater eventual attenuation. The Hamming window provides superior attenuation immediately adjacent to the passband, while the Hanning window provides greater attenuation away from the passband. Other, more exotic window functions are also available, but there is no magic window that does not require some tradeoffs between steepness of cutoff and minimum attenuation in the stopband.

The windowing method provides FIR filters with minimal rippling in the passbands, so they are similar to Butterworth designs. There is, unfortunately, no way to specify the cutoff frequency as part of the design. One can only try various values for a passband edge to find a filter *order-edge-window* combination that is acceptable.

There are two special purpose transfer functions that can be synthesized using an odd symmetric FIR equation. In these applications the magnitude of the transfer function is allowed to carry some of the phase information as well. The first transfer function synthesized using this technique is the *differentiator*.

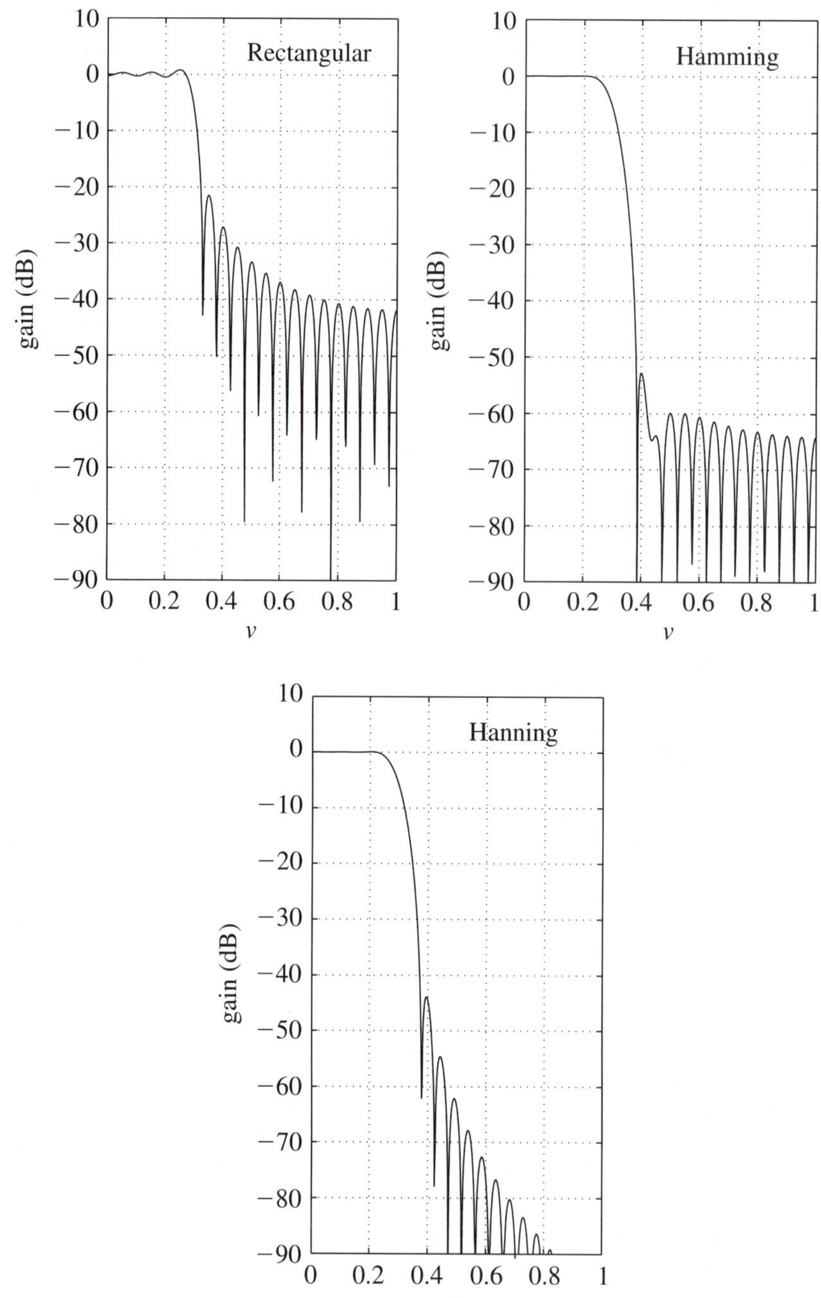

Figure 10.10 Response roll-off for a 40th-order FIR low-pass filter designed for a pass-band edge at $v = 0.3$ using rectangular, Hamming, and Hanning window functions.

FM modulation signals are given a standard pre-emphasis to boost their high-frequency components so that they will be less affected by noise during transmission. The received signal is then provided a matching de-emphasis to restore the original modulation spectrum. The pre-emphasis needs a frequency response like $H(s) = s = j\omega = j(\omega_s/2)v = j(\pi/T_s)v$. The differentiator's high-frequency emphasis also tends to amplify high-frequency noise, so it is best to bring the gain back to zero once the passband has ended.

Figure 10.11 suggests alternative ways of viewing the differentiator characteristic. Both viewpoints have exactly the same net magnitude and phase, and a Fourier series can be used to represent either magnitude function. Ordinarily, we associate negative frequencies with a conjugate phasor and tend to think in terms of spectrum amplitudes having even symmetry and spectrum phases having odd symmetry. Using the Fourier series creates a real function, which has a phase only in the sense that it may change sign. Then we add a linear phase to make the Fourier derived filter equation causal. We could, however, add any desired *constant* phase as well. Implementing Differentiator II is the only way to achieve both the desired magnitude and the desired phase. Differentiator II has the additional advantage that its series representation near the origin will have an easier time matching the straight-line characteristic than it would have had creating the "corner" of Differentiator I.

The derivation for the differentiator coefficients starts with the Fourier definitions

$$|H_a(e^{j\pi v})| = \sum_{m=-M/2}^{M/2} c_m e^{jm\pi v}$$

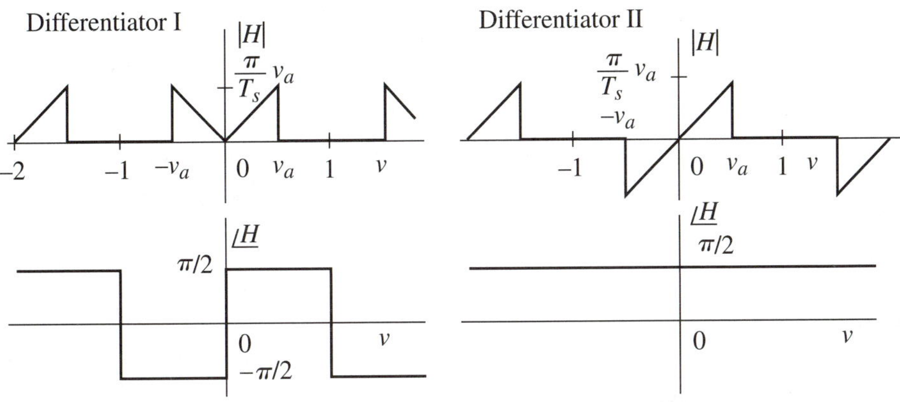

Figure 10.11 Alternate ways of viewing the differentiator characteristic. Differentiator II will have a superior amplitude response at the origin, because it is easier for the Fourier series to approximate a straight line than it is to make a corner. Its constant phase is also easily added to the final transfer function.

where

$$c_m = \frac{1}{2}\int_{-v_a}^{v_a}\left(\frac{\pi}{T_s}v\right)e^{-jm\pi v}\,dv$$

Since Differentiator II is an odd function, the integral may be simplified as

$$c_m = \int_0^{v_a}\left(\frac{\pi}{T_s}v\right)(-j\sin m\pi v)\,dv = \frac{-j}{T_s\pi m^2}[\sin m\pi v - m\pi v\cos m\pi v]_0^{v_a}$$

$$c_m = \begin{cases} \dfrac{-j}{T_s\pi m^2}[\sin m\pi v_a - m\pi v_a \cos m\pi v_a] & m \neq 0 \\ 0 & m = 0 \end{cases}$$

Although these coefficients are imaginary, they produce a purely real H. This is an example of the odd symmetric FIR filter with even M. The $+m$ and $-m$ terms will combine to produce a series of sine functions. To obtain the desired phase, we need to multiply these coefficients by j. Then the differentiator coefficients become

$$d_m = jc_m = \frac{1}{T_s\pi m^2}[\sin m\pi v_a - m\pi v_a\cos m\pi v_a] \qquad m \neq 0 \tag{10.16}$$

and

$$H_c(z) = z^{-M/2}H_a(z) = z^{-M/2}\sum_{-M/2}^{M/2} d_m w_m z^m = \sum_{m=-M/2}^{M/2} d_m w_m z^{m-M/2}$$

$$H_c(z) = \sum_{i=0}^{M} d_{(M/2-i)}w_{(M/2-i)}z^{-i} \tag{10.17}$$

If the notation of the final equation is confusing, remember that all we really need is the set of a coefficients for Equation 10.2. These are generated by windowing d_m and flipping the sequence so that the first coefficient is $d_{M/2}$.

EXAMPLE 10.4

Design a 20th-order FIR differentiator for $v < 0.3$ using a Hamming window. Assume $T_s = 1$.

Solution

The series coefficients are given by (Eq. 10.15)

$$d_m = \frac{1}{\pi m^2}[\sin 0.3m\pi - 0.3m\pi \cos 0.3m\pi]$$

$$|H_a(e^{j\pi v})| = \sum_{m=-10}^{10} d_m w_m e^{jm\pi v}$$

where w_m is the Hamming window.

```
>m=-10:10;
>n=0.3*pi*m;
>dm=(sin(n)-n.*cos(n))./(pi*m.^2);
Warning: Divide by zero.
>dm(11)=0;
>w=hamming(21);
>dmprime=dm.*w';
>a=-dmprime          % one way to reverse the odd symmetric coefficient order
a0-a4      0.0024      0.0023     -0.0012     -0.0104     -0.0182
a5-a9     -0.0069      0.0334      0.0859      0.1113      0.0794
a10-a14    0          -0.0794     -0.1113     -0.0859     -0.0334
a15-a19    0.0069      0.0182      0.0104      0.0012     -0.0023
  a20     -0.0024
```

Figure 10.12 shows the resulting frequency characteristic prior to adding the causal delay.

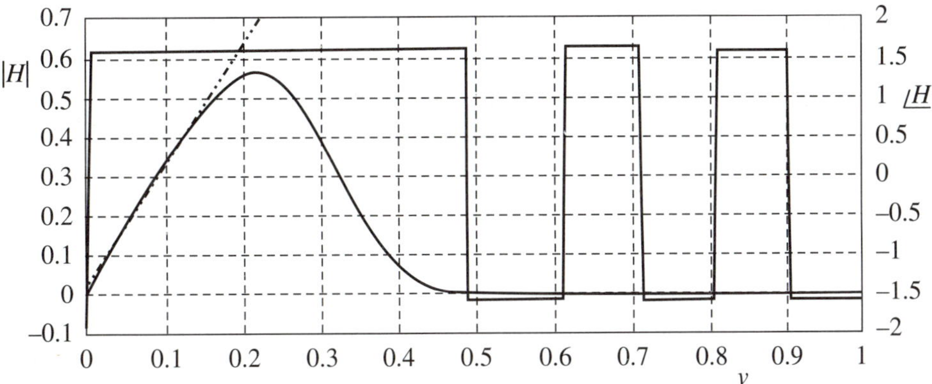

Figure 10.12 A 20th-order FIR differentiator using a Hamming window and $T_s = 1$. Notice that the useful range of the differentiator is only about half that intended. Either more terms are required or the goal value of v_a should be increased. The straight line indicates the slope desired of the differentiator characteristic. Sign changes in the stopband cause the alternating phase activity.

Note: To obtain the frequency response of odd symmetric systems, using H=freqz (a, 1) will provide the response of the causal system. To get the noncausal result we could use the FS.m function developed in Chapter 6, but it must be modified to plot the imaginary part instead of the real part.

Another interesting transfer function that is also handled by making its amplitude carry part of the phase information is the Hilbert transformer. Its idealized characteristic is shown in Figure 10.13. In generating a single sideband-modulated signal by the phase shift method, it is necessary to shift all the modulation signal components by 90°. While it is easy to build a circuit to shift a single frequency by 90°, it is much more difficult to shift an entire audio spectrum by 90°. The Hilbert transformer does exactly that. The Fourier series coefficients are multiplied by $-j$ to give the Hilbert transformer coefficients

$$c_m = \frac{1}{2}\int_{-v_a}^{v_a} |H| e^{-jm\pi v}\, dv = \int_{0}^{v_a} -j \sin m\pi v\, dv = j\frac{\cos m\pi v}{m\pi}\Bigg|_{0}^{v_a}$$

$$h_m = -jc_m$$

$$h_m = \begin{cases} 0 & m = 0 \\ \dfrac{\cos v_a \pi m - 1}{\pi m} & m \neq 0 \end{cases} \tag{10.18}$$

and the corresponding FIR filter coefficients are $a_i = h_{(M/2-i)}$ for $0 \leq i \leq M$.

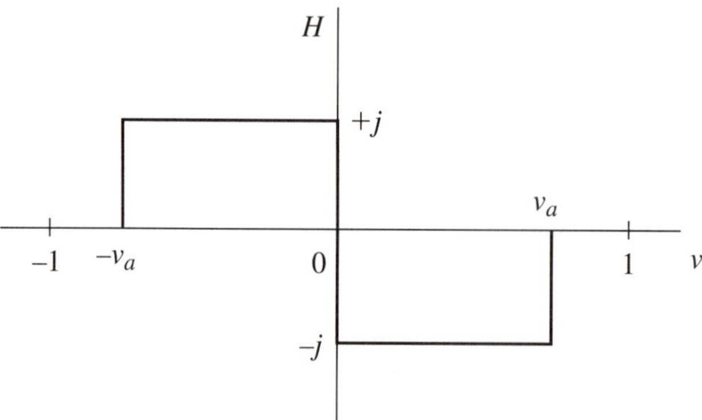

Figure 10.13 The ideal Hilbert transformer.

10.4 FREQUENCY SAMPLING

There are a variety of design techniques that are classified as frequency sampling design approaches. The one we will pursue starts with samples of a desired frequency response. The response will be specified in a manner that is compatible with the DFT. Then we will take the IDFT of these samples to find the impulse response and, therefore, the FIR filter coefficients. After completing this section you will be able to:

- Use the DFT for an empirical frequency sampling design of FIR filters.

Although the analytical development of the frequency sampling approach is rather involved, we have sufficient experience with the DFT to explore this approach empirically as a computer-oriented design procedure.

For an Mth-order FIR filter, we need to describe the desired $H(e^{j\pi v})$ in $M + 1$ points equally spaced around the unit circle of the z domain. If we do it properly, the inverse transform will give real values for the impulse response. Remembering that the DFT provides a frequency spectrum running from $v = 0$ to $v = 2$, we know the H values equal distances either side of the $v = 1$ point must be complex conjugates. If $v = 1$ is a sampling point, which it is for odd M, its phase must be zero. The phase must have a negative slope if plotted versus v in order to have a positive group delay. Using Equation 10.3 for guidance suggests that the phase should vary by about $M\pi$ over the range $0 \le v < 2$.

Using these empirical rules, we start a frequency sampling design by defining the desired frequency response. We will design a 15th-order low-pass filter to demonstrate the process. The desired filter will be described to the IDFT by the data points indicated in Figure 10.14a. Remember that the IDFT starts its next cycle at $v = 2$, so no data is entered for that frequency.

Once the samples of $H(e^{j\pi v})$ are presented to the IDFT, they should result in an entirely real $h(n)$ impulse response.Even if the $h(n)$ appear to be real, it is a good idea to take their real part to eliminate any small misleading effects the imaginary parts may predict for the design. In addition, the impulse response should show an appropriate symmetry about its center point, or the design will not have a linear phase characteristic. The programming and impulse response for this demonstration are shown in Figure 10.14b.

Next, we would like to compare the frequency response actually achieved with our sampling points. To do this we will pad the impulse response with zero fill to obtain more points back in the frequency domain. Remember to make sure $h(n)$ is real before doing this. There is no interest in seeing the frequency response of an FIR filter with complex coefficients.

The frequency response of this demonstration (Fig. 10.14c) matches the sampling points specified, and shows the usual rippling. The rippling will encourage us

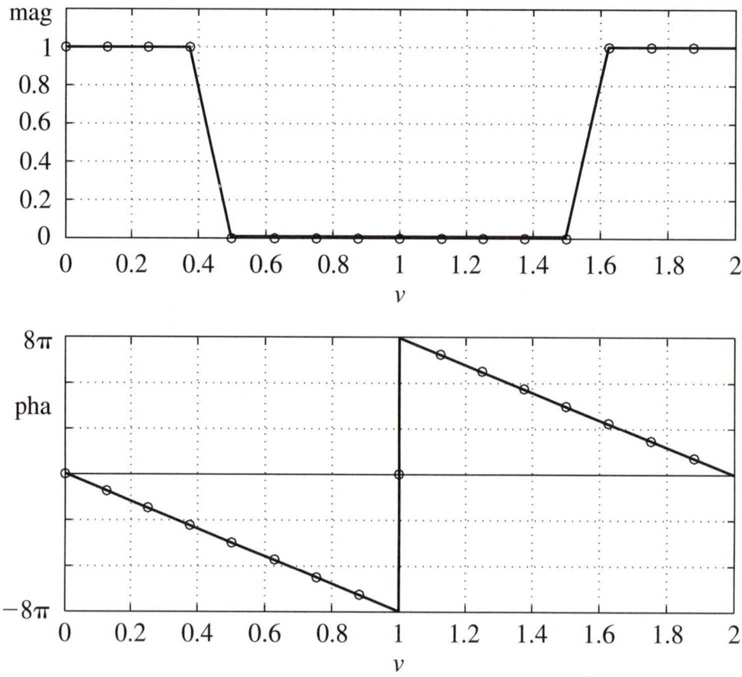

Figure 10.14a Proposed 15th-order low-pass FIR filter (16 samples). Phase should vary *about* $M\pi$ radians over the frequency range.

```
>mag=[1 1 1 1 0 0 0 0 0 0 0 0 0 1 1 1];
>ph1=-(0:7)*pi; ph2=(7:-1:1)*pi;
>pha=[ph1 0 ph2];
>H=mag.*exp(j*pha);
>h=ifft(H);
>h=real(h); % should already be real
>n=0:15;
>stem(n,h)
```

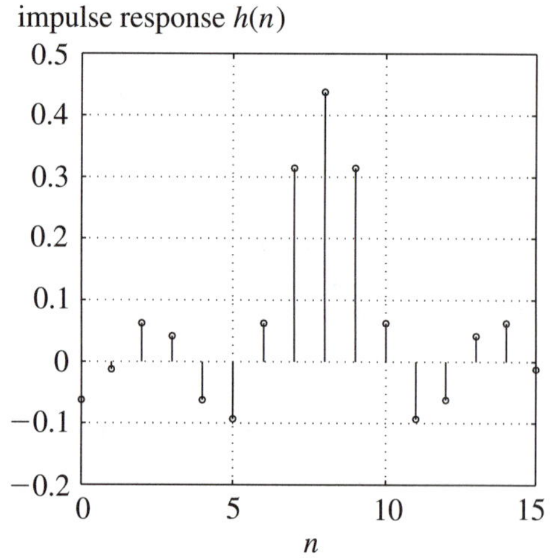

impulse response $h(n)$

Figure 10.14b MATLAB programming and the impulse response provided by the IDFT from the sample points of Figure 10.14a. The impulse response should be entirely real. It should also be *nearly* symmetrical about the center in order to have *nearly* linear phase.

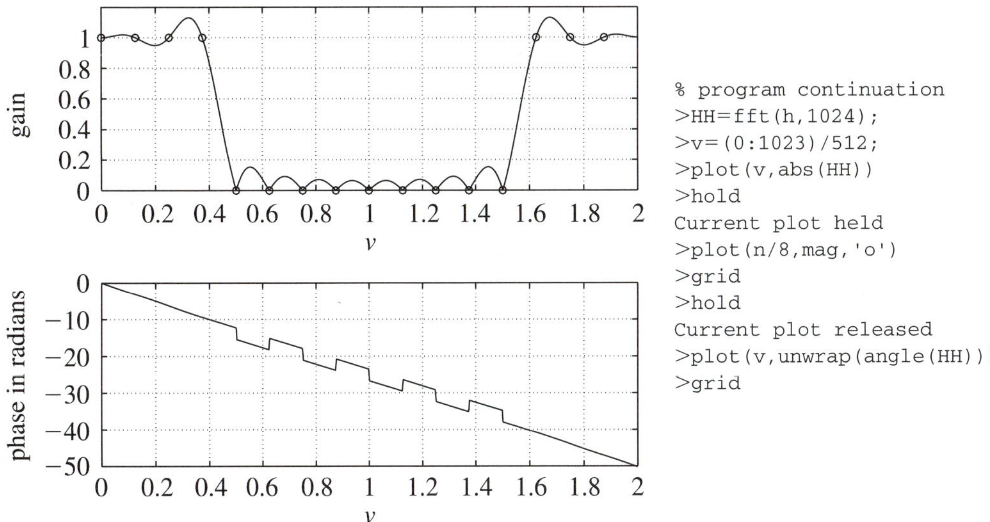


```
% program continuation
>HH=fft(h,1024);
>v=(0:1023)/512;
>plot(v,abs(HH))
>hold
Current plot held
>plot(n/8,mag,'o')
>grid
>hold
Current plot released
>plot(v,unwrap(angle(HH)))
>grid
```

Figure 10.14c The magnitude and phase response of the filter whose impulse response was found in Figure 10.14b is obtained by zero-filling the impulse data record. The original magnitude sampling points are circled.

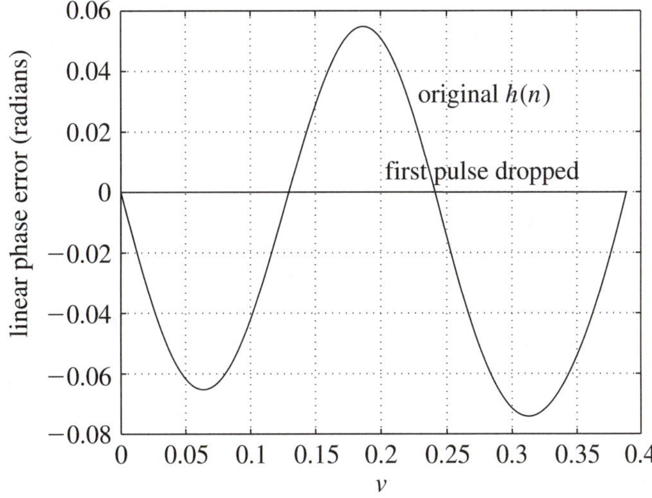

Figure 10.14d Although the phase in Figure 10.14c appears to be linear over the passband, a closer inspection would verify that it is not. One way to look more closely is to pick a phase reading near the end of the passband, and use it to create a straight line equation that can be subtracted from the actual phase. If both are linear, the result should be a difference of zero. In this case the empirical approach has created a phase variation of less than 0.08 radians. No phase variation occurs if the first pulse is dropped.

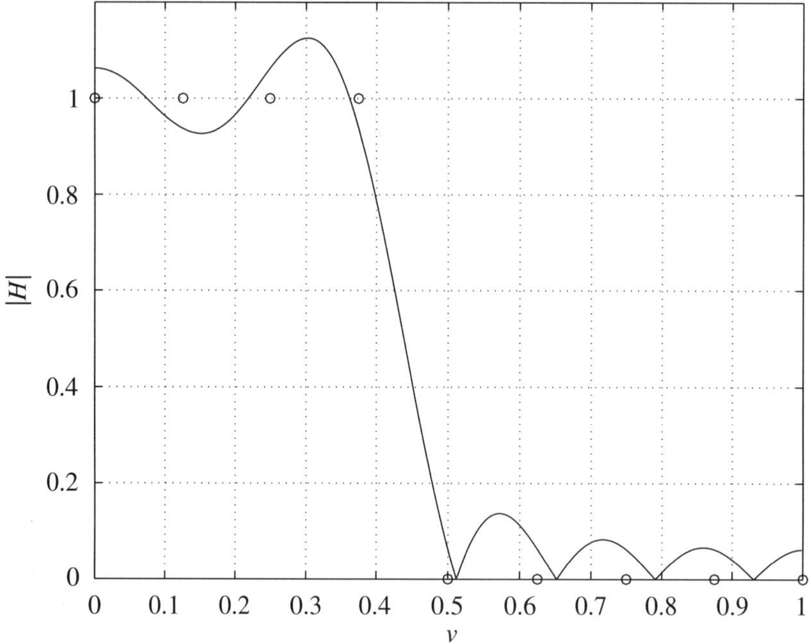

Figure 10.14e Shown is the amplitude response obtained by dropping the first impulse response term to produce a true linear phase. Although the response no longer hits the original sampling points, that may have little impact after windowing.

to use a window function, particularly to improve the stopband attenuation. The phase plot appears quite linear over the passband, but a closer inspection (Fig. 10.14d) shows that it is not ideal. The simple empirical rule we have adopted for selecting the phase samples does not produce a symmetrical impulse response. But Figure 10.14b suggests that the impulse response would be symmetrical if we simply discarded the first $h(n)$ value! This empirical observation seems to hold for any practical M value. With the first impulse term dropped, the resulting amplitude response is shown compared to the original samples in Figure 10.14e.

A more analytical approach to frequency sampled filters would show that a phase *increment* of $M\pi/(M + 1)$ will give a symmetrical impulse response and linear phase from the start. This fact is derived from the phase sample spacing shown in Figure 10.15.

If a truly linear phase is required, this section has demonstrated two techniques of achieving it through frequency sampling: (1) design for a phase increment of $M\pi/(M + 1)$, and (2) design for a total phase change of $M\pi$ and discard the first term in $h(n)$. The resulting filters will be very similar but not identical.

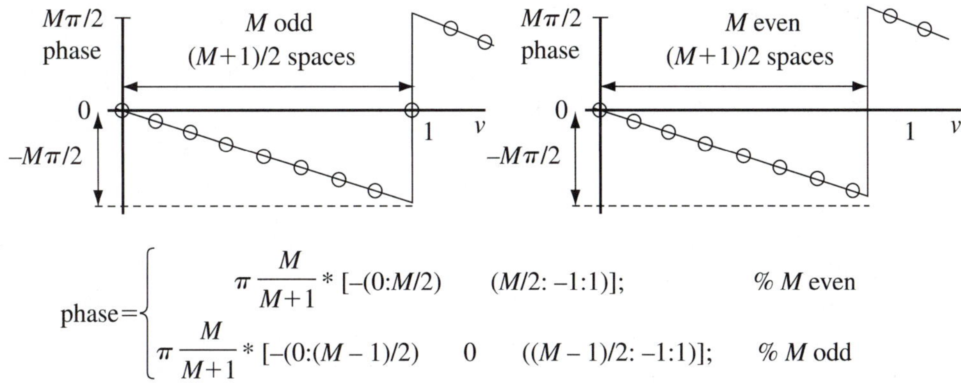

Figure 10.15 Phase sample spacing required for true linear phase, and MATLAB expressions for obtaining it.

10.5 A FINAL MATLAB LESSON

As with other important aspects of systems, MATLAB provides functions that design FIR filters. After completing this section you will be able to:

- Produce a Parks–McClellan equiripple FIR design.
- Use **fir1** to produce window designs.

The window FIR design could use a DFT program to calculate the Fourier coefficients, as was demonstrated repeatedly in Chapter 7. Since the gain is either 1 or 0, a program could deduce the description of $|H|$ from the cutoff frequency information and routinely apply a Hamming window.

MATLAB EXAMPLES

fir1

```
>a=fir1(20,0.2)      % designs a 20th-order FIR filter with vco = 0.2 and a
                       Hamming window
>freqz(a,1)
%       fir1 scales the coefficients to provide a center passband gain of
        exactly 1
>w=[0.4    0.7];               % define the passband of Example 10.3
>a=fir1(6,w, 'noscale')        % do not scale
>w=hamming(7);                 % need to remove the Hamming window
>a=a./w'                       % these are the coefficients of Example 10.3
>help fir1                     % for more options
```

Both the Fourier series and sampled frequency FIR filter designs need to be used with an appropriate window function and consequently have low-ripple passbands most similar to Butterworth filters. The Parks–McClellan algorithm for the design of FIR filters is a well-known and popular alternative to windowed designs. It provides linear-phase filters with an optimal match between the specified and achieved amplitude response. The designs are based on Chebyshev approximation theory, which leads to equiripple passbands and stopbands. The heart of this program is a highly efficient iteration procedure for finding the optimal solution called the *Remez exchange algorithm.*

As with other computer-based design approaches, the design is optimized in the sense that it produces minimum error between the achieved and specified filter magnitude in each frequency band. The error may be given different weightings in each frequency band. Doing so will force the program to reduce the rippling in those bands where the error weighting is higher.

For standard filters, the program user must identify the frequency pass and stopbands, by giving the starting and ending frequency of each, and specify the desired transfer function amplitude (usually 1 or 0) at each of these frequencies. The frequencies specified are *v values* and must start at 0 and end at 1. The original program dates from the early 1970s and punched-card input. The Parks–McClellan program also designs differentiators and Hilbert transformers. The amplitudes specified for the differentiator set its slope.

Remember, an optimum design based on a minimum-error criterion is not necessarily the same as the best design for a particular application. As with other computer-oriented design programs, the success of the resulting design is dependent on how well the programmer is able to describe the desired results to the algorithm. Attempts to force abrupt transitions between bands usually leads to excessive ripple. Since it is easy to make repeated trials, the designer can experiment with different specifications until a suitable result is achieved. This is an essential element of the design process, since ripple and cutoff frequency are not entered as initial design parameters.

remez

```
>f=[0 .1  .2 .6  .7 1];                   % specify 3 frequency bands
>m=[0  0   1  1   0 0];                   % give the desired |H| at each
                                            frequency value (BP filter)
>w=[  2      1      2];                   % optional error weighting per
                                            frequency band
>num=remez(40,f,m,w);                     % a 40th-order bandpass filter
>freqz(num,1)

>f=[0 .3    .6 1];                        % 2 bands
>m=[0 .3*pi  0 0];                        % amplitudes for a differentiator
>h=remez(20, f,m,'differentiator');       % get the impulse response
>stem (real(h))                           % may as well take a look at an odd
                                            symmetric response
>H=freqz(h,1);                            % get the transfer function
```

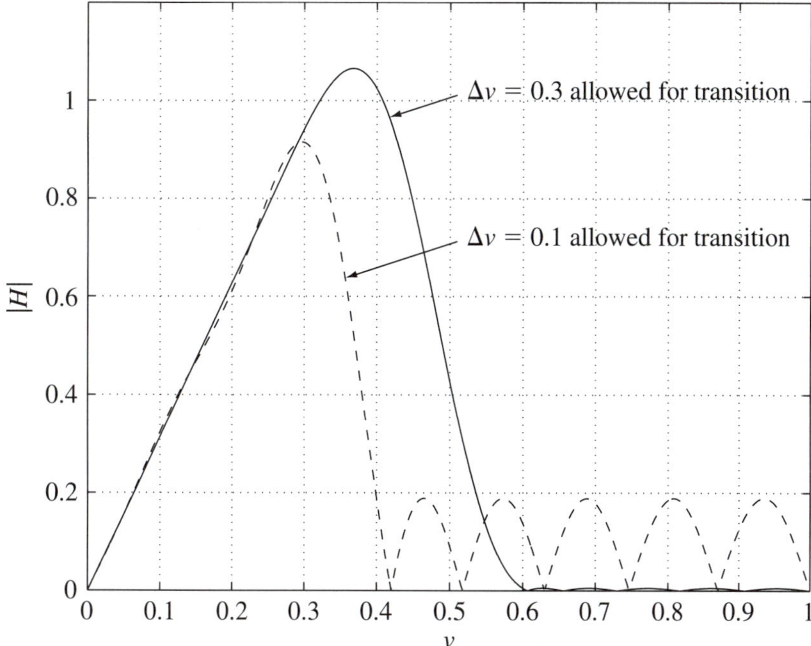

Figure 10.16 Twentieth-order Parks–McClellan designs for a differentiator. The two designs differ only in the way the frequency bands are specified. In both cases, the differentiator region was set for $0 < v < 0.3$. In one case (dashed curve) the stopband was specified as $0.4 < v < 1$, resulting in excessive ripple. In the other case (solid curve) the stopband was specified as $0.6 < v < 1$.

```
>v=(0:511)/512;
>plot(v,abs(H))
% This result is shown as the solid curve in Fig. 10.16
>plot(v,unwrap(angle(H)))          % check the phase
>help remez                        % whenever needed!
```

Windows are not used with Parks–McClellan designs. That would simply undo the optimization procedure inherent in the program. The stopband ripple is controlled by the error weighting given to that band. Weighting the error heavily forces the program to seek solutions that reduce the ripple amplitudes.

This final example of an FIR design technique has also been our final MATLAB lesson. We have concentrated on its numerical tools, and on its system- and signal-related function calls. Additional FIR design tools are available, and *help fir1* will provide a *"see also"* list of most of them.

CHAPTER SUMMARY

FIR filter designs can be obtained from synthetic division of IIR designs, from the Fourier series or the DFT when combined with a window function, and from computer-aided design programs that seek the minimum error between the specified and achieved transfer function amplitudes.

Symmetric or antisymmetric difference equations produce absolutely linear phase in the passband, a feature only approximated by analog filters. This feature, combined with the absence of stability concerns, has made FIR filters very popular. Their main disadvantages are the number of multiply and accumulate operations required and the absence of any way to specify the cutoff frequency as part of the design statement.

PROBLEMS

Section 10.1

1. Find a fourth-order FIR difference equation to approximate the response of the IIR function $H(z) = z/(z - 0.6)$. Demonstrate the extent of their agreement by plotting both transfer functions.
2. Find an eighth-order FIR difference equation to approximate the IIR difference equation $y(n) = x(n) - 0.9y(n - 2)$. Demonstrate the extent of their agreement by plotting both transfer functions.
3. Prepare a series of decibel response curves to compare a fourth-order Butterworth bandpass IIR filter with $\upsilon_L = 0.2$ and $\upsilon_H = 0.40$ to its implementation using a 40th- and an 80th-order FIR filter. Also compare the filters in terms of the number of multiplications required.
4. Prepare a series of decibel response curves to compare a fourth-order Chebyshev high-pass filter with 1-dB ripple and $\upsilon_{co} = 0.40$ to its implementation using a 40th- and an 80th-order FIR filter. Also compare the filters in terms of the number of multiplications required.

Section 10.2

5. Sketch $|H(e^{j\pi\upsilon})|$, and determine the group delay for the difference equation
$$y(n) = x(n) - 2x(n - 1) + 2x(n - 2) - x(n - 3)$$
6. Sketch $|H(e^{j\pi\upsilon})|$ for a 21-term averager. Indicate the frequencies of all zeros, and determine the approximate minimum stopband attenuation.

7. A square wave is input to the difference equation

$$y(n) = x(n) - x(n - 20)$$

   a. Obtain a plot of $H(e^{j\pi v})$, including phase.
   b. Sketch about 100 samples of $y(n)$ given that $x(n)$ is a unit amplitude square wave with a fundamental frequency of $v_o = 1/20$. Repeat for $v_o = 1/15$ and 1/10. Explain how the resulting waveforms could be predicted. (Note: $v_o = 2f_o/f_s = 2$ (cycles/s)/(samples/s) or samples/cycles $= 2/v_o$.)

Section 10.3

8. Design a Fourier-based 40th-order low-pass FIR filter with its band edge at $v = 0.42$. Use a Hanning window. Plot $|H(e^{j\pi v})|$.
9. Plot and compare the magnitude response of a Fourier-based FIR low-pass filter designed for a band edge of $v = 0.3$ using a Hamming window for filter orders of 10, 20, and 40.
10. Derive the formula for the Fourier coefficients of a bandstop filter in terms of band edges of v_L and v_H. Use your results to plot the magnitude response of a 60th-order filter with $v_L = 0.4$ and $v_H = 0.5$ using a Hanning window.
11. Design a 100th-order Fourier-based bandpass filter to pass frequencies in the range $0.3 \le v \le 0.35$. Use a Hamming window, and plot the magnitude response that results.
12. Produce a Fourier-based 60th-order FIR filter to have the response shown in Figure P10.12. Select a window, and plot the amplitude response to show the success of the design.

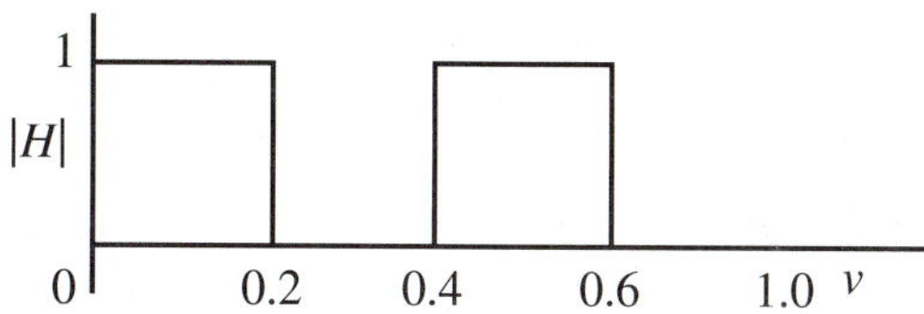

Figure P10.12

13. Design a 40th-order Fourier-based differentiator for frequencies out to $v = 0.7$ using a Hanning window.
   a. Plot the frequency response of your design.
   b. Show the output of the differentiator for the input signal shown in Figure P10.13.

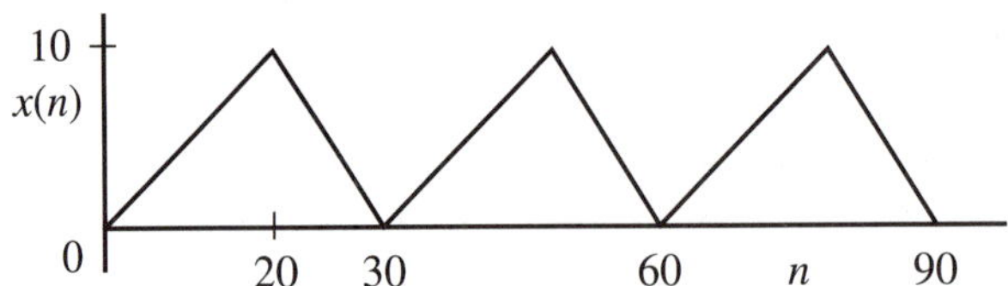

Figure P10.13

14. Design a 40th-order Hilbert transformer and an ordinary low-pass filter, both with a passband range of $0 < v < 0.5$. Use a Hamming window on each. Plot the outputs of the two circuits on the same graph given that their inputs are $x(n) = \sin 0.08\pi n$. What is the relationship between the two outputs for any v in the passband?

Section 10.4

15. Frequency sampling is to be used to create a 20th-order low-pass filter with its band transition at $v = 0.5$. Using the empirical approach provide the following.
    a. A formula giving the frequencies at which the samples will be specified.
    b. A sketch similar to Figure 10.14a showing your magnitude samples.
    c. A plot of the actual phase achieved. Compare it to a linear-phase characteristic.
    d. List the FIR filter coefficients.
16. Use the frequency sampling technique to design a linear-phase 31st-order low-pass filter with a transition to the stopband following the fifth sample. Use the exact-phase formula. Obtain plots of the filter's frequency response with a rectangular window.
17. Repeat Problem 16 by designing for a 32nd-order filter using the empirical-phase design rule. Observe the impulse response of your design; if warranted, reject the first pulse to obtain a symmetrical set of FIR filter coefficients. Obtain plots of the resulting 31st-order filter's frequency response with a rectangular window.
18. Use frequency sampling to design a 50th-order linear-phase filter with a passband of $0.2 \le v \le 0.4$. Use the exact-phase formula. Plot the filter's impulse response and its frequency response before windowing.
19. Use frequency sampling to design a 44th-order Hilbert transformer for use in the range $0.1 < v \le 0.6$. (*Hint:* Add $\pi/2$ to the usual frequency sampling phase, and let the magnitude take care of part of the phase variation.) Demonstrate the frequency response of your design with a rectangular window.

Section 10.5

20. A low-pass filter is needed to reject frequencies above $\upsilon = 0.1$. Compare 32nd-order Remez designs that allow $\Delta\upsilon = 0.02, 0.05$, and 0.10 for the transition to the stopband. Plot the magnitude responses on the same graph. How would you define a cutoff frequency for these filters?
21. Bandpass filters with passbands from $0.2 \le \upsilon \le 0.5$ are to be implemented using 63rd-order designs. Design 1 is to use frequency sampling and a Hamming window. Design 2 is to use the Remez program. The transition regions are to be $\Delta\upsilon = 0.0625$ for both designs. Plot the decibel response curves on the same graph. List the advantages of each design.
22. Design an 80th-order Remez-based FIR differentiator to be used in the range $0 \le \upsilon \le 0.5$. Demonstrate the success of your design by showing its output when the waveform of Figure P10.13 is its input.
23. A 20th-order filter is designed to have a passband for the frequency range $0.2 \le \upsilon \le 0.6$. Given a Remez design, plot the decibel gain that results if the filter is implemented with a processor that uses only two-significant-figure (0.00–0.99) coefficients. Compare with an infinitely precise processor.

11

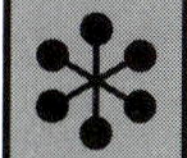

SAMPLING STRATEGIES

OUTLINE

OBJECTIVES

1. Use oversampling to reduce anti-aliasing filter requirements.
2. Use interpolation to simplify reconstruction filtering.
3. Establish sampling requirements for bandpass signals.

INTRODUCTION

While the basic aspects of the sampling interface have already been discussed, special techniques can be used to move between the continuous-time and discrete-time worlds more economically. For the most part this involves shifting more of the burden to the digital arena, where software can be used to reduce the need for elaborate and potentially expensive analog anti-aliasing or reconstruction filters. Most of the improvements to be discussed are achieved by changing the effective sampling rate. In applications where standard sampling rates are required, these changes must and can be made transparent to the user.

The sampling of bandpass signals, such as are often encountered in communications systems, involves special rules. It is usually possible to sample a bandpass signal at well below the Nyquist rate without causing aliasing.

11.1 OVERSAMPLING AND DECIMATION

If the sampling frequency is selected to be much higher than the Nyquist rate, the image frequencies become widely separated, and a simple filter can be used to prevent aliasing. Once in the discrete-time domain, a digital filter can be used to clean up the signal spectrum. Then the excess samples can be discarded to return to a standard sampling rate. After completing this section you will be able to:

- Use oversampling to reduce anti-aliasing filter requirements.
- Select an appropriate-order anti-aliasing filter.
- Describe the properties required in a decimation filter.

The first step in preparation for sampling is to minimize the consequences of aliasing. The typical sampling spectrum of Figure 11.1 demonstrates the problem. Remember that sampling a signal causes its spectrum to be replicated around integer multiples of the sampling frequency. Images of the noise or extraneous signals in the original spectrum may fall into the baseband region and overlap the desired signal spectrum. Aliased noise that does not overlap the desired signal may be eliminated by digital filtering after sampling and is not of immediate concern. The anti-aliasing filter has to achieve its minimum acceptable attenuation in the interval from f_m to $f_s - f_m$. The amount of attenuation it must provide depends on the size of the offending noise and the dynamic range of the analog-to-digital converter (ADC). Ideally, we would like the noise to be too small to change the ADC's least significant bit. Obviously, this requires better anti-aliasing filters as the number of bits in the ADC increase.

In discussing the role of the anti-aliasing filter, it should be understood that all of the standard techniques to keep extraneous noise out of the system must be

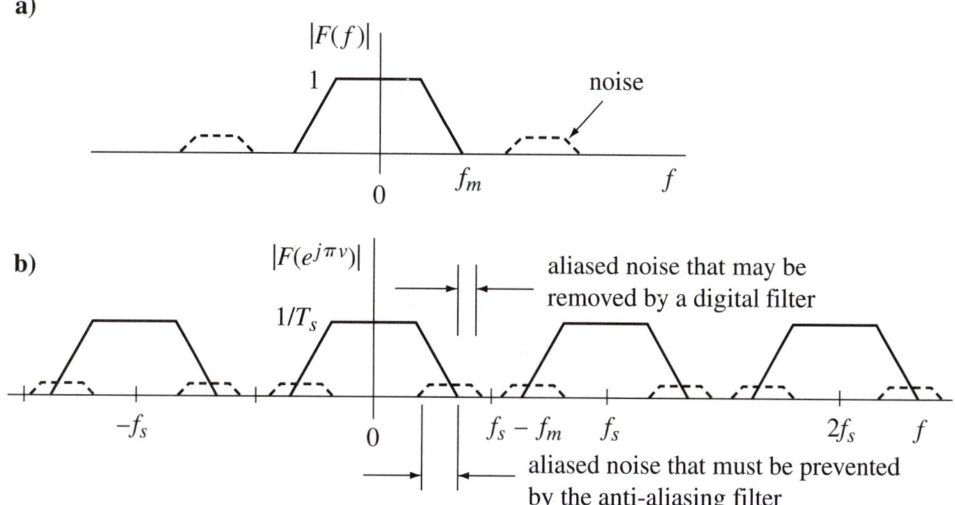

Figure 11.1 **(a)** Continuous-time and **(b)** discrete-time spectra when sampling is slightly above the Nyquist rate. Sampling causes the original spectrum to be replicated around integer multiples of the sampling frequency, including any noise or extraneous signals that are present.

employed. Noise does not necessarily arrive at the sampler by the same path as the desired signal. An elaborate anti-aliasing filter may be ineffective for noise let in by poor shielding, grounding techniques, or circuit board layout.

To review the basics of sampling, consider a digital audiotape (DAT) system with a maximum frequency of $f_m = 20$ kHz and a sampling frequency $f_s = 48$ kHz. The ideal anti-aliasing filter should pass 20 kHz and reject 28 kHz. The specifications defining how well the signal must be passed or rejected will be translated into a filter design. (Audio compact disks (CDs) require even more elaborate anti-aliasing filtering, since they use only a 44.1-kHz sampling rate.) Sampling close to the Nyquist rate requires a formidably elaborate and expensive analog anti-aliasing filter. It would be ironic to have to purchase such a filter just to reach the digital domain, where a high-quality digital filter can easily be implemented.

EXAMPLE 11.1

What order Butterworth filter will provide at least 40-dB attenuation at the first image in a DAT system if its cutoff frequency is 20 kHz?

Solution

Since a DAT system samples at 48 kHz, the low-frequency edge of its first image will be at 48 − 20 = 28 kHz. Then the anti-aliasing filter must have cutoff at 20 kHz and be down 40 dB at 28 kHz:

$$-40 \text{ dB} = 20 \log |H|, \qquad |H| = 0.01$$

From Equation 5.4,

$$10^4 = 1 + (28/20)^{2n}$$
$$2n \log 1.4 \approx 4$$
$$n \geq 13.7$$

A 14th-order Butterworth filter would be required!

An option that reduces the requirements on the antialiasing filter is to oversample. The term *oversampling* is used to denote a sampling rate two or more times the Nyquist rate. Oversampling increases the frequency separation of the baseband and sampling images, so a lower rate of attenuation is acceptable in the anti-aliasing filter. Oversampling by a factor of over 9 is shown in Figure 11.2, but much higher oversampling factors are not unusual. Repeating Example 11.1 with the sampling rate increased to 2 MHz (50 times the Nyquist rate) would allow a one-pole anti-aliasing filter to do the job.

Oversampling is not always a viable option. In very high-frequency applications, the high-speed components required for oversampling may not offer a cost advantage over the analog filter that would otherwise be necessary. In other cases, retention of the excess samples may strain the capacity of the storage medium being used. In many applications, as with the DAT system example, sampling rate standards have been established for compatibility among equipment manufacturers. Such standards do not necessarily preclude the use of oversampling as part of the process for creating a tape, but they do require that the final tape provide samples at the standard rate.

If oversampling is combined with *decimation,* the reduced anti-aliasing filter requirements can be retained without increasing the *apparent* sampling rate. *Decimation* is a term used to indicate reducing the sampling rate by integer factors (originally it meant factors of 10). Decimation by N can be accomplished simply by selecting out every Nth sample and ignoring the others, just as if the samples had been taken at the lower rate to start with. The anti-aliasing analog filter still only has to eliminate the noise above $f_s - f_m$, where f_s is the oversampling rate. An anti-aliasing digital filter must be used to remove any remaining noise in the rest of the

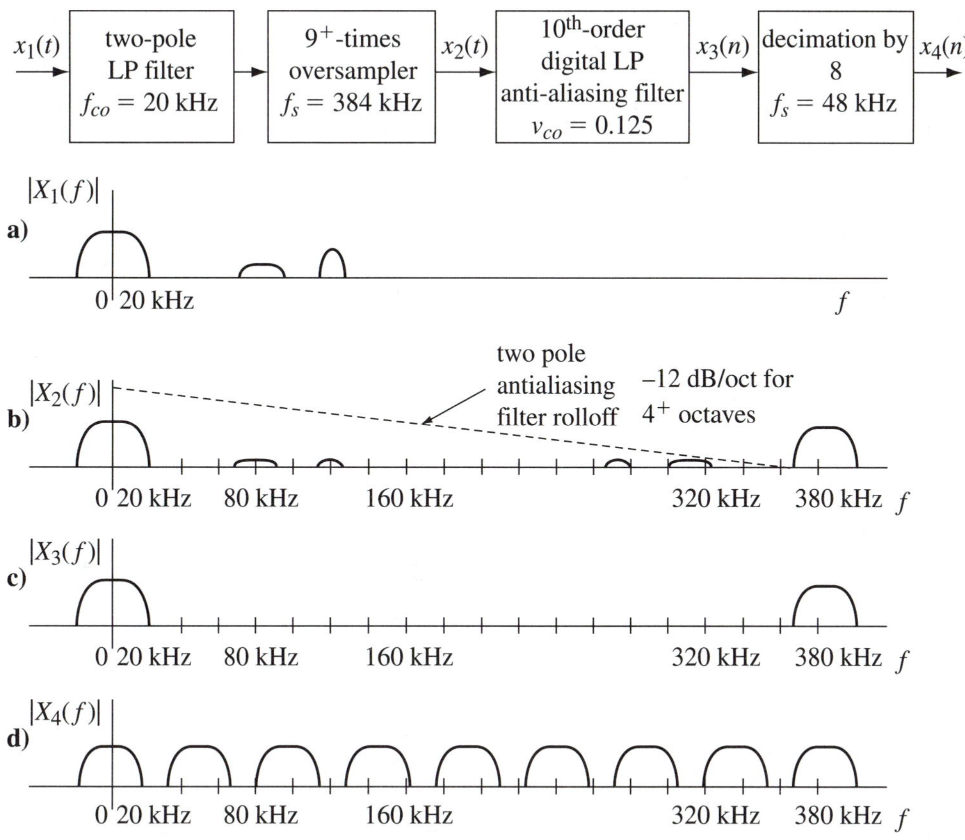

Figure 11.2 DSP input stage using oversampling and decimation to eliminate the need for a high-quality anti-aliasing filter. **(a)** The input signal consists primarily of audio spectrum components, but noise components may also be present. A two-pole anti-aliasing filter provides an attenuation of about 50 dB at 364 kHz. **(b)** The sampled spectrum still contains the remains of some noise sources. **(c)** A low-pass digital filter cleans out everything that remains after sampling except for the audio spectrum and its images. **(d)** Selecting every 8th sample (decimation) provides an effective sampling rate of 48 kHz at the output.

spectrum prior to decimation, since any such noise would otherwise become replicated around multiples of the lower sampling rate created by the decimation.

A *decimation filter* combines the digital filtering to clean out all of the noise signals from the oversampled spectrum with the decimation process. Rather than first filtering at the higher sampling rate and then discarding the unwanted, filtered samples, an FIR filter is usually used as the decimation filter. Since an FIR filter output depends only on current and past inputs, there is no reason to calculate the unwanted output samples. The inputs can be presented to the FIR filter in the usual way, but the outputs need to be calculated only at the reduced output sampling rate.

EXAMPLE 11.2

A line containing voice signals in the frequency range of 0–3.6 kHz is contaminated with noise signals at 5.1 kHz and 15.3 kHz. Given that it is to be oversampled at 24 kHz and decimated by 3, describe the basic features of the analog and digital filters needed.

Solution

A sketch of the spectral situation is shown in Figure 11.3 before decimation. The 15.3-kHz noise harmonic is aliased to 8.7 kHz. The aliased noise does not overlap the signal and can be removed by a digital filter after sampling. No anti-aliasing analog filter is needed *based on the information given.* (If the fifth harmonic of the 5.1-kHz noise source is present, it would be aliased to 1.5 kHz, and that would have to be prevented by an analog filter.) A low-pass digital filter with v_{co} of about 0.3 is needed to clear out everything from 3.6 kHz to $f_s/2$ before decimation.

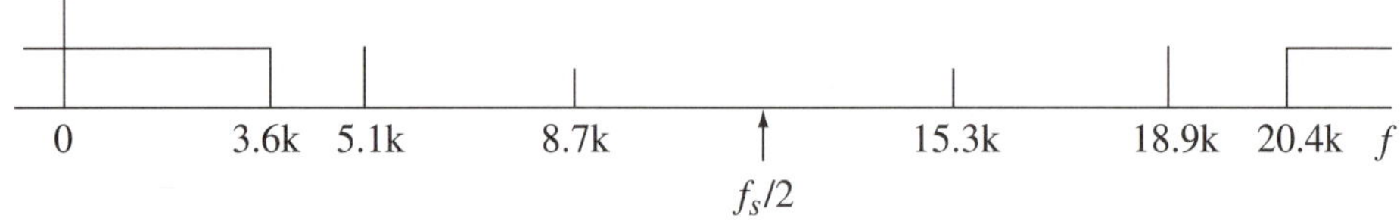

Figure 11.3

11.2 RECONSTRUCTION AND INTERPOLATION

When the digital processor output is to be converted back to the continuous-time domain, the baseband signal must be separated from its images. This requires an expensive reconstruction filter unless the signal has been oversampled. An *interpolation filter* effectively increases the sampling rate of the output to ease the requirements on the reconstruction filter. After completing this section you will be able to:

- Use interpolation to simplify reconstruction filtering.
- Describe what a zero-order hold does to the output spectrum.

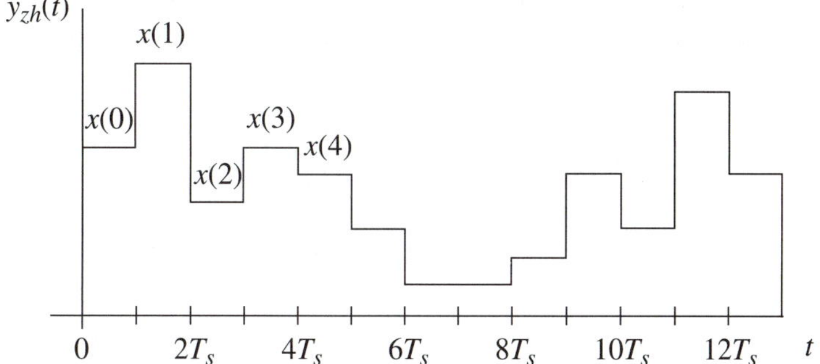

Figure 11.4 Discrete-time signals $x(n)$ are converted to a continuous-time signal by the zero-order hold of the DAC.

- Describe zero padding and what it does to a signal spectrum.
- Explain what a polyphase interpolation filter is.
- Modify sampling rates by any rational fraction.

The procedure for reconstructing an analog signal after the signal has been digitally processed may also involve sampling rate changes. A common way to return a discrete-time signal to the continuous-time domain is with a latched digital-to-analog converter (DAC). The processor delivers the output number to the DAC; the DAC immediately converts it to an analog voltage and holds it constant until the next number is received. This is known as a *zero-order hold.* The continuous-time waveform that results, shown in Figure 11.4, may be described mathematically using a series of unit step functions:

$$y_{zh}(t) = \cdots x(0)[u(t) - u(t - T_s)] + x(T_s)[u(t - T_s) - u(t - 2T_s)] + \cdots$$

$$y_{zh}(t) = \sum_{n=-\infty}^{\infty} x(nT_s)[u(t - nT_s) - u(t - (n + 1)T_s]$$

The spectrum of the zero-order hold signal is found from the Fourier Transform:

$$Y_{zh}(\omega) = \int_{-\infty}^{\infty} y_{zh}(t)e^{-j\omega t}\,dt$$

$$Y_{zh}(\omega) = \int_{-\infty}^{\infty} \sum_{n=-\infty}^{\infty} x(nT_s)[u(t - nT_s) - u(t - (n + 1)T_s)]e^{-j\omega t}\,dt$$

Interchanging the order of summation and integration and noting that the bracketed step function terms are nonzero only from nT_s to $(n + 1)T_s$ gives

$$Y_{zh}(\omega) = \sum_{n=-\infty}^{\infty} x(nT_s) \int_{nTs}^{(n+1)T_s} e^{-j\omega t}\, dt = \sum_{n=-\infty}^{\infty} x(nT_s)\left(\frac{e^{-j\omega t}}{-j\omega}\right]_{nT_s}^{(n+1)T_s}$$

$$Y_{zh}(\omega) = \left[\sum_{n=-\infty}^{\infty} x(nT_s)\, e^{-jn\omega T_s}\right]\left[T_s e^{-j\omega T_s/2}\,\frac{\sin \omega T_s/2}{\omega T_s/2}\right] \qquad \textbf{(11.1)}$$

The left-hand bracket should be recognized as the discrete-time Fourier transform, or spectrum, of the signal delivered to the DAC. The right-hand bracket is evidently the transfer function for the zero-order hold, H_{zh}.

$$Y_{zh}(\omega) = X(e^{j\omega T_s})H_{zh}(j\omega)$$

Notice the T_s multiplier in H_{zh}, which undoes the $1/T_s$ factor introduced by the sampling operation (Eq. 7.11).

Our goal is to recover the $X(e^{j\omega T_s})$ from the baseband region and eliminate all of the images that come with the discrete-time signal. The zero-order hold response creates transmission zeros at integer multiples of the sampling frequency, so it helps to remove the images from the output. Figure 11.5 shows the important features. An analog output filter, called a *reconstruction filter,* is usually also used to further reduce the images.

The inertia of mechanical systems, such as earphones and loudspeakers, tends to make them reject high frequencies naturally, so the reconstruction filter may not be necessary in all applications. Unfortunately, however, those image frequencies most likely to be objectionable are also those that are least attenuated by the zero-order hold. Adding an analog filter to pass signals at f_m but reject those at $f_s - f_m$ is no more attractive an option at the output than it was at the input for a system sampled at near the Nyquist rate.

The other effect of the zero-order hold is to introduce a drooping gain over the baseband region. This distortion of the signal's spectrum can be anticipated and corrected for in the digital processing, but life would be simpler if this were not necessary. Indeed, the beneficial aspects of the zero-order hold can be retained, and the drooping of the baseband gain can be made negligible if the sampling rate can be increased—and it can.

Interpolation is used to increase the apparent sampling rate at the output. Interpolation starts with the insertion of zeros between the existing output samples. To interpolate by a factor of I, $I - 1$ zeros are inserted between the existing samples. It is easy to show that this process, called *zero-padding,* does not change the signal spectrum but does increase the effective sampling frequency. Suppose the spectrum of a sequence $\{x(nT_s)\}$ is

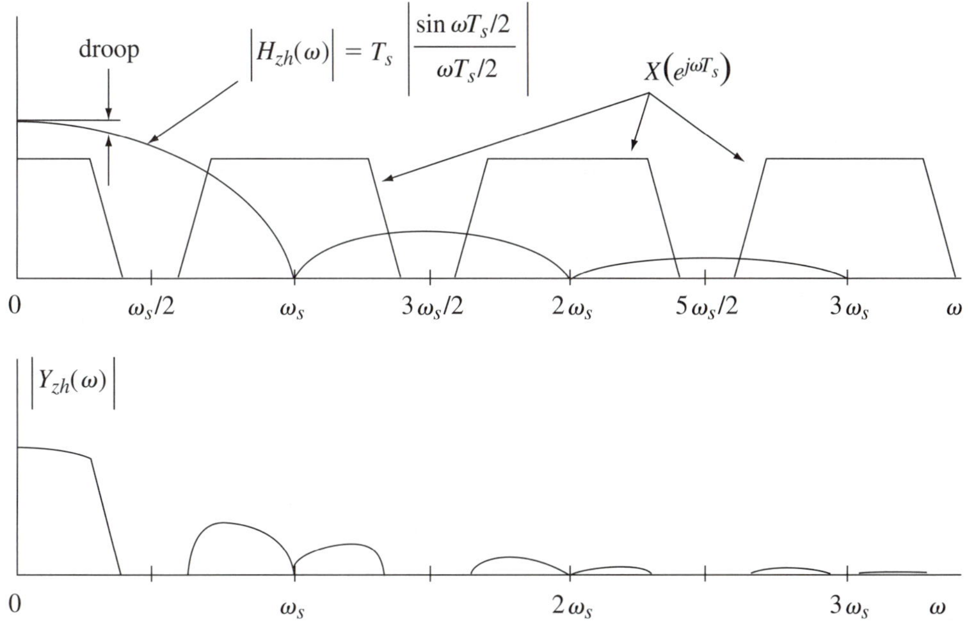

Figure 11.5 The spectrum of the output of a D/A converter is a combination of the (sin *x*)/*x* transfer function of the zero-order hold and the periodic spectrum of the discrete-time signal. The zero-order hold introduces an undesirable droop in the passband but assists in reducing the output image signals.

$$X(e^{j\omega T_s}) = \sum_n x(n)e^{-jn\omega T_s} = \cdots x(0) + x(1)e^{-j\omega T_s} + x(2)e^{-j2\omega T_s} + \cdots$$

If this sequence is padded with, for example, two zeros between each set of actual samples, the new sequence $\{x'\}$ is

$$\{x'(kT_s')\} = \cdots x(0) \;\; 0 \;\; 0 \;\; x(1) \;\; 0 \;\; 0 \;\; x(2) \;\; 0 \cdots$$

where $T_s' = T_s/3$, and the new spectrum is

$$X'(e^{j\omega T'_s}) = \sum_n x'(n)e^{-jn\omega T'_s} = \cdots x(0) + 0e^{-j\omega T'_s} + 0e^{-j2\omega T'_s} + x(1)e^{-j3\omega T'_s} + 0e^{-j4\omega T'_s} + 0e^{-j5\omega T'_s} + x(2)e^{-j6\omega T'_s} + \cdots$$

or

$$X'(e^{j\omega T'_s}) = \cdots x(0) + x(1)e^{-j\omega T_s} + x(2)e^{-j2\omega T_s} + \cdots = X(e^{j\omega T_s})$$

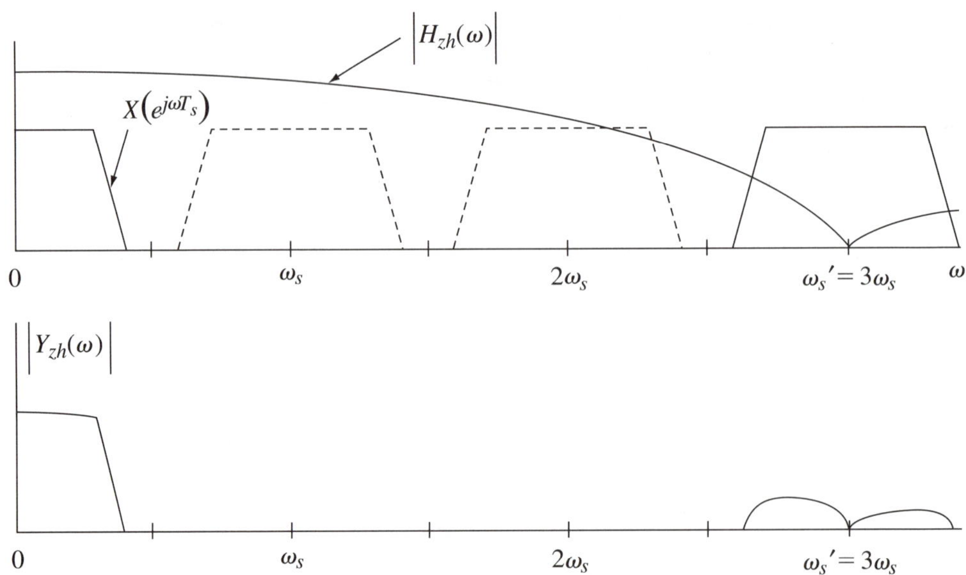

Figure 11.6 Using interpolation, the apparent sampling rate can be increased by an integer factor. The (digital) interpolation filter is responsible for removing the excess images (shown dotted). The widened spacing between the baseband and new first image allows a simpler analog filter to be used to eliminate remaining images at the output. The baseband spectrum takes up a smaller fraction of the distance to the first zero of the zero-order hold function, reducing the amount of droop experienced in the baseband.

Although the spectrum is unchanged by zero-padding, the new sampling rate allows the excess intervening images to be removed with a digital filter. An interpolation filter includes both the zero-padding operation and the low-pass filtering. It may also multiply the output by the interpolation factor to retain the signal energy present before filtering. Figure 11.6 indicates the spectral situation resulting from interpolation. The droop introduced by the zero-order hold is reduced by interpolation, and the requirements on any final analog filter are significantly eased. At high rates of interpolation, a one-pole *RC* filter may be adequate as a reconstruction filter, if one is needed at all.

EXAMPLE 11.3

Audio samples from 0 to 20 kHz are normally delivered to a DAC latch at 48 kHz. What interpolation factor will hold the gain droop experienced by this signal to under 0.1 dB?

Solution

The effective sampling frequency at the output is $f_s = I(48 \text{ kHz})$, where I is the interpolation factor. The relative attenuation of the zero-order hold at 20 kHz is

$$|H_{zh}| \propto \frac{\sin \pi f/f_s}{\pi f/f_s} = \frac{\sin \pi/2.4I}{\pi/2.4I} = \frac{\sin(1.3090/I)}{1.3090/I}$$

Tabulating a few values with MATLAB shows that an interpolation factor of 5 will do the job.

I	$\|H_{zh}\|$ (dB)
1	−2.6399
2	−0.6292
3	−0.2774
4	−0.1556
5	−0.0994

```
>i=1:10;
>h=sin(1.3090./i)./(1.3090./i);
>H=20*log10(h);
>table=[i'  H']
```

An FIR filter is usually used for the interpolation filter. Although the interpolation filter must calculate all of the outputs, the calculations are simpler than normal because a majority of its input samples are zeros. As a consequence, very few of its coefficients are actually used in calculating a particular output. Instead, the coefficients are used in sets, or phases, and the implementation is known as a *polyphase* filter. The result of applying an interpolation filter is very striking in the time domain. The interpolation points fill in the shape of the waveform, making the zero-hold steps so small that very little, if any, additional smoothing is needed in the analog domain.

EXAMPLE 11.4

The signal $x(n) = \sin n\pi/2$ is passed through an 8th order Butterworth interpolation filter with a cutoff frequency of $v_{co} = 0.9$. Show the input and output signals of the filter given that it interpolates by a factor of 3.

Solution

The input signal, $x(n)$, is shown in Figure 11.7a. Since its normalized frequency is $v = 1/2$, only four samples are taken during one cycle of the sinusoidal input.

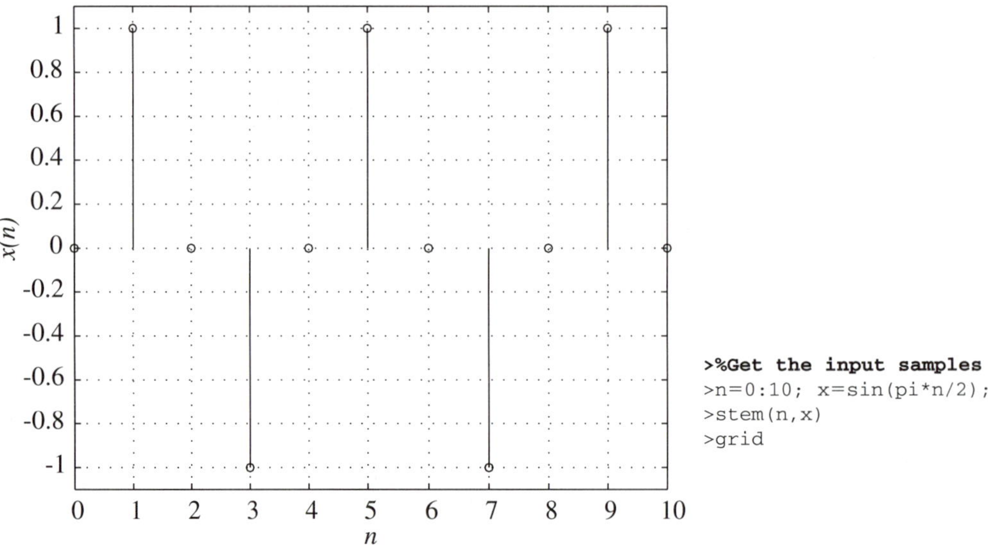

Figure 11.7a A discrete sinusoid with a normalized frequency of $v = 0.5$.

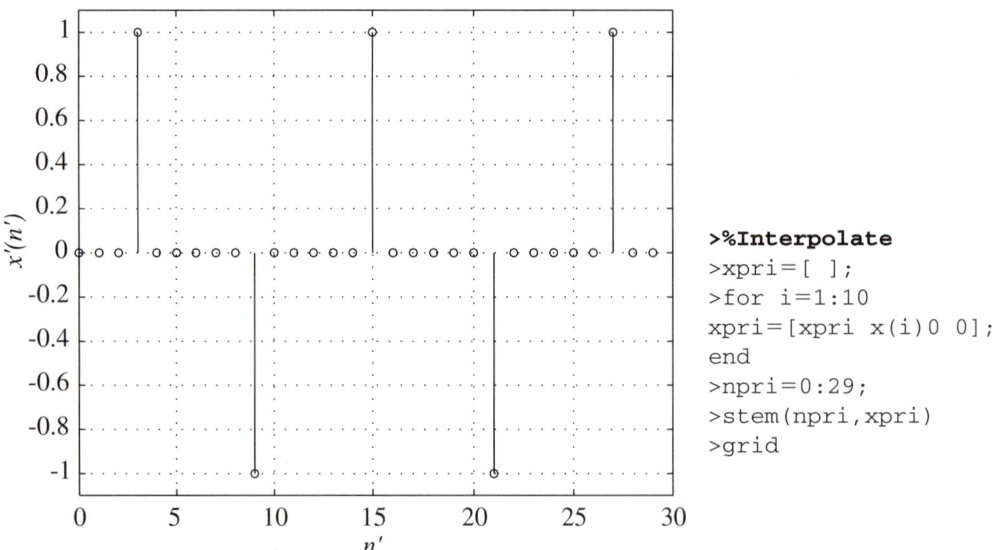

Figure 11.7b A discrete sinusoid after zero padding is added to interpolate by 3. The sinusoid's affective normalized frequency is now $v' = 1/6$.

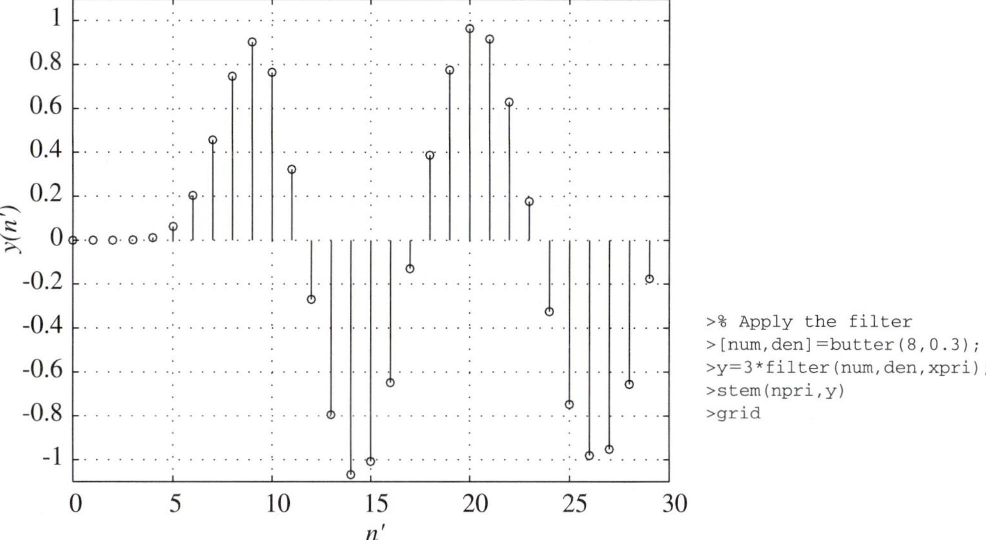

Figure 11.7c The waveform resulting from sampling a sinewave 4 times per cycle, zero padding to increase its effective sampling rate to 12 samples per cycle, and passing it through a low-pass filter.

To interpolate by 3, two zeros are inserted between each of the original samples, producing the waveform of Figure 11.7b. This reduces its normalized frequency to $v' = 1/6$.

Finally, the interpolated signal is sent through the low-pass digital filter specified to eliminate everything that might be present in the spectrum from $0.3 < v' < 1$. This corresponds to the frequency range $0.9 < v < 3$ in the original input spectrum. Specifically, it is removing the signal images at $v = 1.5$ and 2.5 (as well as their negative frequency counterparts). The resulting output, including the startup transient, is shown in Figure 11.7c. Obviously the interpolation process has filled in the shape of the sinusoid.

Using combinations of interpolation followed by decimation, sampling rates may be changed by any rational fraction. One possible application is to convert between CD and DAT data samples having sampling rates of 44.1 kHz and 48 kHz, respectively. Of course, one way to accomplish this is to convert back to the continuous-time domain and then resample at the desired rate. It should, however, be possible to perform the conversions directly while within the digital domain. To convert CD samples to DAT samples requires a conversion ratio of $480/441 = 160/147$, so we could first interpolate by 160 and then decimate by 147. Even after taking advantage of any processing shortcuts, however, it might be difficult for the processor intended for operation at 44 or 48 kHz to do the interpolation filtering at 160(44.1 kHz) = 7.056 MHz. Such considerations argue in favor of achieving the overall conversion by doing it in a series of smaller interpolation/decimation steps, such as (8/7)(10/7)(2/3).

Digital signal processing (DSP) systems having analog input and output require both ADC and DAC units. In telephone systems, where most of the early work in sampled data systems was done, the unit responsible for making these conversions was named a *codec* (coder/decoder). Today, the term *codec* most often refers to a programmable, large-scale integration (LSI) chip, which not only performs the AD/DA conversions but also implements many of the features discussed in this chapter to keep the analog anti-aliasing and reconstruction filter requirements minimal. These chips also provide for serial data exchanges with the main processor and generally have features, such as multiplexing, to handle several signals, programmable gain, selectable sampling rates and word sizes, and standard companding options (nonlinear coding to improve the dynamic range of signals that can be handled using 8-bit words). In some applications, the codec may provide all the processing necessary.

11.3 BANDPASS SAMPLING

The Nyquist rate, $f_s = 2f_m$, is based on the assumption that the signal being sampled has all frequencies present in the range $0 \geq f \geq f_m$. When that is not the case, a lower sampling rate is often possible without aliasing. After completing this section you will be able to:

- Establish sampling requirements for bandpass signals.
- Determine if spectral inversion will occur.

In many communication applications, the signal is confined to a small bandwidth, B, between its lowest, f_L, and highest, f_H, frequencies. Such a signal may be sampled at below the Nyquist rate without alias overlap. The signal must first pass through a bandpass anti-aliasing filter so that the rest of the spectrum is indeed free of extraneous signals. The sampling criteria for the bandpass situation may be deduced from Figure 11.8.

Avoiding overlap between the signal and the image centered on the mth multiple of the sampling frequency requires that $f_s < 2f_L/m$. Then, keeping the next higher replication from overlapping the original spectrum from above requires that $f_s > 2f_H/(m + 1)$. This provides a range of permissible values for f_s:

$$\frac{2f_H}{m+1} < f_s < \frac{2f_L}{m} \tag{11.2}$$

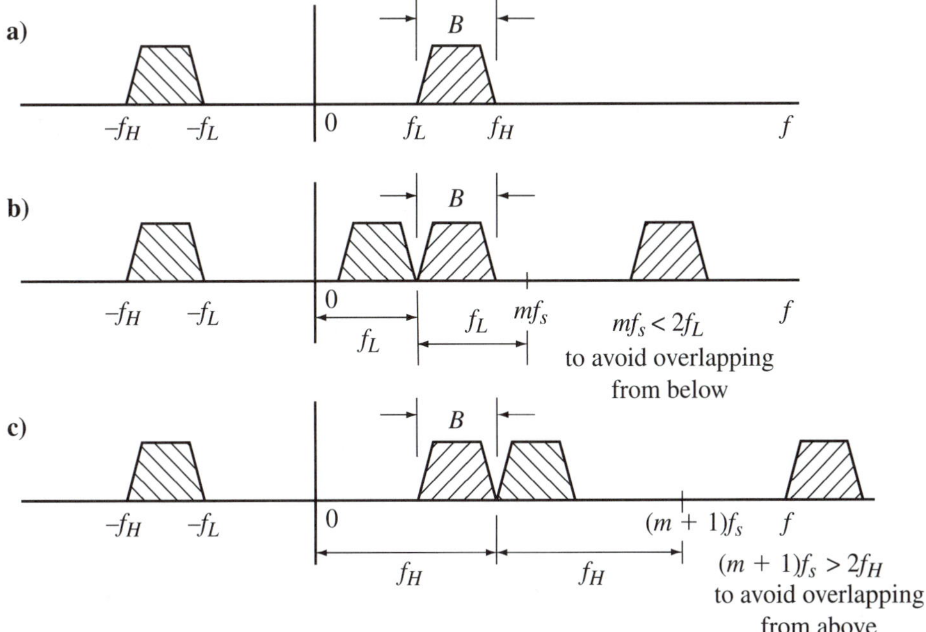

Figure 11.8 If the bandpass signal of **(a)** is sampled, its spectrum will be replicated about integer multiples of the sampling frequency. (Only one replication is shown, for clarity.) Requiring that the replication about the *m*th multiple of f_s not overlap with the original spectrum from below sets an upper bound on f_s, as shown in **(b)**. The next higher replication, about $(m + 1)f_s$, must then avoid overlapping with the original spectrum from above, as shown in **(c)**. This sets a lower bound on f_s.

The range of values allowed by Equation 11.2 converges, and alias overlap becomes unavoidable at higher m values. The ultimate limit is $f_s = 2B$, since that will always fill the entire frequency axis regardless of where the original bandpass signal is located. Generally, for narrow-band systems ($B \ll f_L$), it is possible to approach the $f_s = 2B$ limit quite closely. However, approaching this limit will require better anti-alias filters and will not leave spectral space for any further processing (demodulation) that the signal might need to undergo.

EXAMPLE 11.5

Determine the limiting sampling frequencies to avoid alias overlap for a signal having $f_L = 9$ MHz and $f_H = 11$ MHz, and sketch the sampled spectrum for all cases.

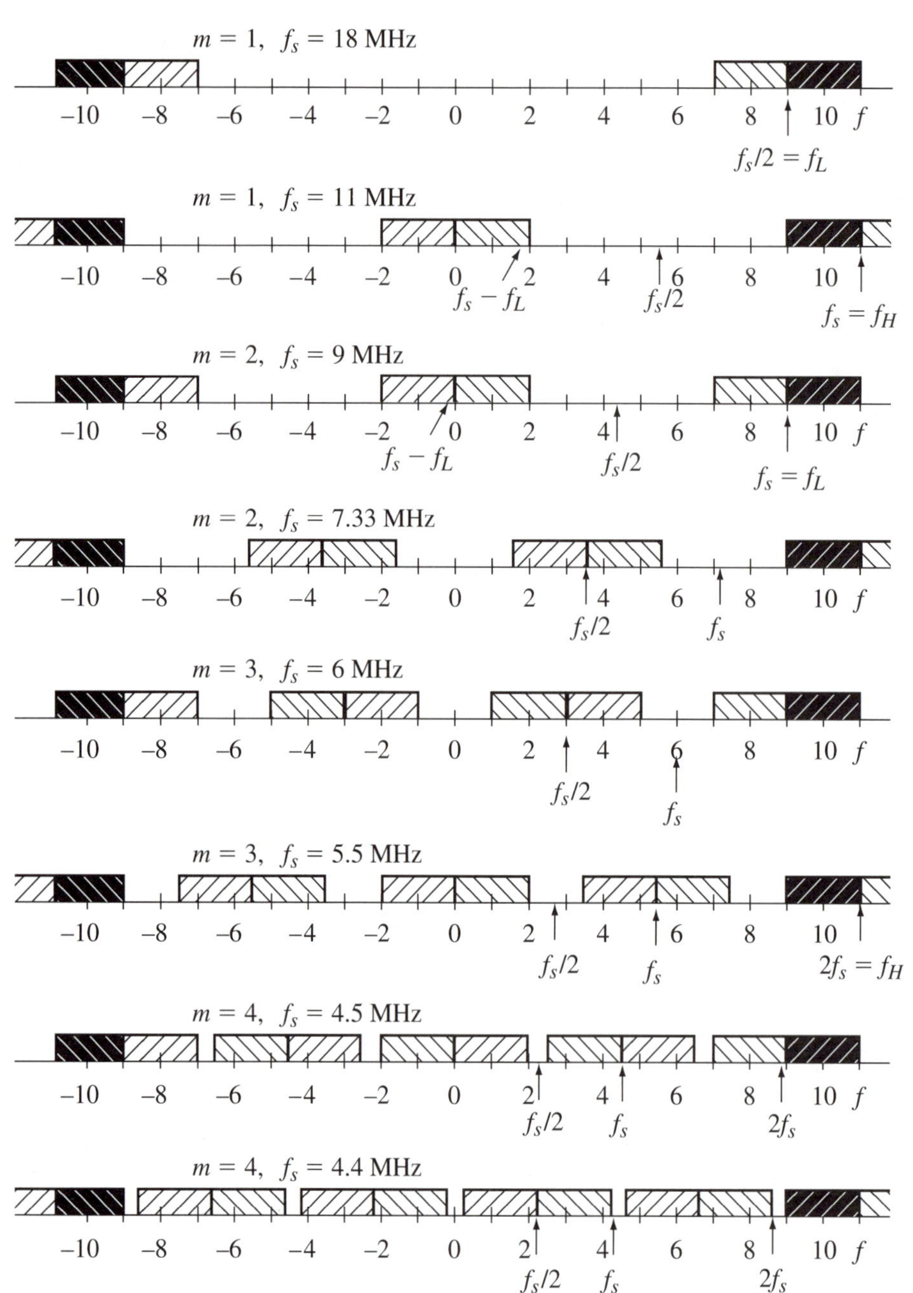

Figure 11.9 Sampled spectra using sampling rates at the extreme limits of each band. The original bandpass signal spectrum is shown dark, and the shading slopes upward from low to high frequency. Notice that some of the sampling options (m = 1, 3) invert the frequency order in the baseband and that the sampled signal's spectrum becomes increasingly crowded as f_s decreases. Each of these facts may influence the selection of a sampling frequency.

Solution

Tabulating values from Equation 11.2 gives

$$\frac{22}{m+1} < f_s \text{ (MHz)} < \frac{18}{m}$$

m	f_{smin}	f_{smax}
1	11.0000	18.0000
2	7.3333	9.0000
3	5.5000	6.0000
4	4.4000	4.5000
5	3.6667	3.6000

At $m = 5$, f_{smax} becomes less than f_{smin}, and both are less than the ultimate limit of $2B = 4$ MHz. Spectra for the four allowed ranges of sampling frequencies are shown in Figure 11.9.

If it were necessary to use the usual sampling theorem, we would have had to sample at twice the highest frequency, or 22 MHz. As this example shows, we can sample at one-fifth that rate without alias overlap. Each specific application of bandpass sampling will impose additional constraints that might further restrict the sampling frequency options.

A thorough examination of the results of Example 11.5 shows that the signal spectrum may be shifted to different positions within the baseband by the choice of f_s and that the baseband spectrum may end up inverted from that of the original bandpass signal. These factors may or may not be important to particular modulation schemes. The following rules apply:

Even m (produces no spectral inversion):
- To position f_L at 0 Hz, sample at the maximum f_s.
- To position f_H at $f_s/2$, sample at the minimum f_s.

Odd m (produces spectral inversion):
- To position f_H at 0 Hz, sample at the minimum f_s.
- To position f_L at $f_s/2$, sample at the maximum f_s.

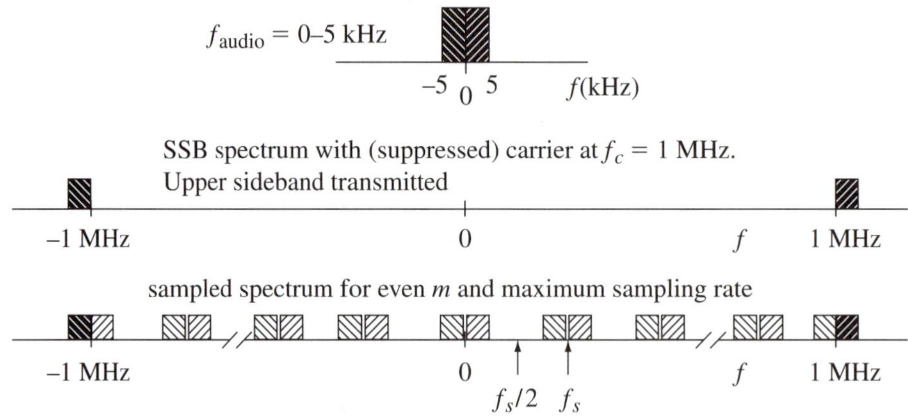

Figure 11.10 Spectra associated with single-sideband modulation in which the upper sideband is transmitted, and its recovery by sampling.

If the nature of the signal requires that it receive further processing within the baseband, it may be necessary to use a sampling rate near the center of the allowed range to leave room for the eventual recovery of the modulation.

In principle, a radio receiver could consist of a tuned circuit selecting out a particular bandpass signal, a sampler, and a filter to recover only the baseband information. In fact, one could argue that the old-fashioned "crystal set" did exactly that. A long wire antenna picked up passing radio waves, and a parallel resonant anti-aliasing filter tuned the desired station (i.e., the strongest station in those days). The crystal (germanium diode) "synchronously sampled" the signal at the carrier frequency rate by allowing only one polarity to pass, and the headphone's inertia rejected everything but the audio information. This technique is still the way to recover amplitude modulation signals, but usually the signal is first amplified to a standard level and further isolated from adjacent noise signals. That may change as chips with radio frequency sampling rates become common.

Perhaps the simplest example of a bandpass sampling application is the recovery of single-sideband (SSB) signal. Figure 11.10 shows the spectrum of a 0–5-kHz audio signal and the corresponding spectrum of an SSB signal modulated by it. In the case shown, the upper sideband is being transmitted. If the transmitted signal is sampled with m even and at the highest rate for the m selected, the original signal will be recovered in the baseband.

EXAMPLE 11.6

A 1-MHz carrier is single-sideband modulated by 0–5-kHz audio. The upper sideband is transmitted with the carrier suppressed. Assume the audio signal is

$$x_{\text{audio}}(t) = \cos 2\pi 1000t + 3 \cos 2\pi 3000t$$

The resulting single-sideband signal is

$$x_{usb}(t) = \cos(2\pi 1{,}001{,}000t) + 3\cos(2\pi 1{,}003{,}000t)$$

Recommend an f_s to recover the audio information, and demonstrate that it works.

Solution

The signal spectrum is the same one shown in Figure 11.10. Its bandwidth is given as 5 kHz, and the recovery process must work for any audio in that range. Then $f_L = 1$ MHz and $f_H = 1005$ kHz:

$$\frac{2f_H}{m+1} = \frac{2010}{m+1} < f_s\ (\text{kHz}) < \frac{2000}{m} = \frac{2f_L}{m}$$

We want an even m, to avoid spectral inversion, so try $m = 100$. This gives

$$19.901 < f_s\ (\text{kHz}) < 20$$

which is a legitimate inequality, implying that this is an acceptable m value. By sampling at 20 kHz (!), we will recover the original signal. We use MATLAB for the demonstration:

```
>t=(2e-3)*(0:200)/200;                        % show about 2 ms worth of
                                                the audio signal
>audio=cos(2*pi*1000*t)+3*cos(2*pi*3000*t);
>plot(t,audio)
>hold

% sampling the modulated signal at 20 kHz means samples are taken every
  0.05ms
>t=(0.05e-3)*(0:39);                          % 40 samples @ 0.05ms
                                                /sample=2ms.
>sideband=cos(2*pi*1001000*t)+3*cos(2*pi*1003000*t);
>plot(t,sideband,'o')
```

The results, shown in Figure 11.11, confirm that the SSB signal has been demodulated by the sampling process.

If 19.901 kHz had been used as the sampling frequency, a double-sideband signal would have resulted, and further processing would have been necessary to recover the original audio. For the m value chosen, there is not sufficient spectral room left to do this, so a higher sampling frequency would have to be used.

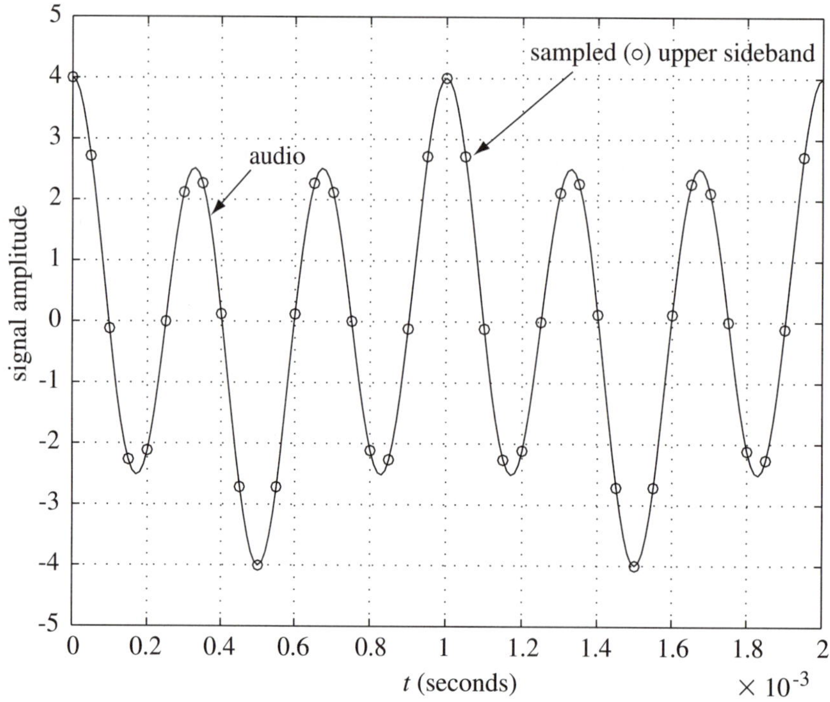

Figure 11.11 An audio signal (solid) used to single sideband modulate a 1 MHz carrier is recovered by sampling (circles) the sideband signal at 20 kHz.

CHAPTER SUMMARY

Oversampling can be combined with a decimation filter to allow an inexpensive low-order analog filter to be used to prevent aliasing. The effective sampling rate at the output of the decimation filter may be very close to the Nyquist rate.

Interpolation involves zero-padding the digital output samples to increase the apparent sampling rate at the output. This allows an inexpensive and low-order analog filter to be used for the reconstruction operation.

Combinations of decimation and interpolation can be used to shift between the standard sampling rates of different systems.

Bandpass signals can be sampled at less than the Nyquist rate without aliasing. At the extreme limit, the sampling rate may have to be only twice the bandwidth of the signal.

PROBLEMS

Section 11.1

1. A compact disk recorder with $f_s = 44.1$ kHz uses a Butterworth anti-aliasing filter with a cutoff frequency of 20 kHz.
   a. Determine the filter order required to provide a minimum of 50-dB attenuation if the edge of the first image is taken as $f_s - f_{co}$.
   b. Determine the filter order required if the signal is oversampled by a factor of 4.
2. A first-order *RC* low-pass anti-aliasing filter is to be used to provide 70-dB attenuation to a signal having $f_m = 4$ kHz. The filter is to have no more than 1-dB attenuation at f_m. Determine the required value of *RC* and the minimum allowed sampling frequency.
3. A telephone voice circuit with $f_m = 3.4$ kHz contains a noise signal of 15.5 kHz due to a nearby TV. The circuit is oversampled at $f_s = 24$ kHz and then passed through a decimation by 3 filter.
   a. Determine the attenuation a second-order Butterworth anti-aliasing filter would provide at the lower edge of the first image if its cutoff frequency is 3.4 kHz.
   b. Sketch the spectrum of the sampled signal from 0 to 30 kHz.
   c. Sketch the spectrum after decimation.
   d. Recommend a value of v_{co} for the decimation filter. Explain the role of this filter.
4. An audio signal in the range 0–3.4 kHz has noise riding on it at 8 kHz, 16 kHz, and 19.8 kHz. The signal and each noise component have equal amplitudes, and are sampled at 32 kHz. Assume a 2-volt peak sinewave at 2 kHz as the signal.
   a. Sketch the sampled spectrum from 0 to 16 kHz given that the analog anti-aliasing filter rolls off at −12 dB/octave starting at 3.5 kHz. Calculate the approximate voltage of each spectral component.
   b. Use a fourth-order low-pass digital bilinear/Butterworth decimation filter with $v_{co} = 0.25$ to clean up the spectrum, and plot one cycle of the decibel spectrum magnitude after decimation.
5. A signal $x(t) = 4 \sin 2\pi t + 2 \sin 4\pi t + \sin 8\pi t$ is sampled at 9 Hz. The 1-Hz component is the desired signal, while the 2- and 4-Hz components are noise.
   a. Use a 1024-point FFT to plot the spectrum of the sampled signal vs. frequency in hertz.
   b. Decimate the signal by 4, and plot its 256-point FFT vs. frequency in hertz.
   c. Use an eighth-order low-pass digital bilinear/Butterworth decimation filter with $v_{co} = 0.33$ to eliminate the signal's 2- and 4-Hz components. Then decimate by 4, and plot the 256-point FFT vs. frequency in hertz.
   d. Explain the frequency components of each spectrum.

Section 11.2

6. A signal $x(n) = 2 \sin \pi n/2$ is delivered to a zero-order hold unit.
   a. Determine the attenuation experienced by the signal due to the zero-order hold.
   b. Sketch two cycles of the output of the zero-order hold.
   c. Take the Fourier series of the waveform in (b) and show that its fundamental is attenuated by the amount calculated in (a).
7. A signal $x(n) = 2 \sin \pi n/3$ is delivered to a zero-order hold unit.
   a. Determine the attenuation provided the signal by the zero-order hold unit.
   b. Sketch two cycles of the output of the zero-order hold.
   c. Take the Fourier series of the waveform in (b) and show that its fundamental is attenuated by the amount calculated in (a).
8. For $x(n) = 2 \sin \pi n/2$, interpolate by a factor of 10 and remove unwanted images with a tenth-order 1-dB bilinear/Chebyshev low-pass filter having a cutoff of $v_{co} = 0.1$. Plot about three cycles of the resulting signal before and after interpolation.
9. For $x(n) = 2 \sin \pi n/3$, interpolate by a factor of 5 and remove unwanted images with a tenth-order bilinear/Butterworth low-pass filter having a cutoff of $v_{co} = 0.1$. Plot about three cycles of the resulting signal before and after interpolation.
10. An eighth-order FIR interpolation filter has the symmetrical coefficients indicated in the following table. The filter interpolates by 3, so most of the input samples are zero. The x are numbered to indicate which nonzero input sample they are. (x_1 is the first nonzero sample, x_2 is the second, etc.) Similarly, y_1 will be taken as the start of an output cycle since a new input sample has entered the filter. Show that this filter could be implemented by holding the input samples constant and calculating the output with three different sets (or phases) of coefficients before moving out the earliest sample and moving in the next new sample.

a_0	a_1	a_2	a_3	a_4	a_3	a_2	a_1	a_0	
0	0	x3	0	0	x2	0	0	x1	$y0 = a_0x1 + a_3x2 + a_2x3$
x4	0	0	x3	0	0	x2	0	0	$y1 =$
0	x4	0	0	x3	0	0	x2	0	$y2 =$

11. For $x(n) = 2 \sin \pi n/4$, interpolate by a factor of 3 and remove unwanted images with a filter of your choice. Plot about three cycles of the signal after interpolation to demonstrate your success. Explain the basis of your choice for v_{co}.

Section 11.3

12. Tabulate all sampling frequencies that can be used with a bandpass signal having $f_L = 5$ MHz and $f_H = 6$ MHz.
13. Tabulate sampling rates that could be used on a bandpass signal having $f_L = 445$ kHz and $f_H = 465$ kHz.
14. A 5-Hz carrier is amplitude modulated by a 1-Hz signal. The equation of the amplitude-modulated wave is $x(t) = (1 + 0.9 \sin 2\pi t) \sin 10\pi t$. One way to demodulate the waveform is to pass it through an ideal full-wave rectifier, which is equivalent to taking its absolute value.
    a. Use an FFT to obtain the spectrum of $x(t)$. Sample at 64 Hz to get 128 points for the FFT.
    b. Use the FFT to obtain the spectrum of $|x(t)|$. Where did all the frequencies come from?
    c. Recover the 1-Hz signal by filtering $|x(t)|$ with a low-pass filter.
15. A 1-MHz carrier is single-sideband modulated by 0–5-kHz audio. The lower sideband is transmitted with the carrier suppressed. If the audio signal is

    $$x_{\text{audio}}(t) = \cos 2\pi 1000t + 3 \cos 2\pi 3000t$$

    the resulting single-sideband signal is

    $$x_{lsb}(t) = \cos(2\pi 999{,}000t) + 3 \cos(2\pi 997{,}000t).$$

    Recommend an f_s to recover the audio information, and show that it works.
16. a. A bandpass signal having $f_L = 15$ MHz and $f_H = 18$ MHz is to be sampled so that it is centered between 0 Hz and $f_s/2$ in the baseband. Determine the f_s that will do this for $m = 1$.
    b. Repeat (a) for $m = 2$.

A

DSP HARDWARE

The electrical hardware of digital signal processing (DSP) consists mostly of integrated circuitry. The builder of a DSP system does not design these units, but rather selects units with the desired specifications from those available. Individual packages containing sampling units, ADC or DAC units, etc. are available and provide sufficient control pins to allow them to be used in a customized system. Increasingly, these individual units are absorbed into a single LSI package and the control options selected by the programming of control registers.

This appendix presents a brief discussion of key elements of DSP hardware and demonstrates a few standard approaches to accomplishing the required D/A and A/D conversions. The example devices presented are generic and represent operating principles rather than actual circuitry. Major suppliers of DSP chips include, among others, Texas Instruments, Motorola, National Semiconductor, and, prophetically, Analog Devices. These companies are in the business of selling silicon real estate, and they compete vigorously to provide the fastest or most complete "bang per buck" in their products.

A1. THE PROCESSOR

While theoretically any general-purpose processor can be used in DSP applications, trying to use software to compensate for hardware limitations would simply eat up too much time unless the signal is in the subaudio range. The processors intended for signal processing are optimized for the types of operations needed to implement the linear difference equation. For the most part they work with a minimum word size of 16 bits and have a hardware multiply. On the other hand, they do not need the elaborate interrupt handling or output port capabilities of a microcontroller.

General-purpose processors use the von Neumann architecture, with one address/data bus combination accessing all of memory. Signal processors generally use the Harvard architecture, which has two or more bus combinations accessing their own dedicated memory areas and registers. For instance, in the FIR difference equation,

$$y(n) = a_0 x(n) + a_1 x(n-1) + a_2 x(n-2) + \cdots + a_m x(n-m) \qquad \textbf{(A.1)}$$

this allows the filter coefficient a_2 and the past input sample $x(n-2)$ to be fetched simultaneously and presented to the multiplier unit in a single processor clock cycle. Next, the multiply-and-accumulate command would be executed, again possibly in a single clock cycle. Other operations can be occurring simultaneously in the background. For instance, pointers identifying the locations of a_2 and $x(n-2)$ may automatically increment to point to the next set of numbers to be fetched.

A variety of hardware features can be used to speed operations. The programming steps to implement the FIR difference equation might look as follows:

1. Fetch the a_i and x terms identified by pointers P_1 and P_2 respectively.
2. Multiply and accumulate.
3. Did an overflow occur?
    If no, continue
    If yes, do something about it
4. Have all the terms been used?
    If no, increment P_1 and P_2
        go to line 1
    If yes, deliver $y(n)$
        reset or adjust P_1 and P_2
        await and store new $x(n)$
        go to line 1

This program contains two decision points, or questions, which must be asked after each multiply-and-accumulate operation. They are equivalent to "Are we done yet?" and "Is the result still valid?" In evaluating one $y(n)$ value for a 100th-order FIR filter, these questions are asked 101 times. A great deal of time could be saved if these questions had to be asked only once for each $y(n)$ calculated.

Arithmetic overflow can be made unlikely if an oversized register is used to accumulate the results. If a system is designed to deliver a 16-bit word to the output, there is usually little that can be done about an overflow other than to set the output to a maximum or minimum value. It is not uncommon, however, for a 2's complement number to overflow the 16 bits temporarily during an extended series of additions but still wind up valid and confined to 16 bits at the end. An overflow flag for the oversized register that can be reset only by instruction, combined with an inspection of the overflow register bits, can be used to ensure the validity of the $y(n)$ calculated.

A *circular addressing* mode allows a pointer to automatically cycle after a programmed number of increments (or decrements). The P_1 pointer to the $\{a_i\}$ coefficients could start a $y(n)$ calculation cycle pointing at a_0, increment automatically in the background, and be pointing at a_0 again to start the next calculation. The P_2 pointer to the $\{x\}$ values would need adjustment only once each $y(n)$ calculation. It too would cycle to its original position, pointing again to the $x(n)$ term used in the just-completed calculation of $y(n)$. Decrementing it would point it at the $x(n - m)$ term. But this $x(n - m)$ value now falls out of the difference equation and can be replaced with the $x(n)$ value for the next calculation. Incrementing P_2 would cause it to point at the $x(n)$ term used in the last calculation, which becomes the $x(n - 1)$ term of the current calculation. In other words, the $\{x(n)\}$ remain where they are initially stored but are associated with the proper filter coefficient by circular addressing and an adjustment of the P_2 pointer's initial setting at the start of each $y(n)$ calculation.

Other hardware enhancements may be present to support the numerical operations required in signal processing. Students who have taken a course in microprocessors will have no trouble appreciating the advantages of most of them.

A.2 *R–2R* LADDER DAC

The R–$2R$ ladder circuitry of Figure A.1 is a practical and common method of converting a multibit binary number into an analog level. Each bit contributes an amount to the output that is proportional to its position down the ladder, which

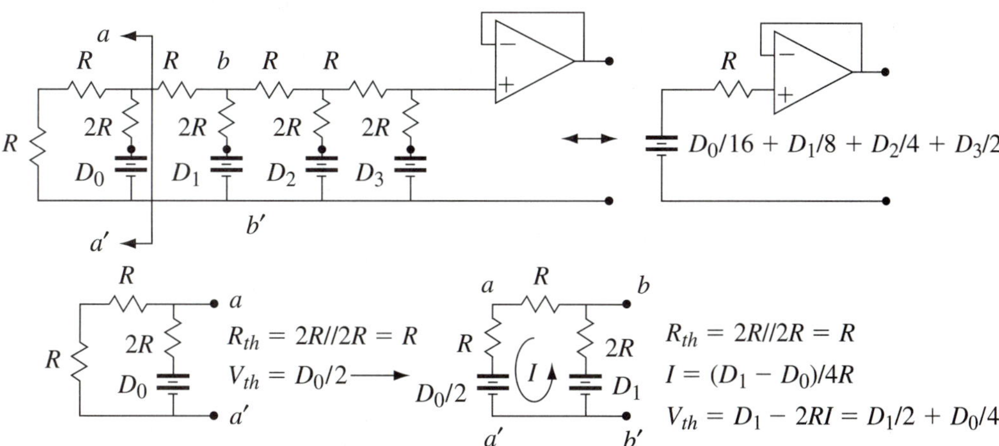

Figure A.1 The *R–2R* ladder D/A converter. The first few applications of Thevenin's theorem show the pattern and that successive bits will have the appropriate weighting in setting the value of the output from the op amp buffer.

corresponds to its positional value in the binary number. Most chips latch the binary numbers, and provide a small queue to the converter. An op amp buffer may also provide offset and gain adjustments at the output. In practice the series $2R$-voltage source combinations are usually implemented as Norton equivalents, since transistors are more ideal current sources than they are voltage sources.

A.3 PULSEWIDTH-MODULATED DACS

Microcontrollers often have timers and output port control tailored to implement a pulsewidth-modulated D/A output. In this scheme, the output period is fixed, and the length of time the output is high is proportional to the amplitude of the digital signal. Since the waveform amplitude is binary, it is an easy control signal for a digital processor to create.

From a simple-minded viewpoint, if the pulsewidth varies slowly, the waveform can be viewed as having a time-varying d-c (average) value that reflects the desired output (Figure A.2). It will also have frequency components at integer multiples of $1/T$, which need to be removed by an analog filter. For mechanical systems, the pulse period may often be set small enough that they would not react to it anyway, so the filter may not need to be fancy.

A critical reader would, of course, recognize that since the waveform has a varying pulsewidth, it is really periodic not in T but in some much longer interval, if at all. We could formulate the equation of a periodic pulsewidth-modulated waveform, with a pulsewidth varying sinusoidally, and carry out an exact Fourier analysis. Doing so would give a d-c component and components at integer multiples of this low repetition rate. One of those harmonics will end up at the frequency of the modulating sinusoid, f_{mod}; another would show up at $1/T$. All of the other harmonics would ideally be negligible in this frequency range. We would want to filter out everything at and above $1/T$, which would yield the sinusoid of frequency f_{mod} riding on a constant d-c level.

In control applications, pulsewidth modulation (PWM) is usually a very satisfactory method of achieving D/A conversion. Some second and third harmonic distortion is predicted by the Fourier analysis outlined, but this diminishes as more pulsewidth cycles are used during a single cycle of the modulation ($1/T \gg f_{\text{mod}}$).

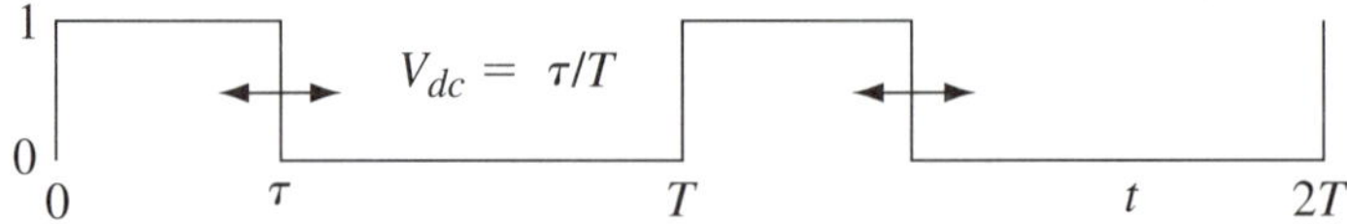

Figure A.2 Pulsewidth modulation produces a "variable d-c level" proportional to the pulse width.

A.4 PULSE-CODE-MODULATED ADCS

Turning to the input side of the processor, there are basically two types of analog-to-digital converters of importance in DSP applications: flash converters and successive-approximation converters. An N-bit flash converter consists of $2^N - 1$ voltage comparators, each set to flip between one of the 2^N possible output number values of the converter (Figure A.3). For a given input voltage, all comparators referenced at levels below the input will have tripped, and all those referenced to levels above the input will still be untripped. A large set of logic gates decode the indications of the comparators into the appropriate N-bit output number. The flash converter does not use a clock, and its conversion speed is limited only by the speed of propagation of the information through the gates.

Successive-approximation converters operate by placing a trial "1" in the most significant bit (MSb) of its initially cleared *successive-approximation register* (SAR) (Figure A.4). The register contents drive an internal DAC whose output is compared to the unknown input voltage. Based on the comparison, the decision is made whether to keep the MSb a "1," or to return it to "0." The trial process is then repeated in successively lower significant-bit positions until the N-bit number has been fully decided. The converter is stepped through its decision process by a clock, but the total number of clock cycles required for the conversion is a constant, independent of the size of the voltage input.

Flash converters involve more analog circuitry than successive-approximation devices, so they are more expensive and used only where their speed is essential. Both types of converters will operate properly only if the input voltage remains constant during the conversion process. If the input voltage were to vary, the flash

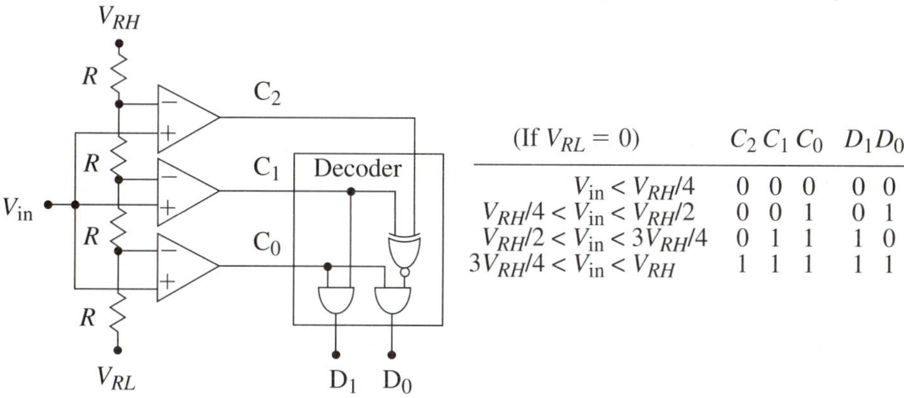

(If $V_{RL} = 0$)	C_2	C_1	C_0	D_1	D_0
$V_{in} < V_{RH}/4$	0	0	0	0	0
$V_{RH}/4 < V_{in} < V_{RH}/2$	0	0	1	0	1
$V_{RH}/2 < V_{in} < 3V_{RH}/4$	0	1	1	1	0
$3V_{RH}/4 < V_{in} < V_{RH}$	1	1	1	1	1

Figure A.3 Flash converters provide an almost instantaneous resolution of an analog amplitude into a digital number, but they require $2^N - 1$ comparators for an N-bit result and a large set of decoding gates.

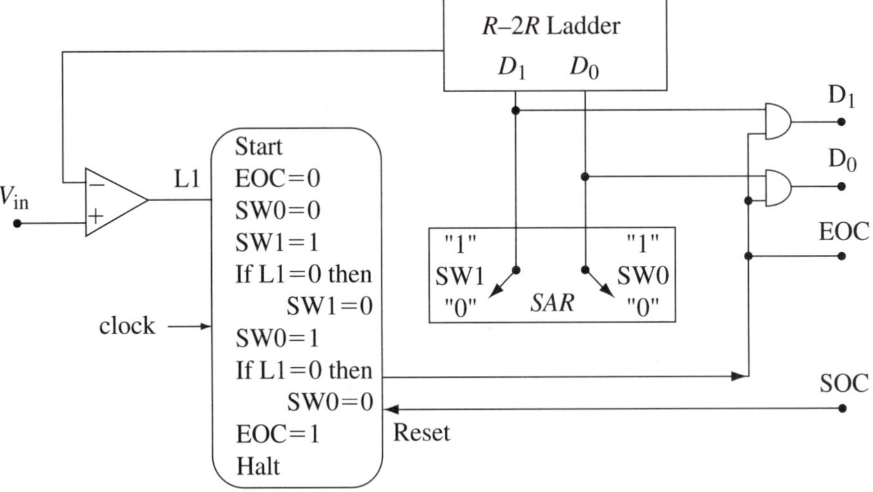

Figure A.4 The successive-approximation converter offers a compromise between cost and speed of conversion that makes it popular in many applications. It is clocked through a programmed sequence and completes its conversion in a fixed number of clock cycles. The start of a conversion (SOC) is under the control of the processor, and the processor is signaled when the conversion is complete (EOC).

converter comparators would keep changing their minds and the output of the decoders would never stabilize. If the input to the successive-approximation unit changes, the initial decision made about the MSb might no longer be valid while the status of a lower bit is being decided. To prevent this type of problem, the A/D units will either include an internal track/hold unit or require that one be supplied by the system designer.

A.5 TRACK/HOLD UNITS

The operation of a track/hold unit is conceptually quite simple (Figure A.5). In the tracking mode, a field-effect transistor (FET) switch is closed, allowing V_{in} to directly charge a storage capacitor. When the hold command is issued, the FET becomes an open switch and the capacitor retains its charge. When the A/D unit signals that it has completed its conversion, the hold command is withdrawn and the capacitor attempts to reacquire the signal and continue tracking it. The storage capacitor must be of high quality (often polystyrene for low leakage and hysteresis effects), and the op amp must have extremely high input resistance, or the capacitor voltage will droop over the holding period.

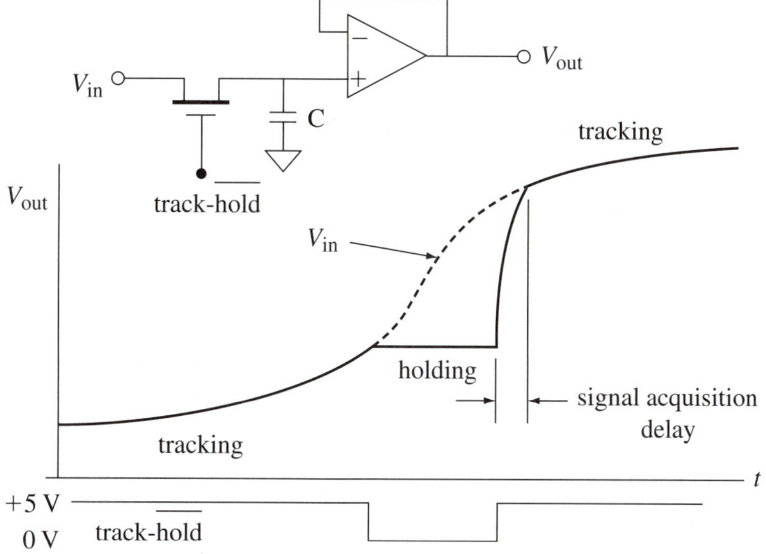

Figure A.5 Track/hold units required for A/D operations.

A.6 SWITCHED CAPACITOR FILTERS

Switched capacitor filters represent sampled-data systems operating at very high rates of oversampling. An analog input voltage is sampled, and voltage information is stored and transferred to the output in the form of charge on capacitors. Switched capacitor circuits are most directly analyzed using discrete-time theory, but their output appears almost continuous due to the high rate of oversampling used. A simple one-pole *RC* filter can be used to eliminate the high-frequency sampling components at the output.

A two-phase sampling clock is generated internally on switched capacitor chips, but the clock itself must be provided externally. Since the filter's frequency characteristics are directly related to the clock frequency, a stable clock is essential. Using switched capacitors instead of resistors as the op amp control elements saves significant silicon real estate. Switched capacitor filters are available in stand-alone packaging, but they are also often used to provide the reconstruction filtering on codec chips.

The switched capacitor integrator is shown in Figures A.6a and b. During the $\phi 1$ portion of the clock cycle, C_1 samples the new input voltage while C_2 holds the result of the previous sampling. During the $\phi 2$ portion of the clock cycle, the new charge on C_1 is transferred to C_2. Assuming that the op amp is ideal, the voltage across C_1 is forced to zero, causing its positive charge to annihilate some of the negative charge

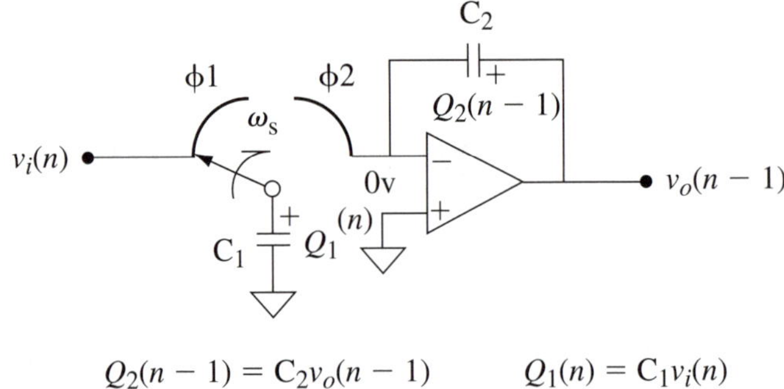

Figure A.6a During the $\phi 1$ portion of the clock cycle, the input capacitor C_1 is sampling the latest input, while the output is held constant at the value resulting from the previous sampling.

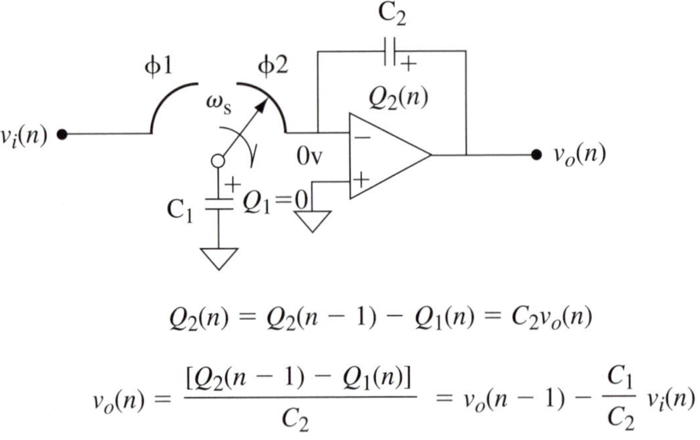

$$Q_2(n) = Q_2(n-1) - Q_1(n) = C_2 v_o(n)$$

$$v_o(n) = \frac{[Q_2(n-1) - Q_1(n)]}{C_2} = v_o(n-1) - \frac{C_1}{C_2} v_i(n)$$

Figure A.6b During the $\phi 2$ portion of the clock cycle, the input capacitor loses its charge to C_2, creating a new output voltage.

on C_2. As a result, a new output voltage is developed. The difference equation and corresponding transfer function for the system are readily shown to be

$$v_o(n) - v_o(n-1) = -\frac{C_1}{C_2} v_i(n)$$

$$H(e^{j\pi v}) = \frac{V_o}{V_i}(e^{j\pi v}) = -\frac{C_1}{C_2} \frac{1}{(1 - e^{-j\pi v})} \qquad \textbf{(A.2)}$$

For the signal frequencies of interest, $v \ll 1$ due to the high degree of oversampling used, so $e^{-j\pi v} \approx 1 - j\pi v$, which reduces the transfer function to

$$\frac{V_o}{V_i}(e^{j\pi v}) = -\frac{C_1}{C_2} \frac{1}{j\pi v} = -\frac{C_1 f_s}{j2\pi f C_2} \qquad \textbf{(A.3)}$$

This is identical to the frequency response of an analog integrator with a resistor of $R = 1/C_1 f_s$.

A.7 DELTA MODULATION CONVERTERS

The R–$2R$ ladder used in many D/A and A/D converters wastes excessive amounts of silicon chip area. The inherent competition among vendors to reduce chip costs has led integrated circuit designers to favor a well-known tracking or servo converter technique for accomplishing the conversions between the analog and digital domains. The delta modulator (aka sigma-delta or delta-sigma) converts an analog signal into a sequence of ones and zeros, where the percentage of ones is proportional to the analog level. It is inherently an oversampling modulator, and lends itself to switched capacitor implementation for its analog sections.

A basic delta modulator is shown in Figure A.7. The output of the integrator is the sum of the latest signal out of the summing junction and its own previous output. The DAC is just a comparator, and the summing junction is a difference amplifier.

To observe the circuit operation, select a constant input signal of $A = 1$ V and reference voltages of ± 5 V.

If we assume the initial states of the integrator and D-type flip-flop are 0, then this is the initial output:

For the first cycle (second output):
- If $Q = 0$, the output of the DAC is -5 V.
- The output of the summing junction is $E = 1 - (-5) = 6$ V.
- The output of the integrator is $6 + 0$.
- The comparator and FF will output a "1."

For the second cycle:
- Since $Q = 1$, the output of the DAC goes to $+5$ V.
- The output of the summing junction is $E = 1 - (5) = -4$ V.
- The output of the integrator goes to $-4 + 6 = 2$ V.
- The comparator and FF will output a "1."

For the third cycle:
- Since $Q = 1$, the output of the DAC stays at $+5$ V.
- The output of the summing junction is $E = 1 - (5) = -4$ V.

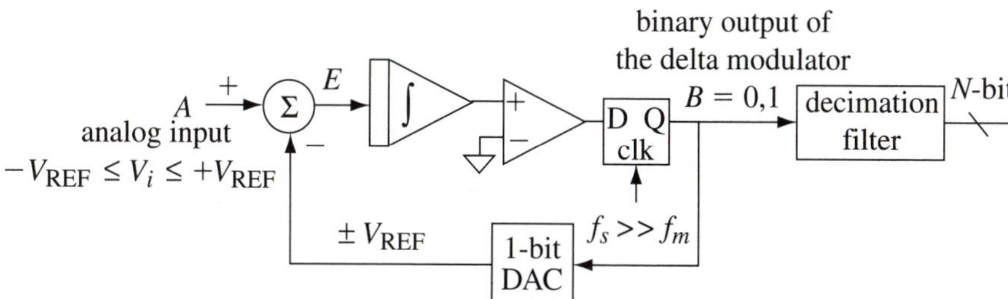

Figure A.7 A basic delta modulator ADC.

The output of the integrator goes to $-4 + 2 = -2$ V.
The comparator and FF will output a "0."
and so on.

Trying this procedure for an input of $-V_{REF}$ will quickly show that the B signal remains all zeros, while an input of $+V_{REF}$ will switch the output permanently to all ones. A simple MATLAB program to continue this cycling and count the number of ones is given in Figure A.8. The number of ones in 2^N successive samples can easily be converted to an N-bit number by this process. In practice, an FIR decimation filter accomplishes the same thing while also performing needed filtering. Since this filter needs to handle only one-bit $x(n)$ when used with the delta modulator, it is particularly easy to implement in hardware. The multibit filter coefficients are either added or ignored, depending on whether their corresponding $x(n)$ is one or zero. The output of the FIR filter is consequently a multibit number.

Exactly the same circuit arrangement can be used for the DAC. The output of the delta modulator may be converted to a binary number by counting, as we have noted. However, the single-bit, high-frequency data stream leaving the delta modulator also has an average value proportional to the input number, so its amplitude is recovered once the reconstruction filter discards all the high-frequency switching components. This is entirely equivalent to the recovery process used in PWM, but the high level of oversampling used in the delta modulator gives superior results.

In the DAC unit, the input is already in the form of a multibit binary number coming from the output of the interpolation filter, so the delta modulator DAC construction is entirely digital.

delta.m

```
% Enter with -5 ≤ A ≤5
% Calculates the binary output sequence B from the delta modulator
%              for 256 cycles, and counts the number of 1s.
1=0; B=[];
for I = 1:256
          if I>0
                B=[B 1];
                E=A-5;
          else
                B=[B 0]
                E=A+5;
          end
I=E+I;
end
Count = B*ones(1,256)
```

A	A+5	Count	
+5 V	10	255	Count = 25.5(A+5)
+1 V	6	153	
−5 V	0	0	

Figure A.8 A MATLAB m-file to determine the B output for 256 clock cycles in the delta modulator of Figure A.7.

B

MATLAB INDEX

This index identifies MATLAB commands introduced in the text, and cites some pages where their variations were demonstrated. It also includes some special functions developed in the text, and identifies them as *.m files.* Some commands, especially common mathematical functions, have been listed but may not have been demonstrated. Other undemonstrated functions have been listed with demonstrated functions having a similar purpose and syntax.

If you know the function name, *help function_name* will provide full information on the function options in the MATLAB command window.

C

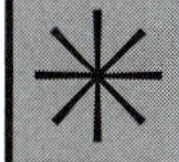

ANSWERS TO THE ODD-NUMBERED PROBLEMS

Chapter 1

1.1 a) $2.2361 \angle 153.4349°$ b) $4.4721 \angle -63.4349°$ c) $3.1623 \angle -71.5651°$

1.3 a) $-1.3074 + 2.8566i$ b) $-1.3073 + 1.5136i$ c) $3.3016 - 3.7549i$

1.5 a) $3 \angle \pi$ b) $2 \angle -1.1416$ c) $0.25 \angle -5$

1.7 a) $4\cos(1) - 6\cos(2)$ b) $2i \sin \pi/2 - 8i \sin 3 = 2i(1 - 4 \sin 3)$

1.9 a) $6.0000 + 22.0000i$ b) $0 + 1.5000i$ c) $1.3580 + 0.5940i$

1.11 a) $0.3245 - 0.0712i$ b) $0.0009 - 0.0125i$ c) $-0.0258 - 1.3389i$

1.13 a) $|F| = 2$ if $(x - 1)^2 + y^2 = \left(\frac{1}{2}\right)^2$. This is a circle centered at $z = 1$ with radius 1/2.

b) $F = 0 + j2$ only at the point $z = 1 - j0.5$.

1.15 a)

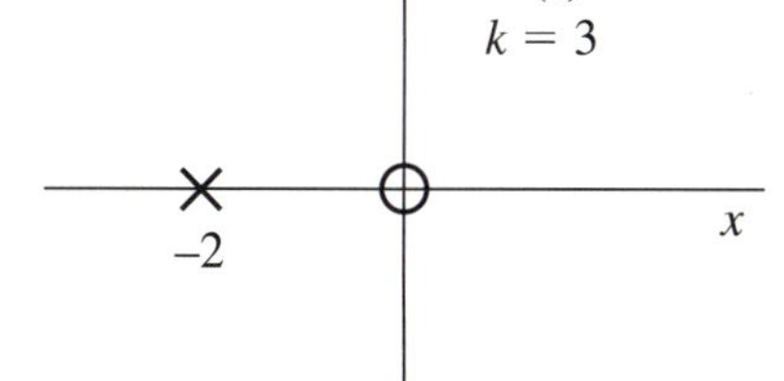

b) $F(jy) = \dfrac{3jy}{2 + jy} = \dfrac{3y}{\sqrt{4 + y^2}} \angle \dfrac{\pi}{2} - \arctan\left(\dfrac{y}{2}\right)$

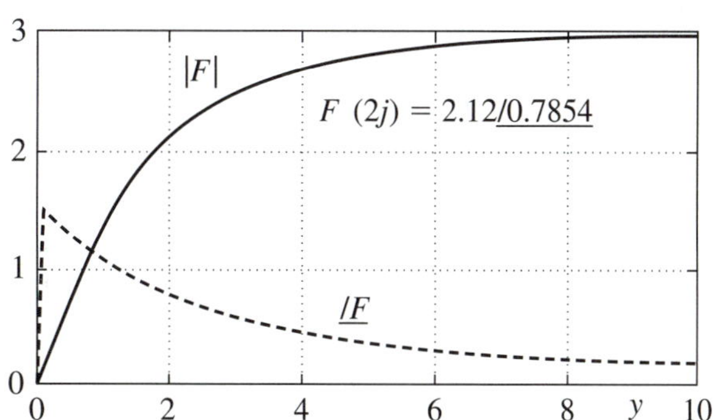

1.17 a) $F(0) = 0, F(2) = \infty, F(10) = 17.50$ b) $G(2) = 2/5$

1.19 `x=(0:12)*pi/12 %colon notation    x=linspace(0,pi,13) %MATLAB function`

1.21 The horizontal axis will run from 0 to $\pi/2$. The graph will show a half-cycle of a sinewave calculated using 123 equally spaced points and plotted in a solid red line.

1.23 The following program shows that Real(F) = 3 except for the singular point at the origin.

```
theta=linspace(0,2*pi,20);
z=1+exp(j*theta);
F=6./z
```

Chapter 2

2.1 $\frac{1}{4}\frac{di_1}{dt} + i_1 = i_g$ 2.3 $v_g(t) = \frac{1}{2}\int_0^t i(t)dt + v_C(0) + i(t)$ or $\frac{di}{dt} + \frac{1}{2}i = \frac{dv_g}{dt}$

2.5 a) $y(t) = -e^{-t}$ b) $y(t) = (3/14)e^{2t}$

2.7 a) $\tilde{X} = 1.5\angle 12°, s = 2j$ b) $\tilde{X} = 0.5\angle 0°, s = -3 + j$
c) $\tilde{X} = 6\angle -70°, s = 2j$ d) $\tilde{X} = 1\angle 0°, s = 1 + 3j$

2.9 $3e^{-2t} \sin t \leftrightarrow 3\angle 0°, s = -2 + j,$
$\tilde{Y} = -1.896\angle 108.4° \leftrightarrow y(t) = 1.896 \sin(t - 71.6°)$

2.11 a) $\frac{\tilde{Y}}{\tilde{X}} = \frac{4(s+1)}{(s+1)^3} = \frac{4}{(s+1)^2}$ b) $\frac{\tilde{Y}}{\tilde{X}} = \frac{s}{s+4}$

2.13 a)

$$\frac{\hat{Y}}{\hat{X}} = \frac{s^2 + 6s + 4}{(s+2)(s+4)}$$

$j\omega$ $\frac{\hat{Y}}{\hat{X}}$ $k = 1$ σ

−5.2 −4 −2 −0.76

b)

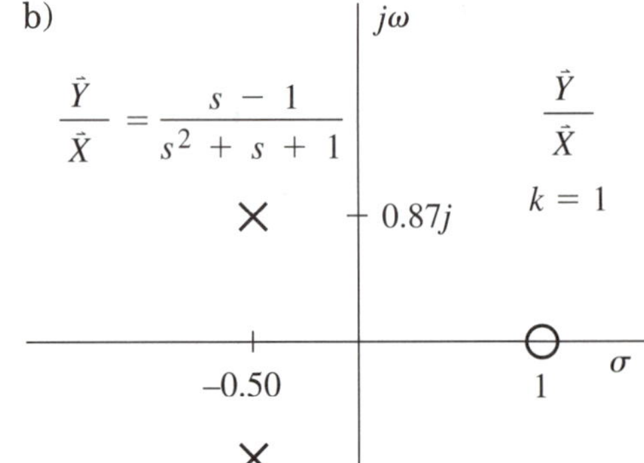

2.15 a) $y(t) = Ke^{-2t}\cos(2t + \phi)$; stable
b) $y(t) = K_1e^{-4t} + K_2\cos(2.828t + \phi)$; marginally stable
c) $y(t) = K_1e^{-8t} + K_2e^{t/2}\cos(1.937t + \phi)$; unstable
d) $y(t) = K_1e^{-t} + K_2e^{-t/2}\cos(1.323t + \phi)$; stable

2.17 `>y = [1 0 0 2 1 -4]; x=[-2 3j 4]; polyval(y,x)`

2.19 `>y(2)=0    % identifies the indeterminate value at x = 1`

2.21
```
>num=[1 -2 1]; den=[1 2 16 18 2];    % define the transfer function
>x = [2 4 6]; s = [0 -3 3j];         % define phasors and frequencies
>Y = polyval(num,s)./polyval(den,s) % calculate forced responses
Y =      0.5000        0.1345      0.1311 + 0.0984i
>abs(Y(3))
ans = 0.1639
>angle(Y(3))*180/pi
ans = 36.8699
```
$\therefore\ y_f(t) = 0.5 + 0.1345e^{-3t} + 0.1639\cos(3t + 36.87°)$

```
>roots(den)
ans = -0.3998 + 38401i   -0.3998 - 3.8401i   -1.0756   -0.1247
```
$\therefore\ y_n(t) = K_1e^{-0.3998t}\cos(3.8401t + \phi) + K_2e^{-1.077t} + K_3e^{-0.1247t}$

Chapter 3

3.1 $Z(s) = 1 + \dfrac{2/s}{2 + 1/s} = \dfrac{2 + 3/s}{2 + 1/s} = \dfrac{s + 3/2}{s + 1/2}$ 3.3 $Y_T = \dfrac{0.5(s + 2)}{(s + 1)^2}$

3.5 a) $\dfrac{\vec{V}_o}{\vec{I}_g}$, $k = 0.4$; $j\omega$, $j0.894$, 3, σ

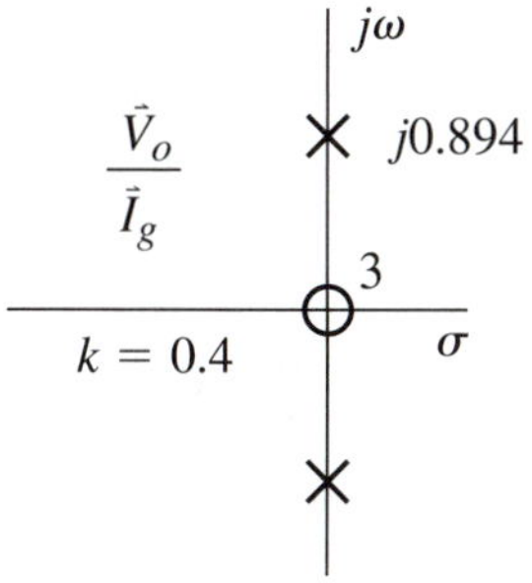

b) Z, $k = 0.4$; $j\omega$, $j1$, $j0.894$, σ

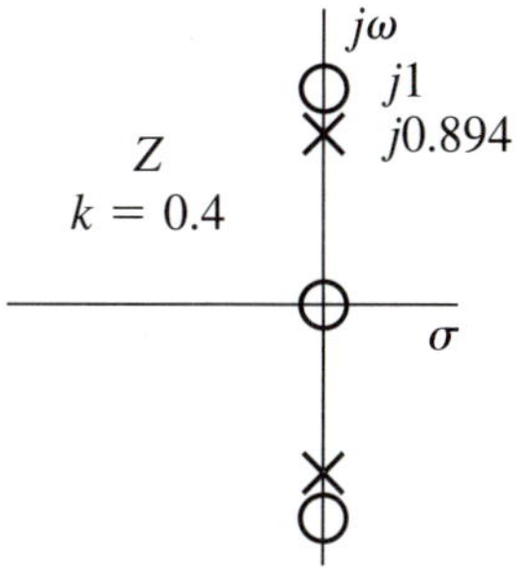

3.7 $\dfrac{\vec{I}_x}{\vec{I}_g} = \dfrac{4}{s^2 + 2s + 4}$ 3.9 $\dfrac{\vec{I}_o}{\vec{V}_g} = \dfrac{s^3}{4(s^2 + 1)}$

3.11 $L_2 = \dfrac{1}{a_2}, L_1 = \dfrac{a_1}{a_0} - \dfrac{1}{a_2}, C = \dfrac{a_2^2}{a_1a_2 - a_0}$

3.13 $g_m = \dfrac{1}{R}\left(\dfrac{C_1}{C_2}\right), \omega_o^2 = \dfrac{C_1 + C_2}{C_1C_2L}$ 3.15 $G > 99/H$

3.17 a) 1000 RPM   b) 12 V   c) 1092 RPM

3.19 $\dfrac{\vec{C}}{\vec{R}} = \dfrac{G_2(G_1 - 1)}{1 + G_1H}$ 3.21 $\vec{\omega} = \dfrac{60K\vec{V}_i - 20\vec{T}_L}{s + 10 + 6K}$

a) $V_i = 6$ V, $s = 0$, $\omega_{NL} = 45$ rad/s
b) $T_L = 20$, $\omega_{FL} = 45 - 10 = 35$ rad/s
c) $V_i = 4.650$ V for 45 rad/s, $\omega_{FL} = 43.71$ rad/s

3.23 a) $K > \sqrt{12}$ b) $K > 2.4$ c) $0.4174 < K < 9.583$
d) $0 < K < 19$ e) $10.26 < K < 149.7$

3.25
```
>subplot(2,1,1); plot(w,abs(H))
>title('|H| vs. w')
>subplot(2,1,2); plot(w,angle(H))
>title('angle of H vs w')
```

3.27 a) x = 0 1 0 2 0 3 0 4 0 5 0 6 0 7 0 8 0 9 0 10
b) x = 2 1 0 −1 −2

Chapter 4

4.1 a) $Z(s) = \dfrac{2s(s \pm j1.4142)}{3(s \pm j0.8165)}$ b) $Z(s) = \dfrac{3(s \pm j1.1547)}{s(s \pm j1.4142)}$

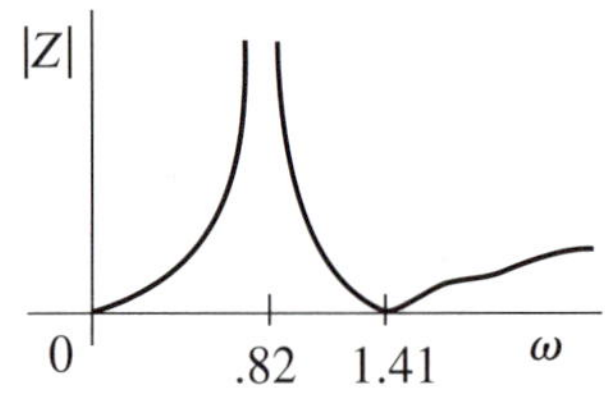

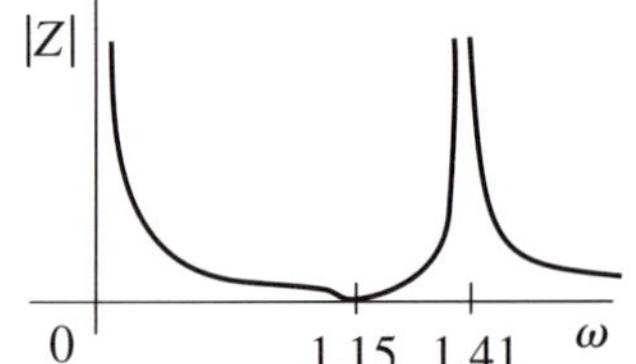

4.3 a) $\dfrac{\vec{V}_o}{\vec{V}_g} = \dfrac{2}{s^2 + 8/3}$ b) $Z(s) = \dfrac{s(s \pm j1.633)}{(s \pm j0.8165)}$

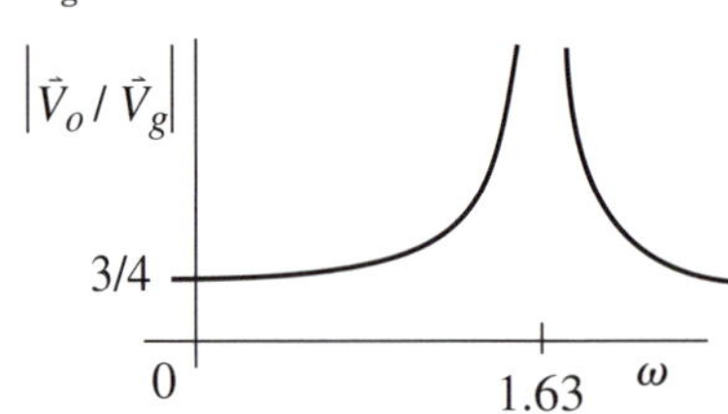

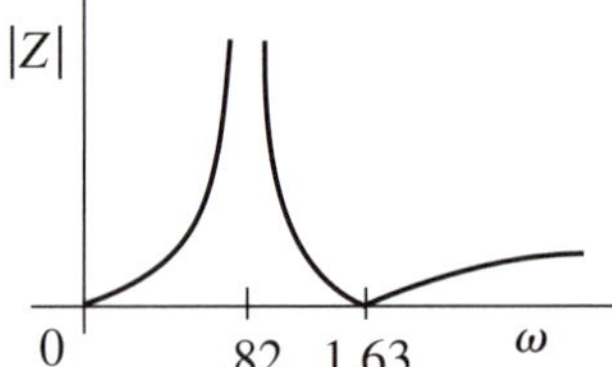

4.5 $Z(s) = \dfrac{2(s + 0.0167 \pm j0.577)}{s(s + 0.0167 \pm j)}$

a) $|Z(j1)| = \dfrac{2(0.423)(1.577)}{1(2)(.0167)} = 39.9\ \Omega;$

$B \approx 2(0.0167) = 0.0334$ rad/s

b) $|Z(j0.577)| = \dfrac{2(0.0167)(1.154)}{0.423(1.577)(0.577)} = 0.10\ \Omega$

$B \approx 0.0334$ rad/s

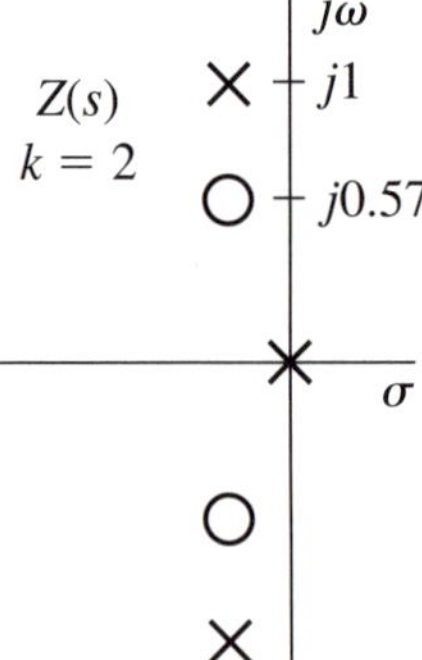

4.7 a)

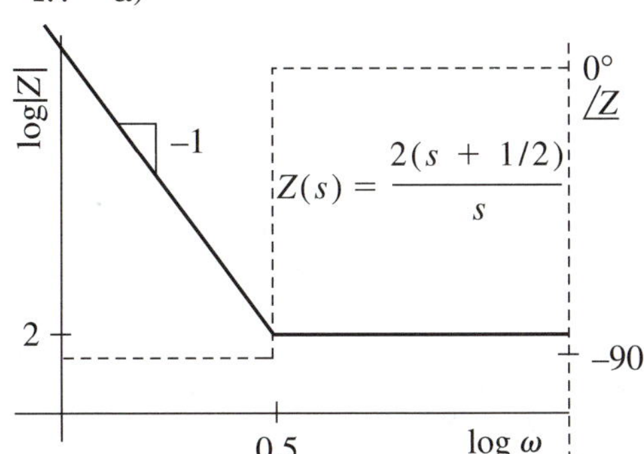

b)

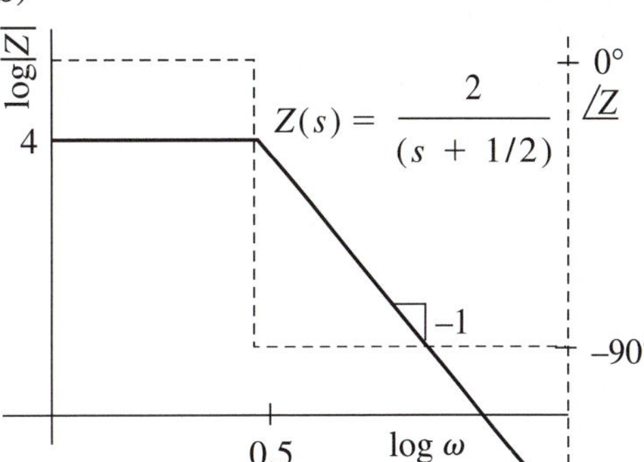

4.9 a)

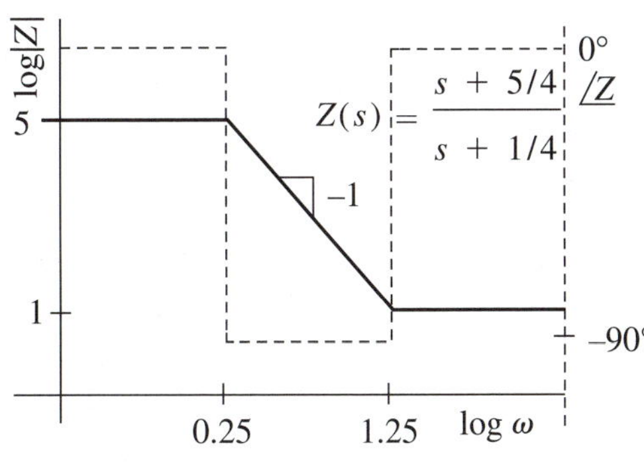

b)

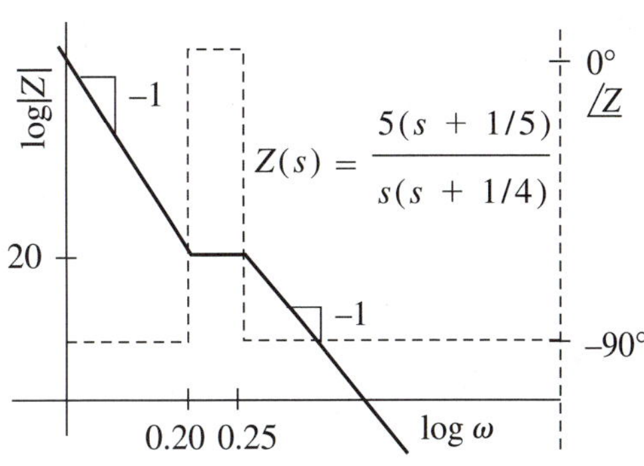

$\omega = \sqrt{0.25(1.25)} = 0.559$ rad/s for max capacitance

$\omega = \sqrt{0.20(0.25)} = 0.224$ rad/s for min capacitance

4.11 38.45 dBm = 8.451 dBW

4.13

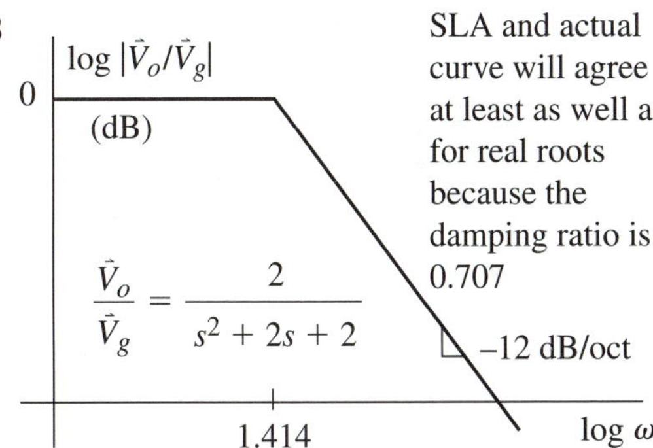

SLA and actual curve will agree at least as well as for real roots because the damping ratio is 0.707

4.15

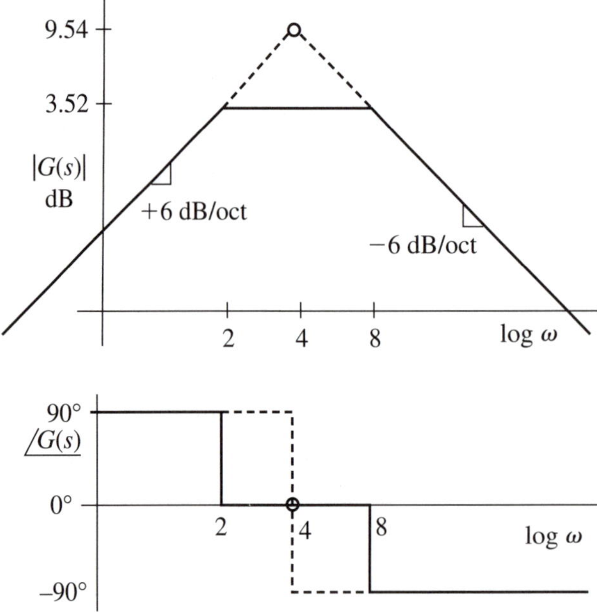

G of (a) has the solid line SLA, (b) and (c) have the same dotted SLA. Phase crosses zero at $\omega = 4$ rad/s in all cases. Maximum gain also occurs at $\omega = 4$ rad/s.

a) $G(j4) = 1.2 = 1.58$ dB b) $G(j4) = 6 = 15.6$ dB c) $G(j4) = 1.5 = 3.52$ dB

4.17

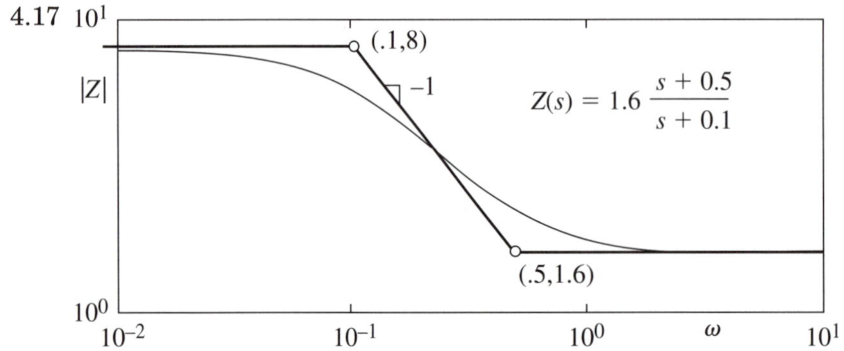

4.19

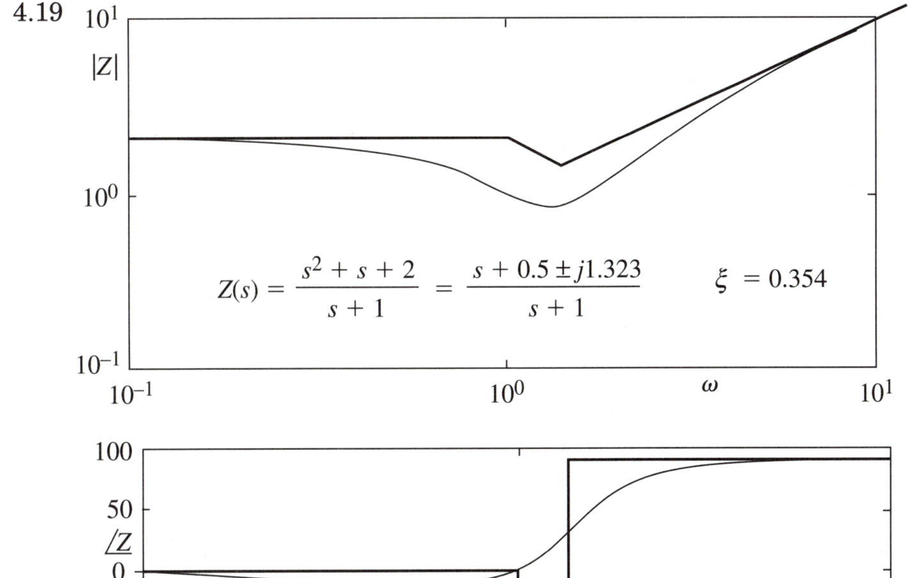

4.21

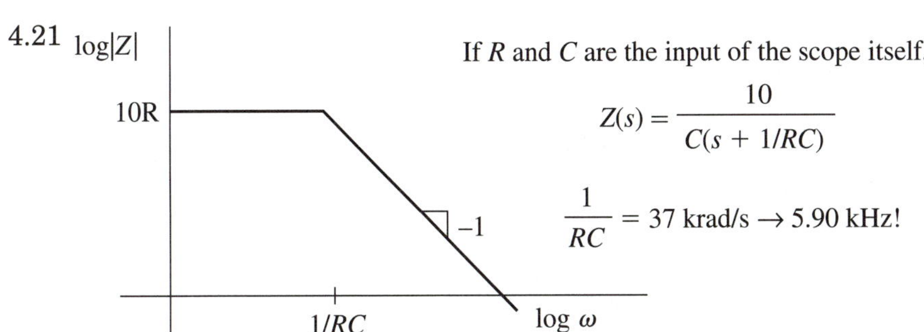

If R and C are the input of the scope itself,

$$Z(s) = \frac{10}{C(s + 1/RC)}$$

$$\frac{1}{RC} = 37 \text{ krad/s} \rightarrow 5.90 \text{ kHz!}$$

Chapter 5

5.1 The output will be amplified by a factor of 3 and delayed 1.5 seconds but otherwise will be identical to the input.

5.3 a) If you want to see it, do as instructed.

5.3 b) Use plot(t,g,t,x,t,y,t,z) in MATLAB for different-colored curves.

5.5 $H(s) = \dfrac{1}{s^3 + 2s^2 + 2s + 1}$

$$|H(j\omega)| = \left|\frac{1}{1 - 2\omega^2 + j\omega(2s - \omega^2)}\right| = \frac{1}{\sqrt{(1 - 2\omega^2)^2 + \omega^2(2 - \omega^2)^2}}$$

$$|H(j\omega)| = \frac{1}{\sqrt{(1 - 4\omega^2 + 4\omega^4) + \omega^2(4 - 4\omega^2 + \omega^4)}} = \frac{1}{\sqrt{1 + \omega^6}}$$

5.7 $n = 5.65 \rightarrow 6$

5.9 $p \rightarrow \dfrac{s}{2\pi 4000}$ $\quad H(s) = \dfrac{15.88 \times 10^{12}}{s^3 + 50.27 \times 10^3\, s^2 + 1.263 \times 10^9\, s + 15.88 \times 10^{12}}$

5.11

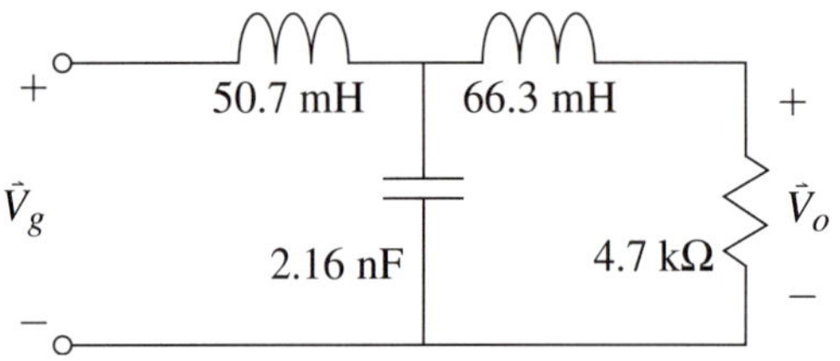

+ $\hat{V}_g$ − | 50.7 mH | 66.3 mH | 2.16 nF | 4.7 kΩ | + $\hat{V}_o$ −

$$L_1 \rightarrow 1.355\,\frac{4700}{40{,}000\pi} = 50.68 \text{ mH}$$

$$C \rightarrow \frac{1.274}{4700(40{,}000\pi)} = 2.157 \text{ nF}$$

5.13

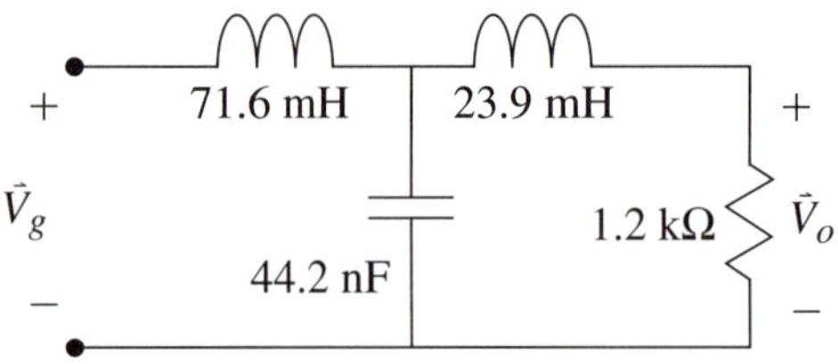

+ $\hat{V}_g$ − | 71.6 mH | 23.9 mH | 44.2 nF | 1.2 kΩ | + $\hat{V}_o$ −

$$R' \rightarrow 150R = 1200\ \Omega$$

$$L'' \rightarrow \frac{150}{2000\pi}L$$

$$C'' \rightarrow \frac{C}{2000\pi(150)}$$

5.15 $p \rightarrow \dfrac{s^2 + 15}{2s}$ $\quad H(s) = \dfrac{4s^2}{s^4 + 2.828s^3 + 34s^2 + 42.43s + 225}$

5.17 $p \rightarrow \dfrac{3s}{s^2 + 4}$ $\quad H_{BS}(s) = \dfrac{0.7943s^4 + 6.3546s^2 + 12.709}{s^4 + 2.9299s^3 + 18.935s^2 + 11.719s + 16}$

5.19

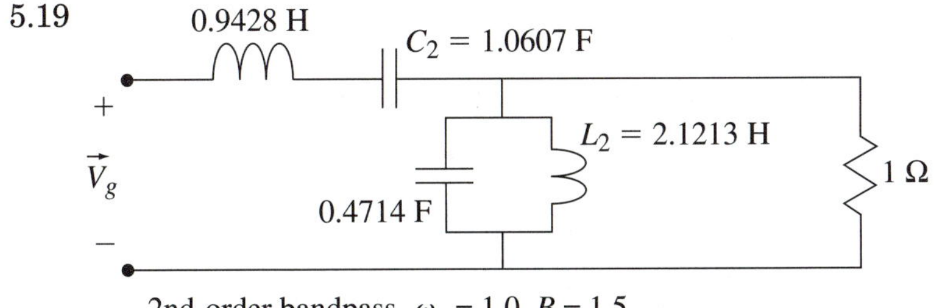

2nd-order bandpass, $\omega_c = 1.0$, $B = 1.5$

5.21 We need a low-pass filter with three inductors or capacitors. One such circuit is

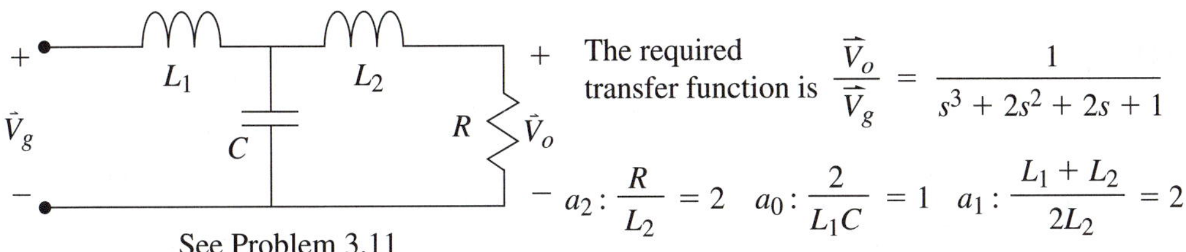

See Problem 3.11

The required transfer function is $\dfrac{\vec{V}_o}{\vec{V}_g} = \dfrac{1}{s^3 + 2s^2 + 2s + 1}$

$a_2 : \dfrac{R}{L_2} = 2 \quad a_0 : \dfrac{2}{L_1 C} = 1 \quad a_1 : \dfrac{L_1 + L_2}{2L_2} = 2$

If $R = 1\ \Omega$, then $L_2 = 1/2$ H, $L_1 = 3/2$ H, and $C = 4/3$ F.

5.23

```
num = 6.1529e + 28
den = 1.0
2.4276e + 05
2.9467e + 10
2.2676e + 15
1.1633e + 20
3.7836e + 24
6.1529e + 28
```

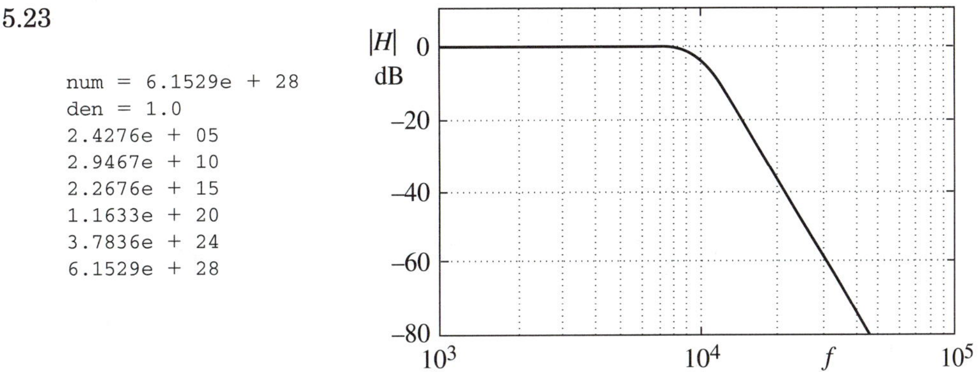

5.25 Use the same *ez1.m* file used for Example 5.7, but change the first instruction to [z,p,k]=cheb1ap(n,2); and use **plot** instead of **semilogx** in the last instruction.

Chapter 6

6.1 $c_m = \dfrac{1}{T}\displaystyle\int_{-\tau/2}^{\tau/2} Ae^{-jm\omega_o t}\,dt = \dfrac{A\tau}{T}\dfrac{\sin m\pi\tau/T}{m\pi\tau/T}$

6.3 a) $f(t) = \cdots -0.2387je^{-j4\pi t/3} - 0.4775je^{-j2\pi t/3} + 0 + 0.4775je^{+j2\pi t/3} + 0.2387je^{+j4\pi t/3} + \cdots$

b) $f(t) = -0.9550 \sin 2\pi t/3 - 0.4774 \sin 4\pi t/3 - 0.2388 \sin 8\pi t/3 - \cdots$

6.5 a) $c_m = \dfrac{1}{8}\displaystyle\int_0^2 \sin\left(\frac{\pi}{4}t\right)e^{-jm\frac{\pi}{4}t}\,dt$

b) $c_m = \dfrac{1}{9}\left[\displaystyle\int_{-2}^{-1} 3\,e^{-jm\frac{2\pi}{9}t}\,dt + \int_1^2 3\,e^{-jm\frac{2\pi}{9}t}\,dt\right] = \dfrac{2}{3}\displaystyle\int_1^2 \cos m\frac{2\pi}{9}t\,dt$

6.7 $c_m = \dfrac{\cos 0.25\pi m - \cos 0.75\pi m}{j\pi m}$

6.9 a)

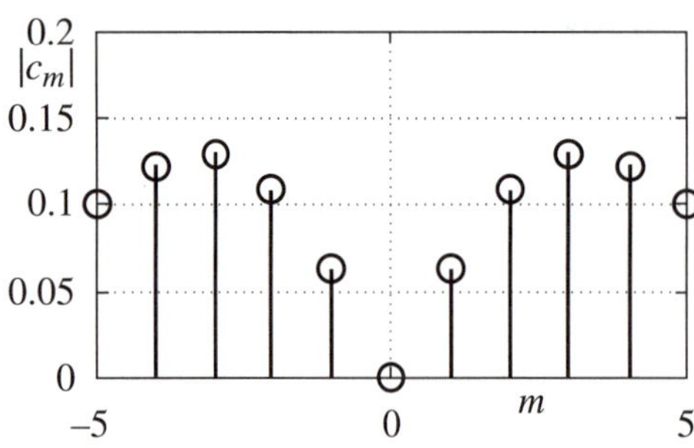

b)

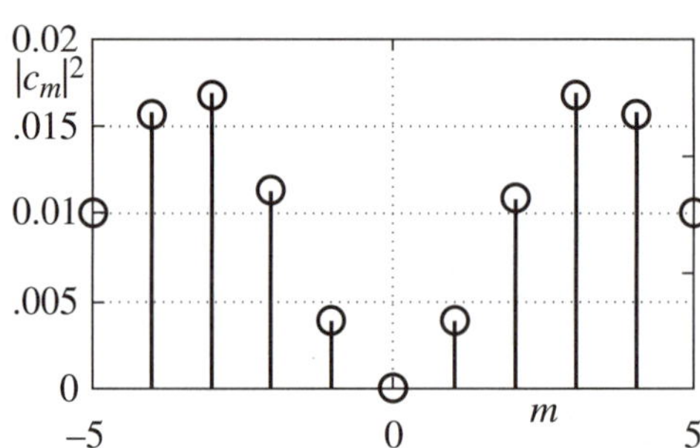

6.11 a) $F(\omega) = \displaystyle\int_0^2 e^{-j\omega t}\,dt - \int_2^3 2e^{-j\omega t}\,dt = \frac{1 - 3e^{-j\omega 2} + 2e^{-j\omega 3}}{j\omega}$

b) $f(t) = \cdots 0.4135e^{-j(4\pi t/3-\pi/3)} + 0.8270e^{-j(2\pi t/3-2\pi/3)} + 0 + 0.8270e^{+j(2\pi t/3-2\pi/3)} + 0.4135e^{+j(4\pi t/3-\pi/3)} + \cdots$

6.13 $X(\omega) = \displaystyle\int_{-3ms}^{0} -2e^{-j\omega t}\,dt + \int_0^{3ms} 2e^{-j\omega t}\,dt = \frac{4j}{\omega}(\cos(3\times 10^{-3}\omega) - 1)$

6.15

c_{-5}	c_{-4}	c_{-3}	c_{-2}	c_{-1}	c_0	c_1	c_2
−0.0320	−0.0839	−0.2652	−0.6821	−1.8243	1.0000	1.8243	0.6821

c_3	c_4	c_5
0.2652	0.0839	0.0320

6.17 Waveforms c and d are discontinuous and need a window.

6.19

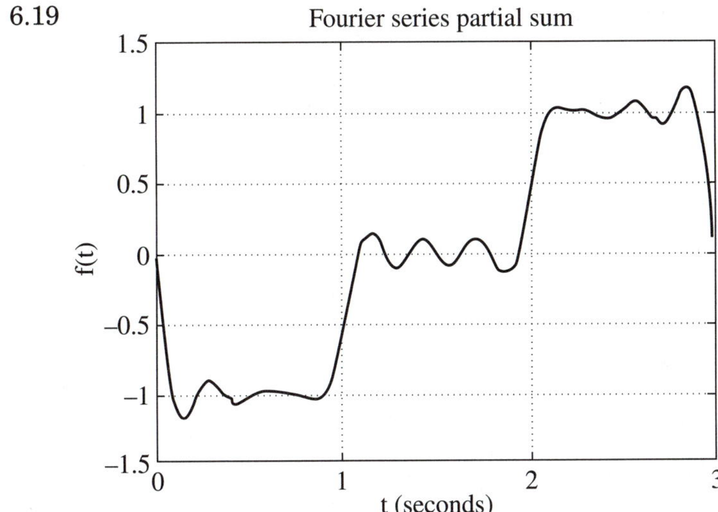

6.21 a) b)

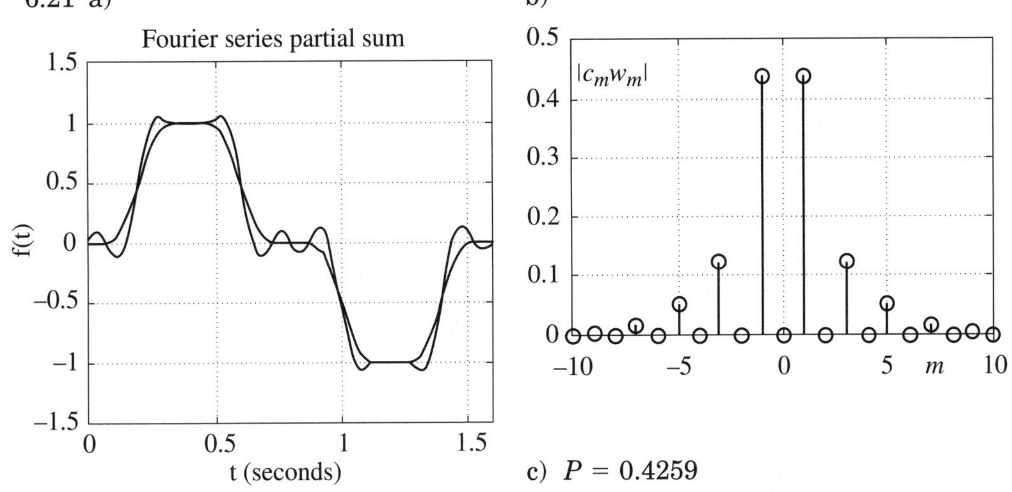

c) $P = 0.4259$

6.23 a = 7 b = 0 – 2.0000i 4.000 –1.0000 c = Error d = 3.0000 – 2.0000i

6.25 Fourier series partial sum

f(t)

t (seconds)

$$\frac{\vec{V}_o}{\vec{V}_i} = \frac{3}{4 + jm\pi}$$

Chapter 7

7.1 $F(m) = \sum_{n=0}^{3} f(n)e^{-j\pi mn/2} = 1$, all m; $f(n) = \frac{1}{4}[1 + j^n + (-1)^n + (-j)^n]$

7.3 $F(m) = 1 - e^{-jm\pi} = 0\ \ 2\ \ 0\ \ 2$; $f(n) = \frac{1}{2}[(j)^n + (-j)^n] = 1\ \ 0\ \ -1\ \ 0$

7.5 $f_s = 1/40\mu = 25$ kHz. $f_m = m(f_s/37)$;
$f_{30} = f_{(30-37)} = f_{-7} = -7(25\text{ k})/37 = -4.73$ kHz

7.7 a) $x(t) = \cos(0.20n + 56°)$; $x(5) = \cos(1 + 0.9774) = -0.395$
b) $x(t) = \cos((2\pi m + 0.20)n + 56°) = \cos(6.4832n + 0.9774)$ @ $m = 1$;
$x(5) = -0.395$
c) $x(n) = \cos((2\pi m - 0.20)n - 56°) = \cos(6.0832n - 0.9774)$ @ $m = 1$;
$x(5) = -0.395$

7.9 The baseband frequencies are $\omega = 2$, 2.472, and 0.292 rad/s.

7.11

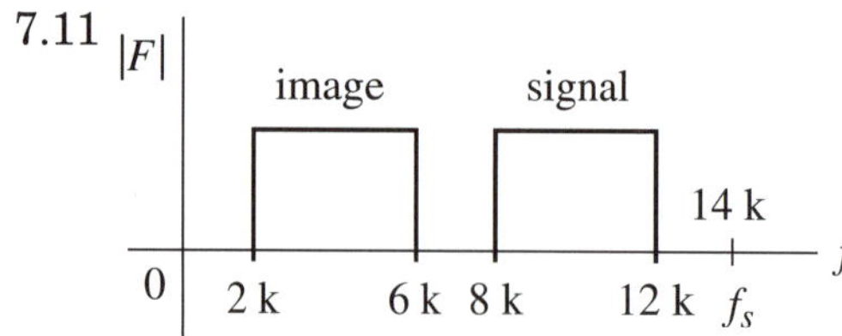

The signal and its image do not overlap, so a bandpass filter with $f_L = 8$kHz and $f_H = 12$kHz could still recover the original sampled signal.

7.13 $X(e^{j\omega}) = \sum_{n=0}^{\infty} x(n)e^{-j\omega n} = 1 + e^{-j8\omega} = 2e^{-j4\omega}\cos 4\omega$

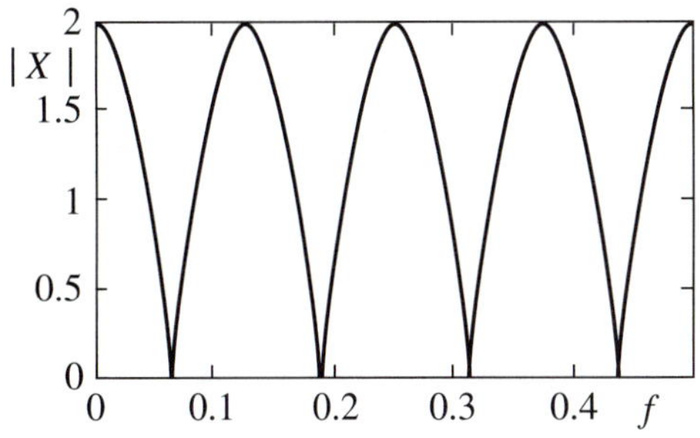

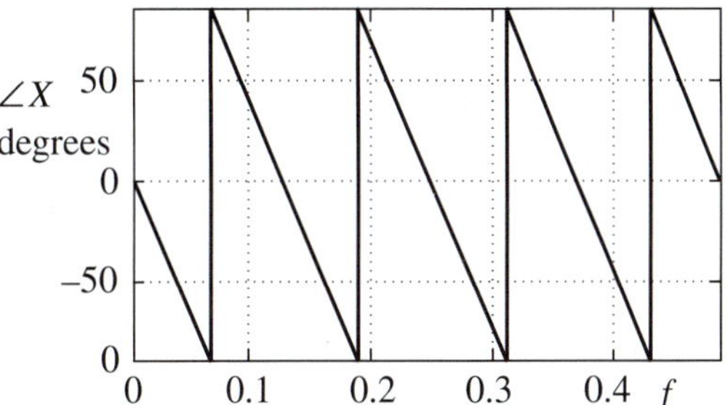

7.15 a) $X(e^{j\omega}) = \sum_{n=0}^{\infty} (0.9)^n e^{-j\omega n} = \sum_{0}^{\infty} (0.9e^{-j\omega})^n = \dfrac{1}{1 - 0.9e^{-j\omega}}$

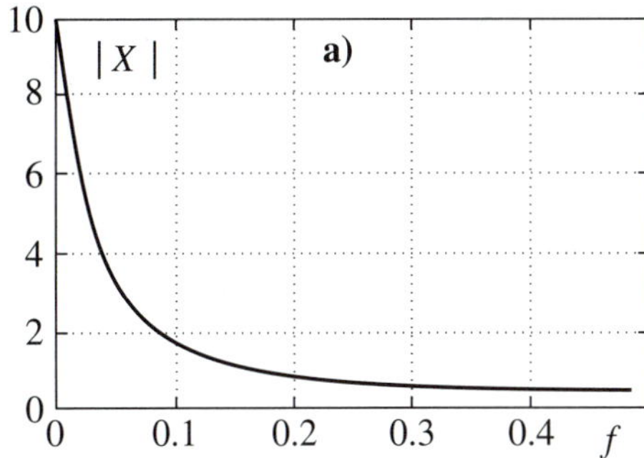

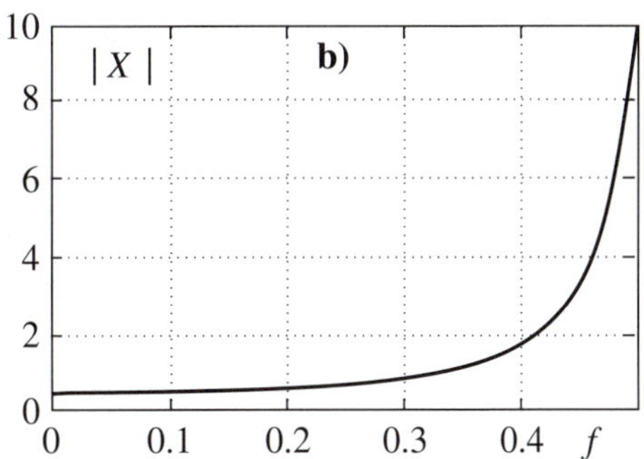

b) $X(e^{j\omega}) = \sum_{n=0}^{\infty} (-0.9)^n e^{-j\omega n} = \sum_{0}^{\infty} (-0.9e^{-j\omega})^n = \dfrac{1}{1 + 0.9e^{-j\omega}}$

Signal (a) is a steady decay, so it is mostly d-c. Signal (b) changes sign every T_S, so it has most of its spectrum amplitude at $f_S/2$.

7.17 a) `f=[cos((0:204)*pi/410) zeros(1,614) sin((0:204)*pi/410)]`

b) `f=[2*(340:-1:0)/340 2*(1:683)/683]`

c) `f=[sin((0:255)*pi/512) zeros(1,768)]`

d) `f=[zeros(1,114) ones(1,114) zeros(1,568) ones(1,114) zeros(1,114)]`

7.19
```
>f=[zeros(1,114) -3*ones(1,114) zeros(1,569)];
>f=[f 3*ones(1,114) zeros(1,113)];
>plot(f);
>subplot(2,1,1)
>partsum(f,15,2,9)
>title('15 harmonics, Hanning window')
>subplot(2,1,2)
>dispect(f)
```

7.21
```
>f=[sin(pi*(0:575)/128) zeros(1,448)]; plot(f)
>subplot(2,1,1)
>partsum(f,20,1,8)
>title('20 harmonics, Hamming window')
>subplot(2,1,2)
>powspect(f)
```

7.23

```
>n=0:99;
>x=sin(.4*pi*n/10);
>dtft(x,10)
>n=0:49;
>x=sin(.4*pi*n/10);
>hold
Current plot held
>dtft(x,10)
>grid
>gtext('1 cycle')
>gtext('2 cycles')
```

As more cycles are taken the spectral density @ 0.2 Hz increases proportionally, and the spectrum narrows in on that frequency.

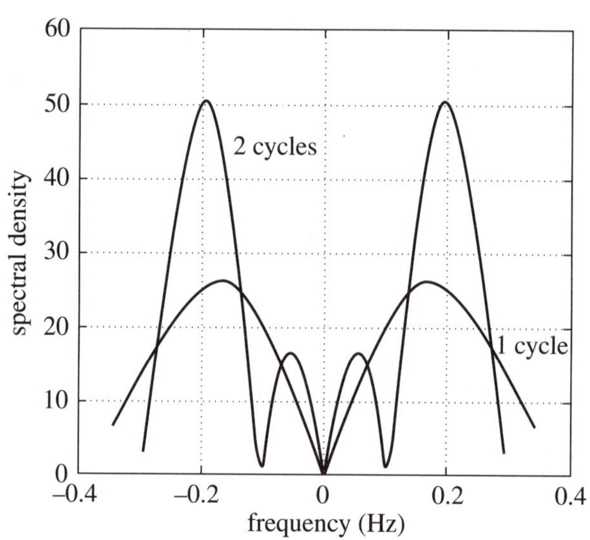

Chapter 8

8.1 a) $\vec{X} = 1; z = e^{0.1} = 1.1052$ b) $\vec{X} = 2\angle -45°; z = e^{j\pi} = -1$
c) $\vec{X} = 6\angle 14°; z = e^{-1}e^{j0.2}$ d) $\vec{X} = 3\angle -115°; z = 1.2e^{j3}$

8.3 $v = 2f/f_s = 5/12 = 2f'/9$ kHz; $f' = 1.875$ kHz is the new notch frequency.

8.5 a)

n	$x(n)$	$-x(n-1)/2$	$y(n-1)/2$	$y(n)$
0	1	0	0	1
1	1	–1/2	1/2	1
2	1	–1/2	1/2	1
3	1	–1/2	1/2	1
4	1	–1/2	1/2	1
5	1	–1/2	1/2	1

8.5 b)

n	$x(n)$	$-x(n-1)/2$	$-y(n-2)/2$	$y(n)$
0	1	0	0	1
1	1	–1/2	0	1/2
2	1	–1/2	–1/2	0
3	1	–1/2	–1/4	1/4
4	1	–1/2	0	1/2
5	1	–1/2	–1/8	3/8

8.7 a) $H(z) = \dfrac{z^2}{z^2 + 0.2z - 0.1}$ b) $H(z) = \dfrac{z^2(z^2 - 0.5)}{z^4 - 0.5z^2 + 0.5}$

8.9 $H(e^{j\pi 0.2}) = 0.8840\angle 0.0198$ $y(n) = 10.61\cos(0.2\pi n + 0.0198)$

8.11 $H(e^{j\pi 0.3}) = 1 - e^{-j1.2\pi} = 1.902\angle -18.0°$ $y(n) = 1.902\cos(0.3\pi n - 18.0°)$

8.13 a) $y(n) = x(n-1) - x(n-2) + x(n-3) + 0.25y(n-2)$
b) $y(n) = x(n) - x(n-3) + 0.2y(n-1) - 0.2y(n-2) + y(n-3)$

8.15 The system is noncausal, since the next input, $x(n+1)$, is needed to calculate the current $y(n)$.

8.17

n	$0.25\,x(n)$	$+0.25y(n-4)$	$y(n)$
0	0.25	0	0.25
1	0	0	0
2	0	0	0
3	0	0	0
4	0	$(.25)^2$	$(.25)^2$
8	0	$(.25)^3$	$(.25)^3$
12	0	$(.25)^4$	$(.25)^4$

Stable: $y(4m) = 0.25(1/4)^m$, m integers, all other, $y = 0$.

8.19 a) Poles at $z = 1.02$, ∴ unstable. b) Repeated poles at $z = -1$, ∴ unstable.

8.21 `>h=freqz(num,den); v=(0:511)/512; plot(v,abs(h))`

a) num = [.1373 .1373]; den = [1 −.7254];
Low-pass filter with cutoff @ $v = 0.1$

b) num = [.0301 0 −.0301]; den = [1 −1.8439 .9398];
Bandpass filter with $v_c = 0.1$

c) num = [.9288 −1.7682 .9288]; den = [1 −1.7682 .8576];
Bandpass filter with $v_c = 0.1$

d) num = [.8627 −.8627]; den = [1 − .7254];
High-pass filter with cutoff @ $v = 0.1$

8.23 a)

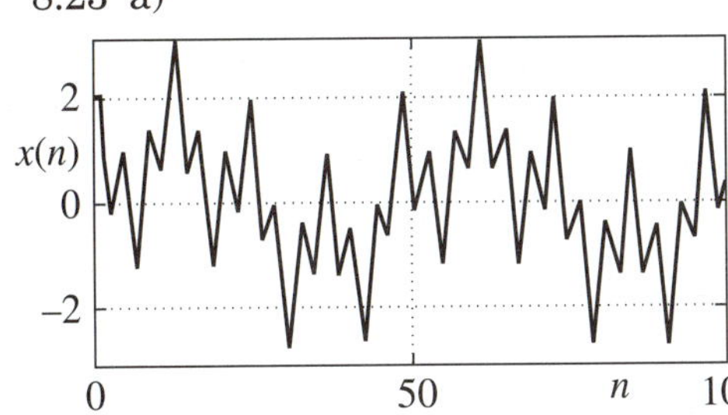

b)

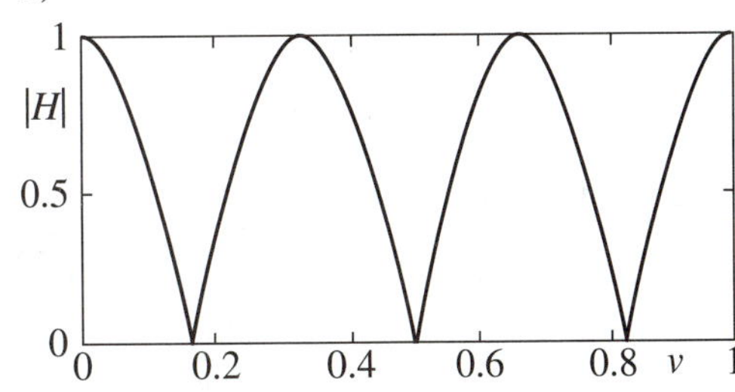

c)

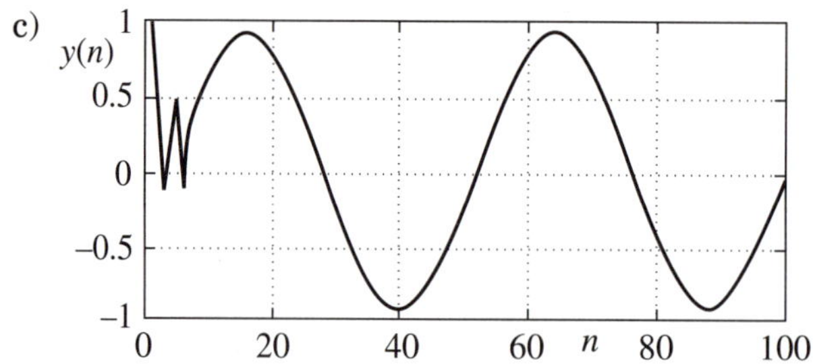

d) Input consists of equal-sized components at $v = 0.042$, 0.167, and 0.50. The components at 0.167 and 0.5 are not passed by the filter, so only the lowest-frequency component emerges. Output freq. = 0.042(3000) = 126 Hz. (Check: T ≈ 48 samples; $f = 6000/48 = 125$ Hz.)

8.25

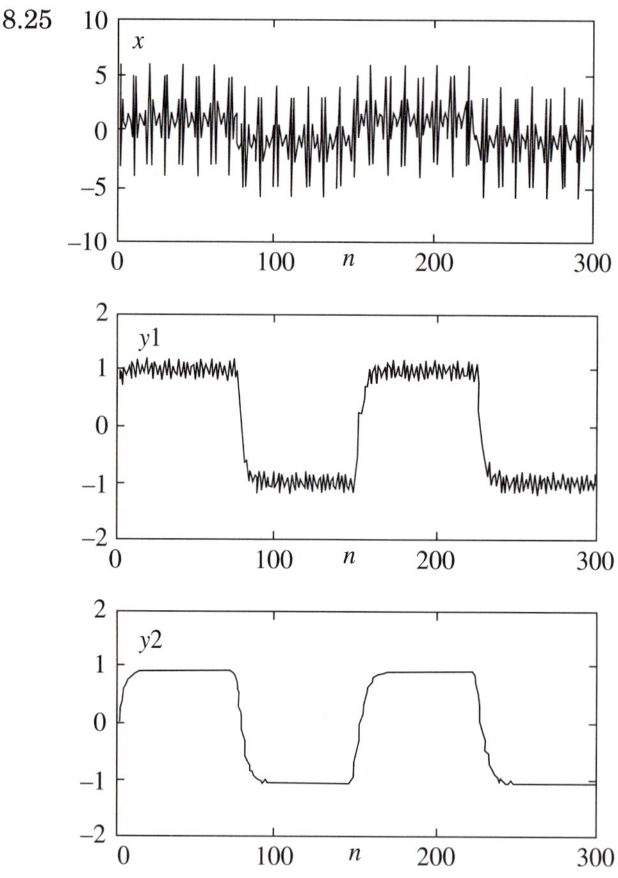

Chapter 9

9.1 $H(z) = \dfrac{1}{\left[C\dfrac{z-1}{z+1}\right]+1} = \dfrac{z+1}{(C+1)z-(C-1)} =$

$$\frac{z+1}{[1+\cot(\pi v_{co}/2)]z+[1-\cot(\pi v_{co}/2)]}$$

9.3 $H_p(s) = \dfrac{1.4314}{s^2+1.4256s+1.5162}$, $C = 1.6319$, $H(z) = \dfrac{0.2200(z^2+2z+1)}{z^2-0.3525z+0.2848}$

9.5 $H_p(p) = \dfrac{1}{p^2+1.4142p+1}$, $H_{HP}(s) = \dfrac{s^2}{s^2+1.4142s+1}$, $C = 1$,

$$H(z) = \frac{0.2929(z^2-2z+1)}{z^2+0.1716}$$

9.7 $C = 1.5158$, $B = 0.3446C = 0.5224$

9.9 $C = 3.5200$, $B = 1.2359$, $H_{BS}(z) = \dfrac{0.8981 - 1.5279z^{-1} + 0.8981z^{-2}}{1 - 1.5279z^{-1} + 0.7961z^{-2}}$

9.11 a)

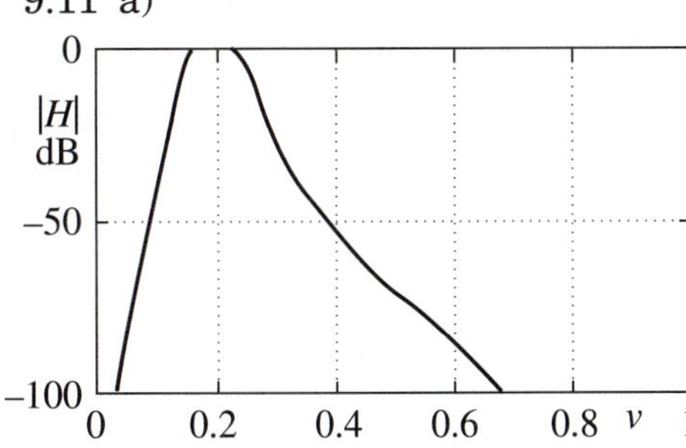

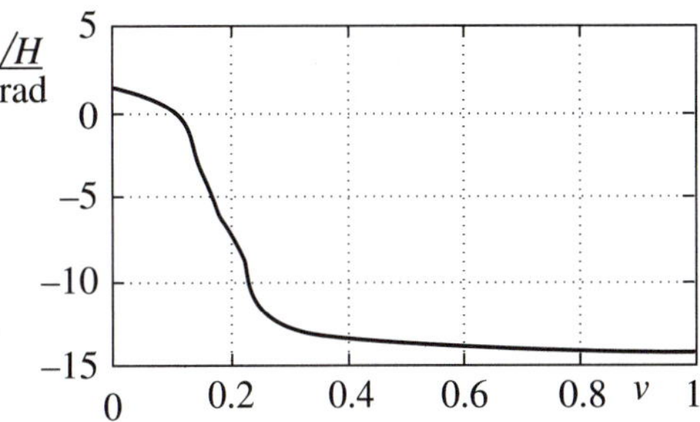

b)

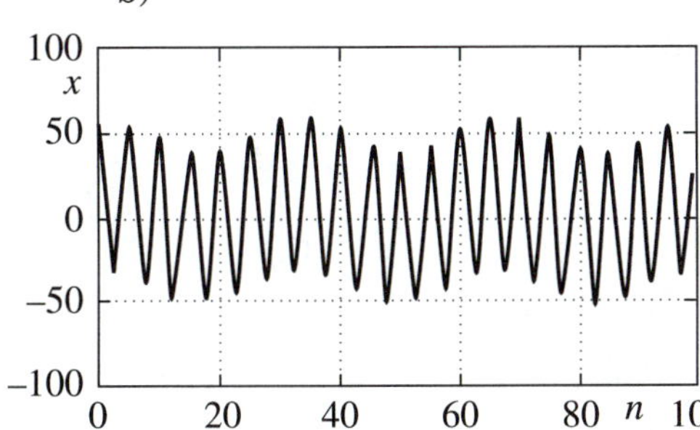

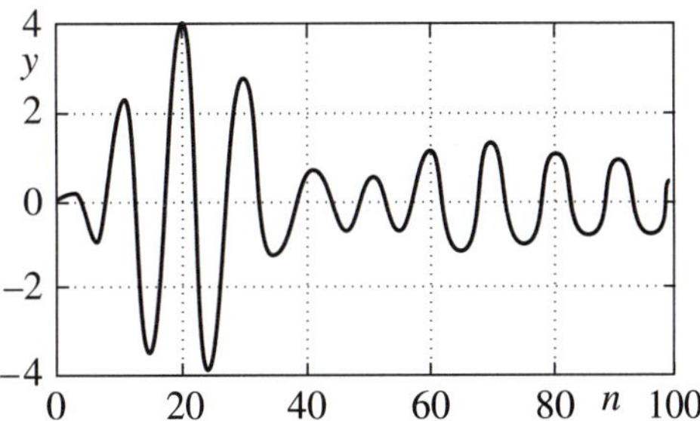

After about 50 transient samples, the output is essentially the $v = 0.2$ signal (10 samples/cycle).

9.13 a)

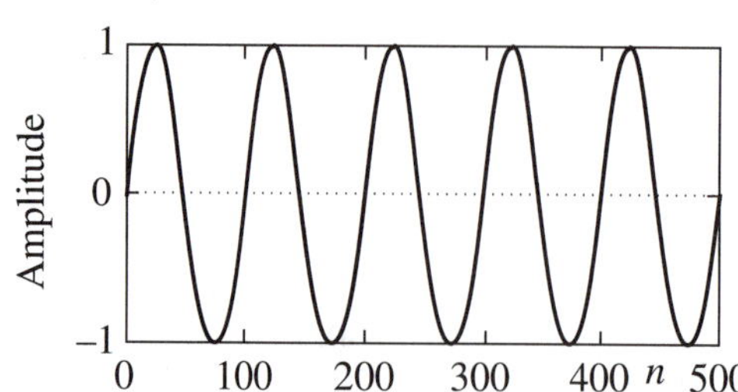

b)

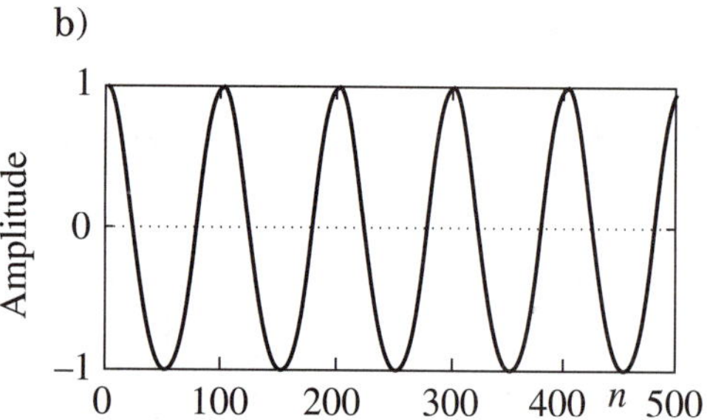

9.15 One solution is:

v = [0 .1 .20 .20 1];
m = [1 1 1.5 0 0];

The Yule–Walker design does not compare well with a Chebyshev type 1 filter (solid). Other vectors may do better or worse. Seven multiplications are required in each case.

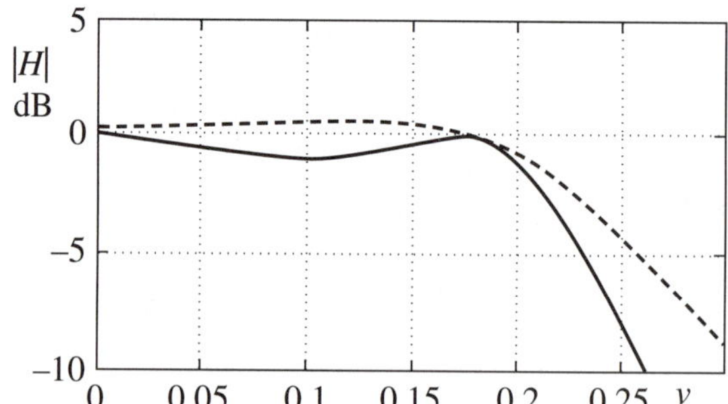

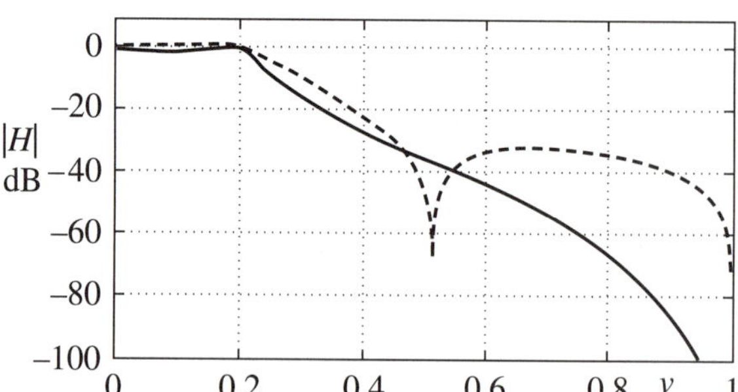

9.17 a) Using two significant figures puts poles on the unit circle—conditionally stable.

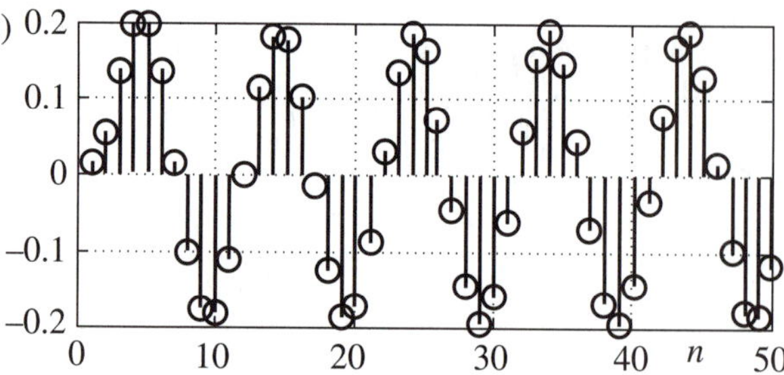

b) $H(z) = \left(\dfrac{0.0115(z + 1)}{z - 0.7233}\right)\left(\dfrac{z^2 + 2z + 1}{z^2 - 1.4145z + 0.7462}\right)$

Two-significant-figure cascade implementation is stable and resembles the desired Chebyshev response.

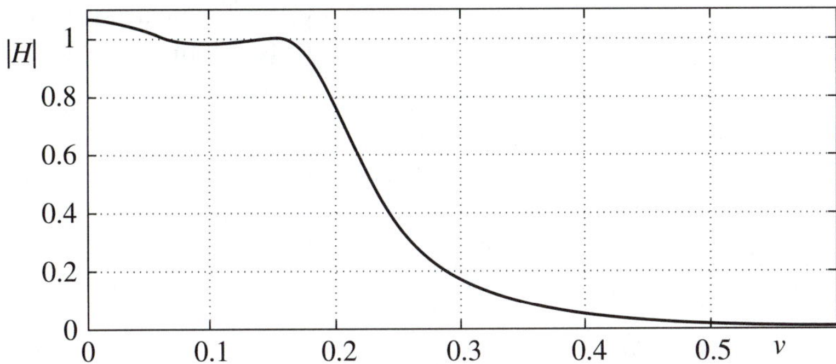

9.19

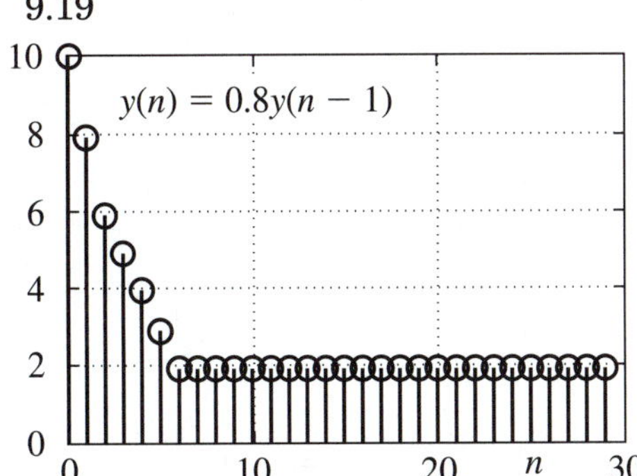

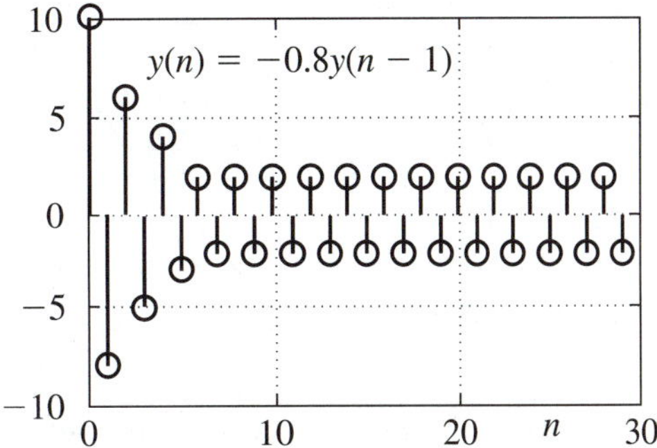

Chapter 10

10.1 $y(n) = x(n) + 0.6000x(n - 1) + 0.3600x(n - 2) + 0.2160x(n - 3) + 0.1296x(n - 4)$ IIR curve is smooth, FIR curve is wiggly.

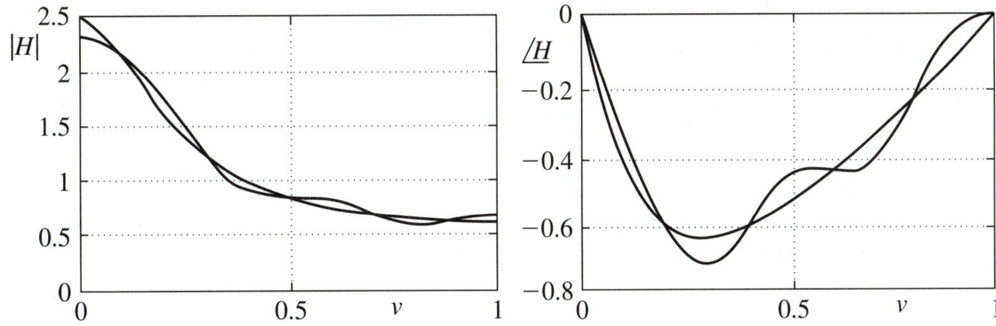

10.3 40th-order FIR agrees well to −40 dB; 80th-order agrees well to −70 dB.

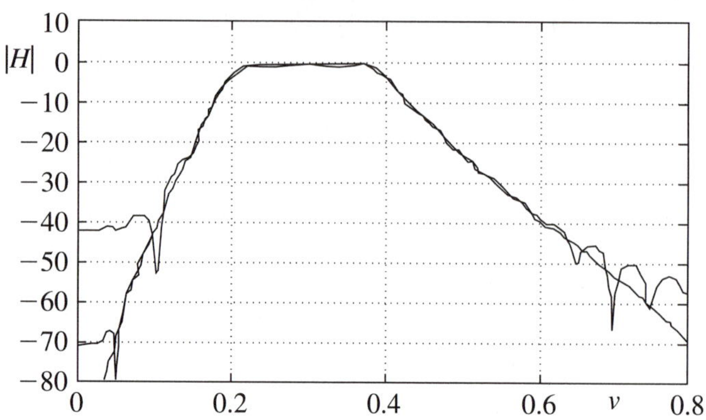

The IIR filter requires 17 multiplications compared to 40 and 80, respectively.

10.5 $y(n) = x(n) - 2x(n-1) + 2x(n-2) - x(n-3)$
$H(z) = 1 - 2z^{-1} + 2z^{-2} - z^{-3}$
$H(e^{j\pi v}) = e^{-j(\pi v 3/2 - \pi/2)}\,[2 \sin 3\pi v/2 - 4 \sin \pi v/2]$

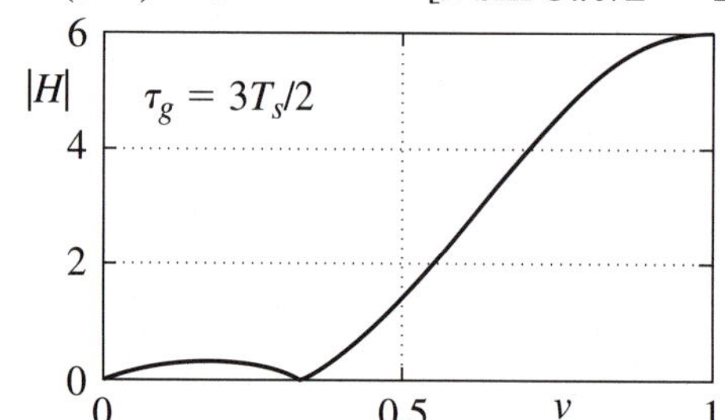

10.7 a) $v_o = 2f/f_s = 2$ cycles/samples $= 1/20$; $T = 40$ samples per cycle

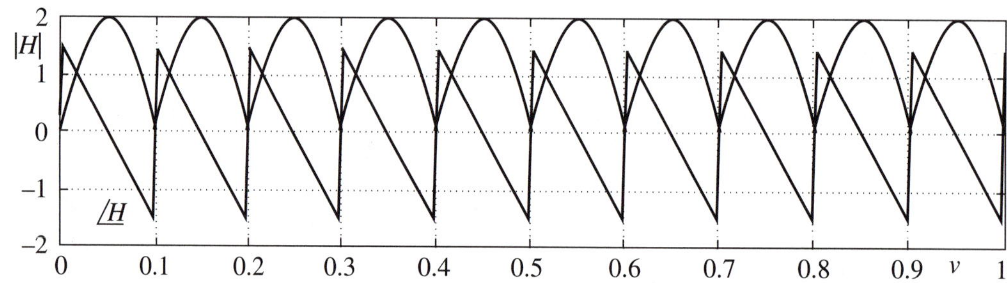

b) The results are of two identical square waves displaced by 20 samples and subtracted.

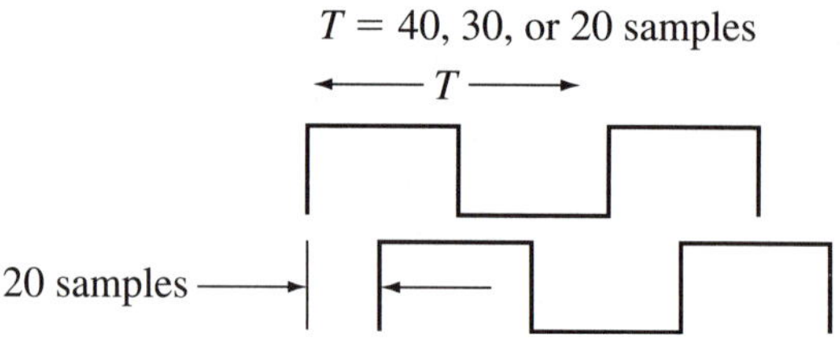

10.9

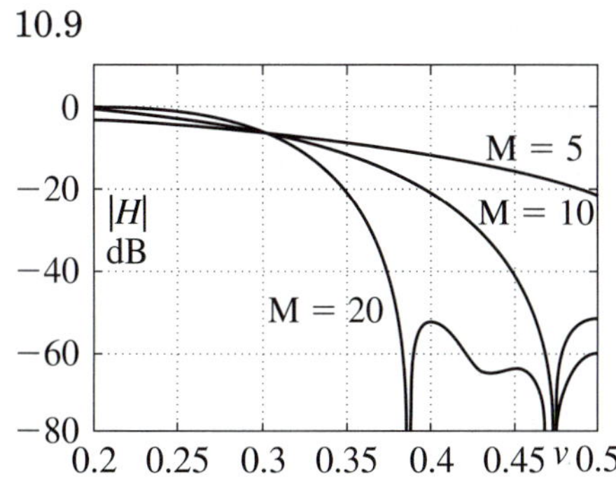

10.11

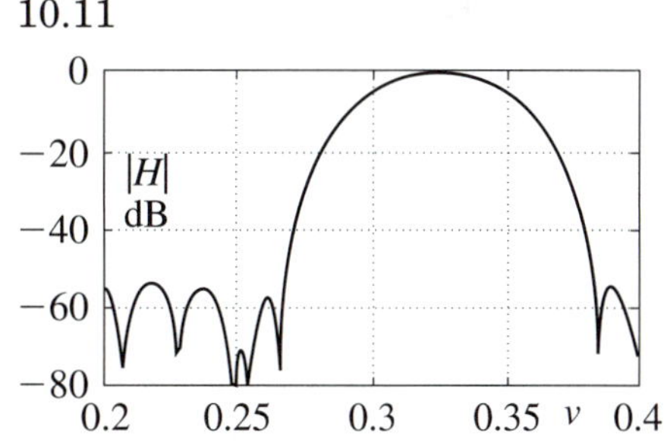

10.13 a)

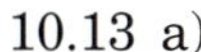

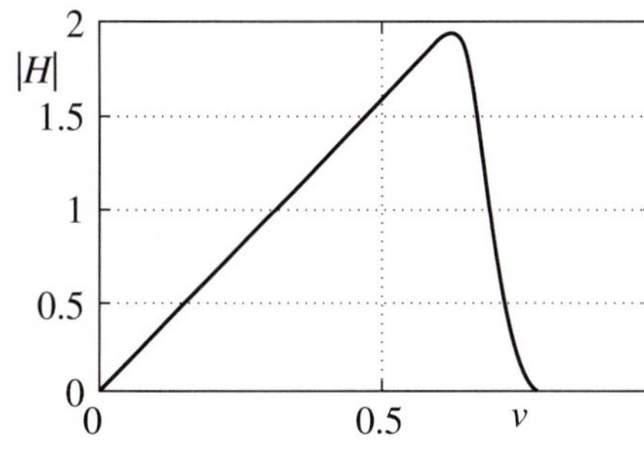

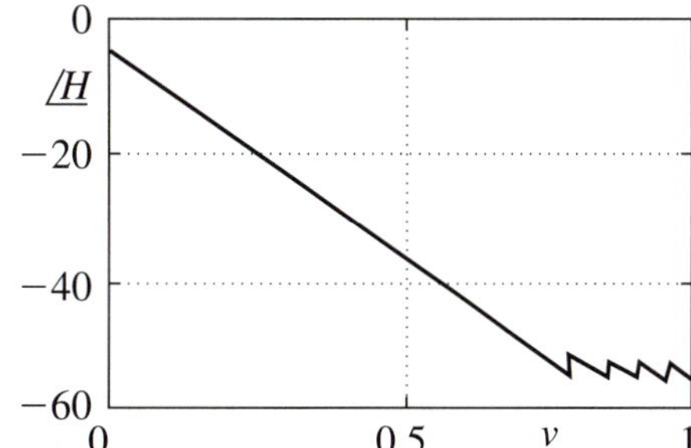

b)

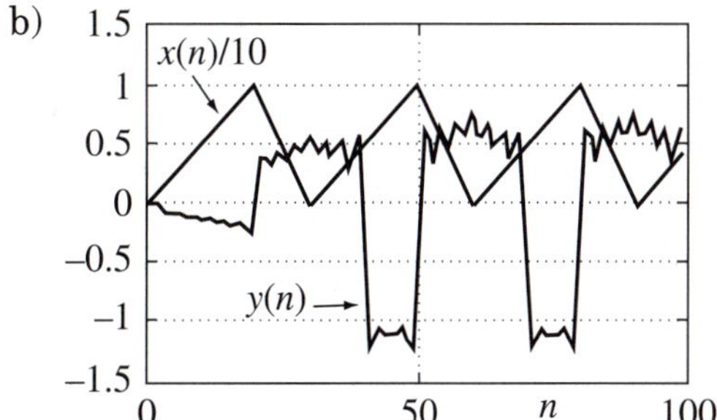

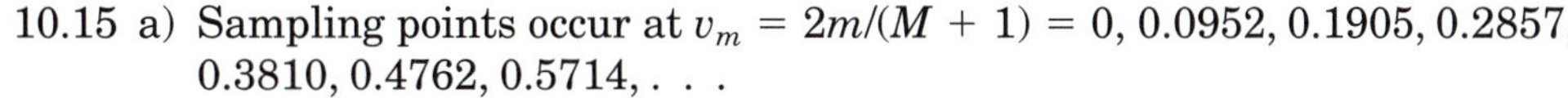

10.15 a) Sampling points occur at $v_m = 2m/(M + 1) = 0, 0.0952, 0.1905, 0.2857, 0.3810, 0.4762, 0.5714, \ldots$

b)

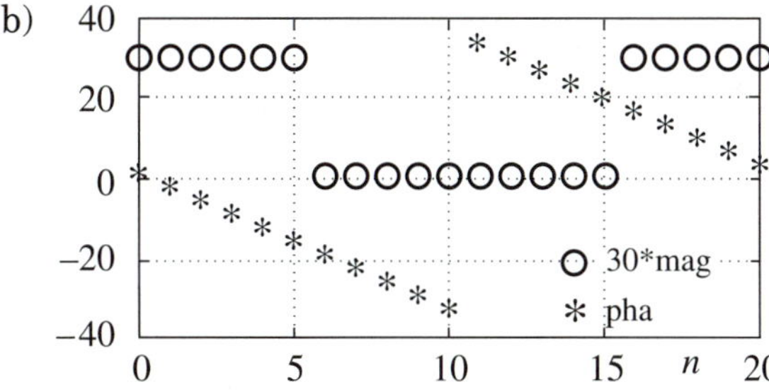

c)

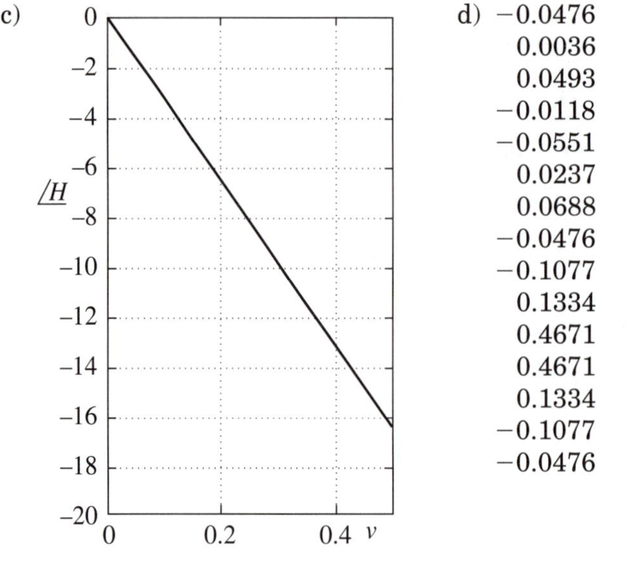

d)

−0.0476
0.0036
0.0493
−0.0118
−0.0551
0.0237
0.0688
−0.0476
−0.1077
0.1334
0.4671
0.4671
0.1334
−0.1077
−0.0476

0.0688
0.0237
−0.0551
−0.0118
0.0493
0.0036

10.17
```
>mag=[ones(1,6) zeros(1,22) ones(1,5)]; pha=pi*[-(0:16) (16:-1:1)];
```

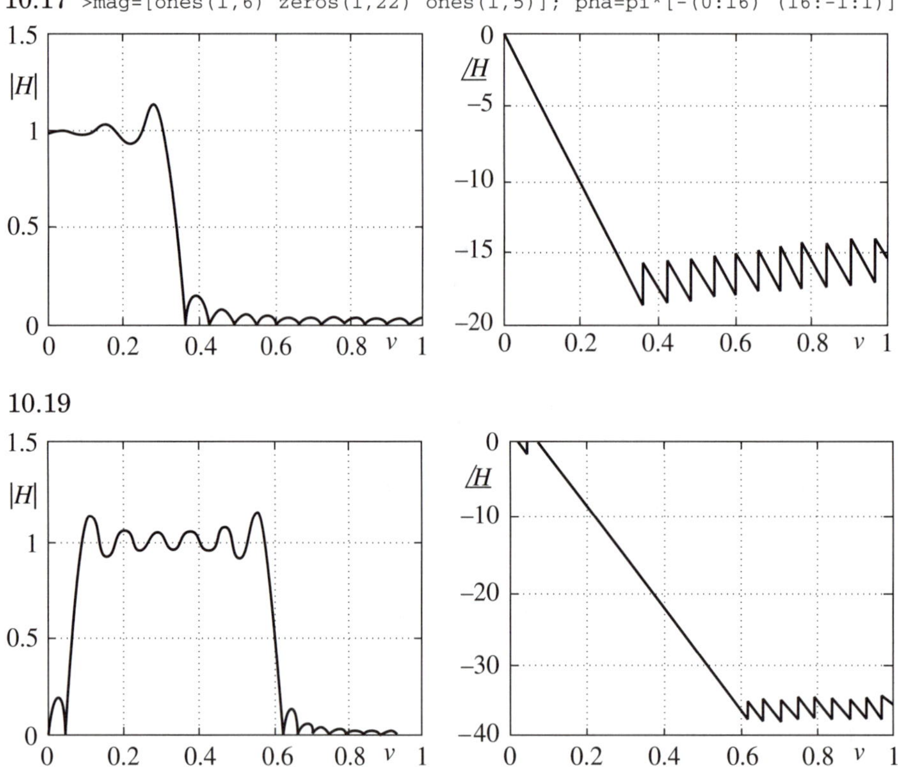

10.19

```
>mag=[0 0 -ones(1,12) zeros(1,18) ones(1,12) 0];      % 45 pts, M=44
>pha=(pi*44/45)*[-(0:22) (22:-1:1)]; pha=pha+pi/2;
```

10.21 Remez sets the passband precisely and has steeper rolloff. Frequency sampling/Hamming window has superior stopband attenuation. Both are good filter options.

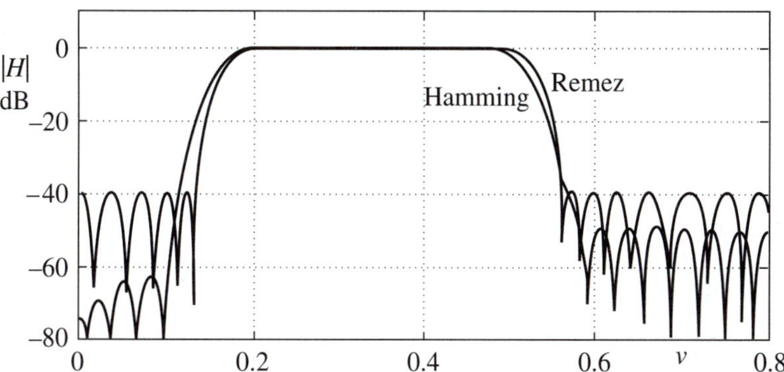


```
% start Remez design
>f=[0 0.1375 .2 .5 .5625 1];
>m=[0    0   1   1   0    0];
% start freq sampling design (2nd trial to adjust passband)
>mag=[zeros(1,5) 0.1 ones(1,11) .1 zeros(1,29) .1 ones(1,11) .1
 zeros(1.4)];
>pha=(pi*63/64)*[-(0:31) 0 (31:-1:1)];
```

10.23 `>m=[0 0 1 1 0 0]; f=[0 .3 .4 .6 .7 1]; num=remez(20,f,m)`

Not much difference between exact and 2-significant-figure designs!

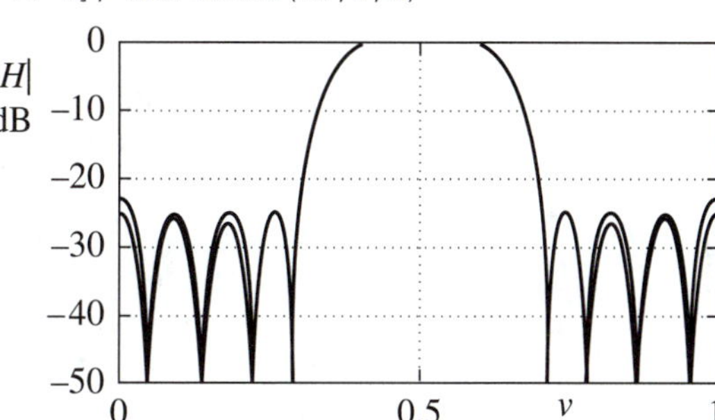

Chapter 11

11.1 a) $n = 30.8$ b) $n = 2.8$

11.3 a) $H = -31.3$ dB @ 20.6 kHz

b)

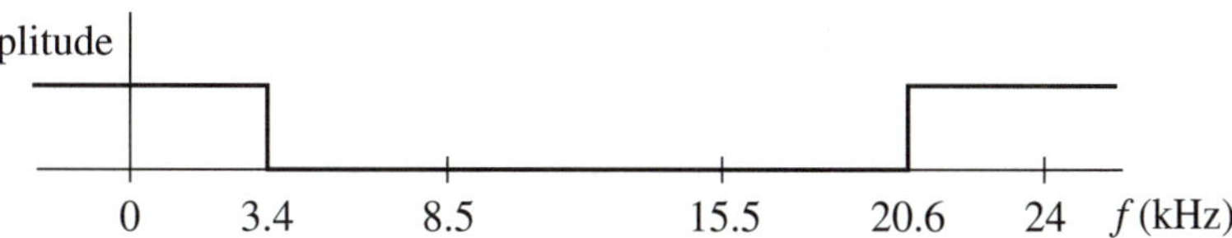

c)

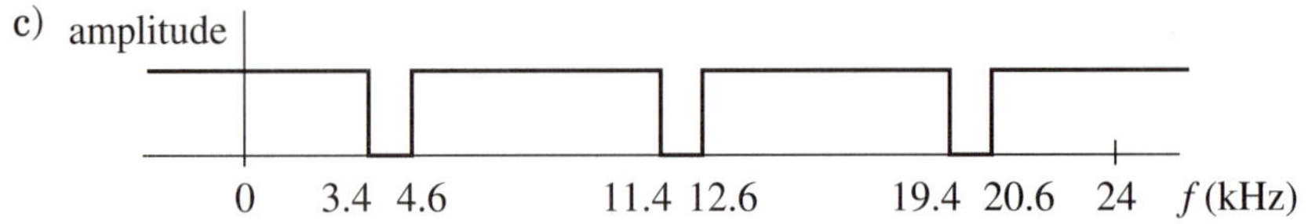

d) The decimation filter must eliminate everything from 3.4 kHz to 20.6 kHz, so it should be a low-pass filter with $\upsilon_{co} = 2(3.4)/24 = 0.283$.

11.5 a) b)

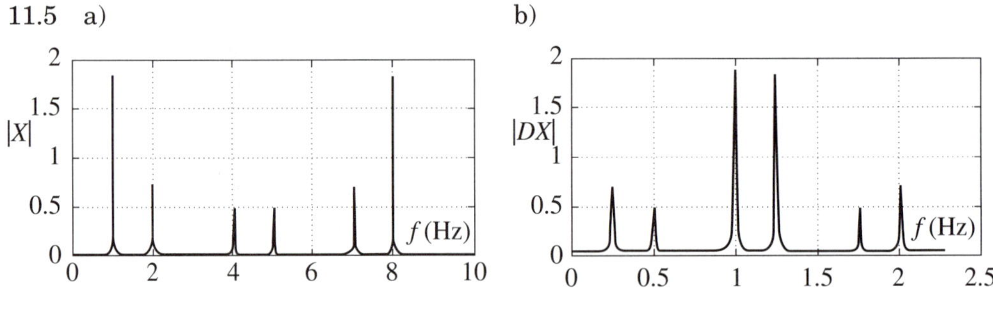

c)

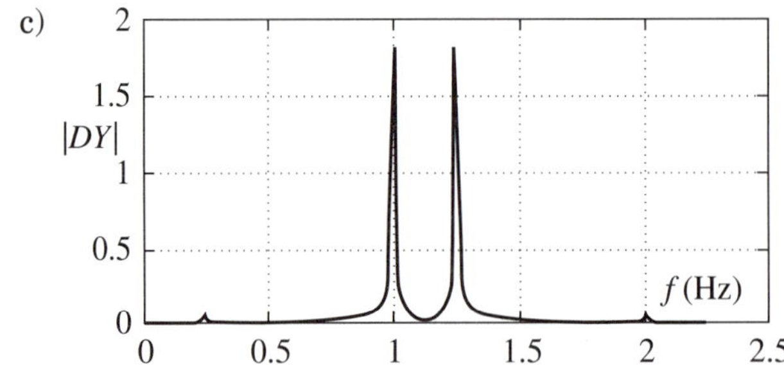

d) The signals appear at their correct frequencies when sampled at 9 Hz. If decimated by 4, the effective sampling frequency becomes 2.25 Hz. Without filtering, the 2- and 4-Hz signals still appear but are aliased to 0.25Hz and 0.5 Hz, respectively. If the noise is first eliminated by the low-pass filter, only the desired signal remains after decimation.

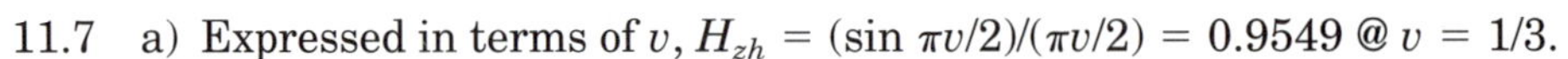

11.7 a) Expressed in terms of υ, $H_{zh} = (\sin \pi\upsilon/2)/(\pi\upsilon/2) = 0.9549$ @ $\upsilon = 1/3$.

b)

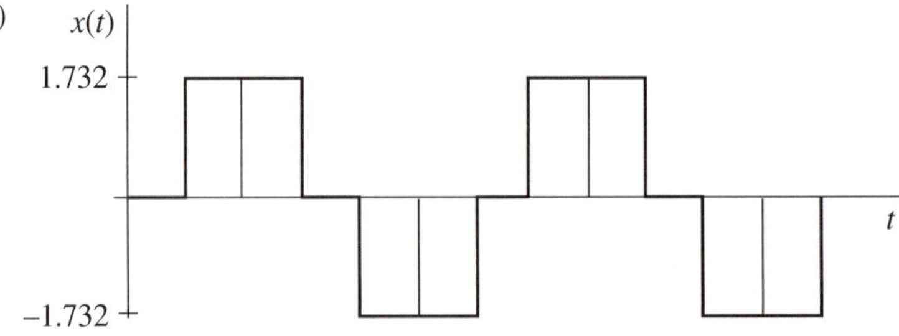

c)
```
>x=[zeros(1,171) 1.732*ones(1,341) zeros(1,171) -1.732*ones(1,341)];
>X=fft(x)/1024;
>abs(X(1:4))
ans = 0 0.9543 0 0.0011      %fundamental = 0.9543
```

11.9

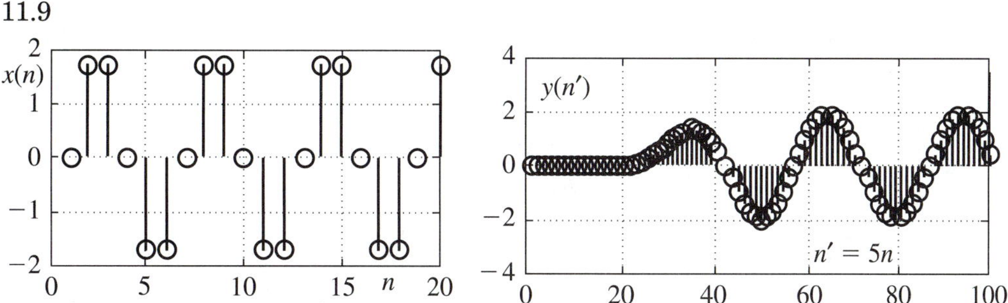

11.11

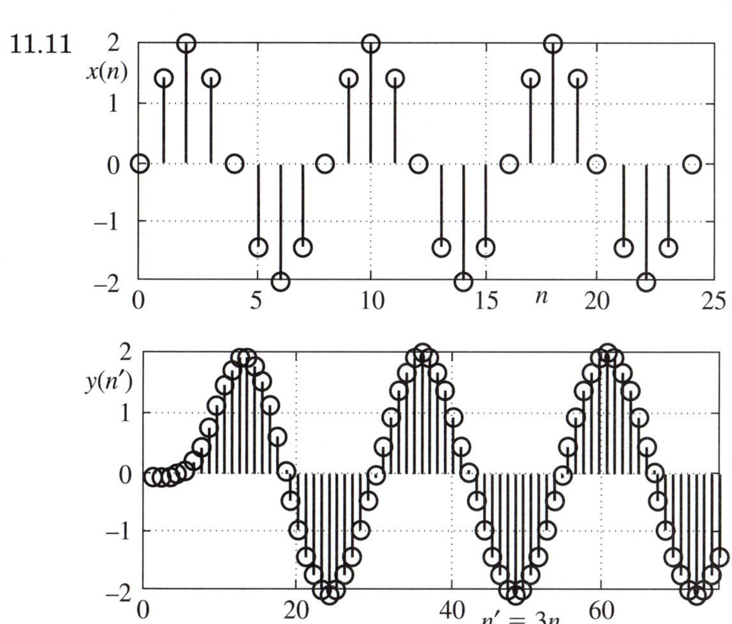

Selection of $v_{co} = 1/6$ is based on the interpolated signal having $v = 1/12$, so any higher value would be adequate for this simple case.

11.13

```
>m=1:25;
>min=930./(m+1);
>max=890./m;
>a=[m' min' max']
```

There are 22 possible ranges for f_s. Some are shown here. The lowest is only 1/23 of the Nyquist rate (930 kHz).

m	f_{smin}	f_{smax}
1	465.0000	890.0000
3	232.5000	296.6667
5	155.0000	178.0000
7	116.2500	127.1429
9	93.0000	98.8889
11	77.5000	80.9091
13	66.4286	68.4615
15	58.1250	59.3333
17	51.6667	52.3529
19	46.5000	46.8421
21	42.2727	42.3810
22	40.4348	40.4545

11.15 Since the lower sideband is to be detected, we want a spectral inversion. We need to sample at the minimum f_s in an m = odd range. From Example 11.6, select $m = 99$, then $f_H = 1$ MHz and $f_L = 995$ kHz. $2/100 = 0.02 < f_s(\text{MHz}) < 0.0202 = 1.99/99$

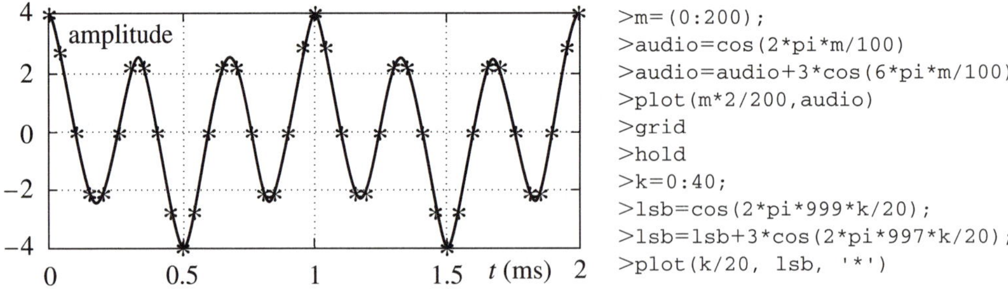


```
>m=(0:200);
>audio=cos(2*pi*m/100)
>audio=audio+3*cos(6*pi*m/100);
>plot(m*2/200,audio)
>grid
>hold
>k=0:40;
>lsb=cos(2*pi*999*k/20);
>lsb=lsb+3*cos(2*pi*997*k/20);
>plot(k/20, lsb, '*')
```

INDEX

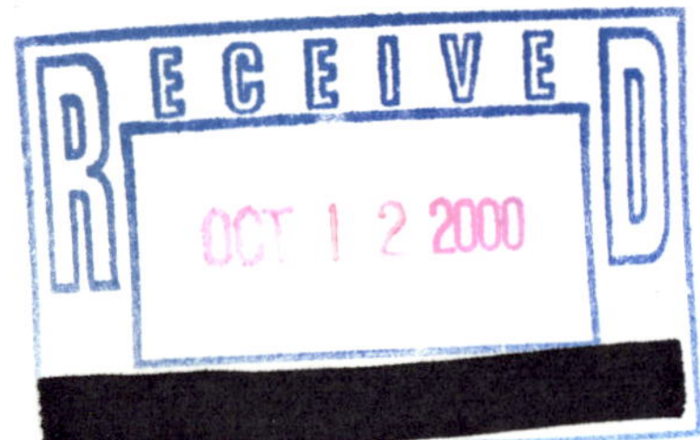
RECEIVED
OCT 12 2000